Annals of Mathematics Studies

Number 143

Cycles, Transfers, and Motivic Homology Theories

by

Vladimir Voevodsky, Andrei Suslin,
and Eric M. Friedlander

PRINCETON UNIVERSITY PRESS

PRINCETON, NEW JERSEY

2000

The Annals of Mathematics Studies are edited by
John N. Mather and Elias M. Stein

Library of Congress Catalog Card Number 00-100291

ISBN 0-691-04814-2 (cloth)
ISBN 0-691-04815-0 (pbk.)

The publisher would like to acknowledge the authors for
providing the electronic files that were reformatted to
produce the camera-ready copy from which this book was printed

The paper used in this publication meets the minimum requirements of
ANSI/NISO Z39.48-1992 (R 1997) (*Permanence of Paper*)

www.pup.princeton.edu

1 3 5 7 9 10 8 6 4 2

3 5 7 9 10 8 6 4 2
(Pbk.)

Contents

Chapter 1   Introduction   3
             Eric M. Friedlander, A. Suslin, and V. Voevodsky

Chapter 2   Relative Cycles and Chow Sheaves   10
             Andrei Suslin and Vladimir Voevodsky

Chapter 3   Cohomological Theory of Presheaves with Transfers   87
             Vladimir Voevodsky

Chapter 4   Bivariant Cycle Cohomology   138
             Eric M. Friedlander and Vladimir Voevodsky

Chapter 5   Triangulated Categories of Motives Over a Field   188
             Vladimir Voevodsky

Chapter 6   Higher Chow Groups and Etale Cohomology   239
             Andrei A. Suslin

Cycles, Transfers, and
Motivic Homology Theories

1

Introduction

Eric M. Friedlander, A. Suslin, and V. Voevodsky

Our original goal which finally led to this volume was the construction of "motivic cohomology theory," whose existence was conjectured by A. Beilinson and S. Lichtenbaum ([2], [3], [17], [18]). Even though this would seem to be achieved at the end of the third paper, our motivation evolved into a quest for a deeper understanding of various properties of algebraic cycles. Thus, several of the papers presented here do not deal directly with motivic cohomology but rather with basic questions about algebraic cycles.

In this introduction, we shall begin with a short reminder of A. Beilinson's formulation of motivic cohomology theory. We then proceed to briefly summarize the topic and contents of individual papers in the volume.

Let k be a field and Sm/k denote the category of smooth schemes over k. A. Beilinson conjectured that there should exist certain complexes $\mathbf{Z}(n)$ ($n \geq 0$) of sheaves in the Zariski topology on Sm/k which have the following properties:

1. $\mathbf{Z}(0)$ is the constant sheaf $\mathbf{Z}$.

2. $\mathbf{Z}(1)$ is the sheaf $\mathcal{O}^*$ placed in cohomological degree 1.

3. For a field F over k, one has

$$H^n(Spec(F), \mathbf{Z}(n)) = K_n^M(F)$$

where $K_n^M(F)$ is the n-th Milnor K-group of F.

4. For a smooth scheme X over k, one has

$$\mathbf{H}_{Zar}^{2n}(X, \mathbf{Z}(n)) = A^n(X)$$

where $A^n(X)$ is the Chow group of cycles of codimension n on X modulo rational equivalence.

5. For any smooth scheme X over k, there is a natural spectral sequence with the E_2-term of the form

$$E_2^{p,q} = \mathbf{H}_{Zar}^p(X, \mathbf{Z}(q))$$

and differentials $d_r : E_r^{p,q} \to E_r^{p+r-1,q+2r-1}$ which converges to Quillen's K-groups $K_{2q-p}(X)$.

After tensoring with $\mathbf{Q}$ this spectral sequence degenerates and one has

$$\mathbf{H}_{Zar}^i(X, \mathbf{Z}(n)) \otimes \mathbf{Q} = gr_\gamma^n K_{2n-i}(X) \otimes \mathbf{Q}$$

where the groups on the right hand side are quotients of the γ-filtration in Quillen's K-theory of X.

Observe that the complexes $\mathbf{Z}(n)$ are determined rationally by the last of above properties.

The hypercohomology groups $\mathbf{H}_{Zar}^i(X, \mathbf{Z}(n))$ are usually denoted by $H_{\mathcal{M}}^i(X, \mathbf{Z}(n))$ and called *motivic cohomology* groups of X.

This definition of motivic cohomology is not "topology free." In particular one may consider the corresponding hypercohomology groups in the etale topology instead of the Zariski topology. S. Lichtenbaum ([17], [18]) has in fact suggested axioms for the etale analog of Beilinson's motivic cohomology. We emphasize that everywhere in this volume "motivic (co-)homology" mean motivic (co-)homology in the Zariski topology unless the etale topology is explicitly specified.

In addition to the axioms given above, Beilinson's original list contained two further axioms. These we state below in the form of conjectures.

Beilinson-Lichtenbaum Conjecture. For a field F over k and a prime l not equal to $char(k)$, one has

$$\mathbf{H}_{Zar}^i(Spec(F), \mathbf{Z}(n) \otimes \mathbf{Z}/l) = \begin{cases} H_{et}^i(F, \mu_l^{\otimes n}) & \text{for } i \leq n \\ 0 & \text{for } i > n \end{cases}$$

Beilinson-Soule Vanishing Conjecture. For a smooth scheme X over k, one has

$$\mathbf{H}_{Zar}^i(X, \mathbf{Z}(n)) = 0$$

for $i < 0$.

In conjunction with the spectral sequence relating motivic cohomology to algebraic K-theory, these two "axioms" imply the validity of highly nontrivial conjectures in algebraic K-theory.

All the approaches to motivic (co-)homology suggested in the last several years can be roughly divided into two types depending on which of property 3 and 4 was considered to be "more fundamental":

1. To construct motivic cohomology (usually of a field) as cohomology groups of certain complexes with terms being given by explicit generators and relations generalizing Milnor's definition of K_M^n ([3],[4], [13]).

2. To construct motivic cohomology of a scheme as cohomology of a complex defined in terms of algebraic cycles thus generalizing the classical definition of Chow groups ([5],[6],[8],[14][1]).

The basic problem with the first approach is that it is very difficult to prove functorial properties of theories constructed in such a way; for example, the proof that Milnor K-theory has transfer maps is quite non-trivial. On the other hand, an important advantage of this first approach is that from this point of view it is possible to construct natural looking complexes (see [13]) satisfying the Beilinson-Soule vanishing property (saying that $H_{\mathcal{M}}^i(X, \mathbf{Z}(n)) = 0$ for $i < 0$).

The second approach was pioneered by S. Bloch in [5], who introduced "higher" Chow groups with many good properties. In the papers of this volume, we develop an alternative (and independent) theory of motivic homology and cohomology based upon algebraic cycles. (The reader should be forewarned that many of our results apply only to varieties over a field for which resolution of singularities is valid.)

A major difficulty which must be confronted in this approach is that standard moving techniques used in the classical theory of algebraic cycles are not sufficient to prove basic properties of such a theory. S. Bloch recently solved this problem for his higher Chow groups by introducing an ingenious but very complicated moving technique based on blow-ups and M. Spivakovsky 's solution of Hironaka's polyhedron game ([7]). A similar problem was encountered by A. Suslin when attempting to prove some properties of "algebraic singular homology theory."

Our theory culminates in the fourth paper of this volume "Triangulated categories of motives over a field": this applies the results of the other papers to construct a consistent triangulated theory of mixed motives over

---

[1]The construction of motivic cohomology given by D. Grayson does not refer explicitly to algebraic cycles, but uses instead some version of algebraic K-theory with supports.

a field[2]. We construct there a certain rigid tensor triangulated category $DM_{gm}(k)$ together with an invertible object $\mathbf{Z}(1)$ called the Tate object and two functors

$$M : Sch/k \to DM_{gm}(k)$$

$$M_c : (Sch/k)_{prop} \to DM_{gm}(k)$$

(where $(Sch/k)_{prop}$ is the category of schemes of finite type over k and proper morphisms) which satisfy triangulated analogs of functorial properties of homology and Borel-Moore homology respectively. Aside from further "motivic" applications, the value of this theory is that it provides a natural categorical framework for different kinds of "algebraic cycle (co-)homology" type theories.

In particular Bloch's higher Chow groups $CH^i(X, j)$ and Suslin's algebraic singular homology $H_i^{alg}(X, \mathbf{Z})$ admit the following descriptions in terms of $DM_{gm}(k)$:

$$CH^i(X, j) = Hom_{DM}(\mathbf{Z}(d - i)[2d - 2i + j], M_c(X))$$

(where $d = dim(X)$)

$$H_i^{alg}(X, \mathbf{Z}) = Hom_{DM}(\mathbf{Z}[i], M(X)).$$

Similarly, one defines motivic cohomology by

$$H_{\mathcal{M}}^i(X, \mathbf{Z}(j)) \equiv Hom_{DM}(M(X), \mathbf{Z}(j)[i]).$$

In the case of a smooth scheme X, the motivic Poincare duality theorem asserts that cohomology is canonically isomorphic to Borel-Moore homology and thus isomorphic to higher Chow groups.

We prove that our motivic cohomology theory satisfies properties (1)-(4) listed above. The comparison with higher Chow groups together with the recent construction of a "motivic spectral sequence" by S. Bloch and S. Lichtenbaum [9] and work in progress by Eric M. Friedlander, M. Levine, A. Suslin and M. Walker gives us the fifth property[3].

As was mentioned above, attempting to develop such a theory one encounters certain difficulties related to the fact that we do not yet know how to work efficiently with algebraic cycles up to "higher homotopies".

There are four main technical tools which we use to overcome these difficulties:

---

[2]Two other constructions of triangulated categories of mixed motives have been proposed by M. Hanamura [15] and M. Levine [16].

[3]We would like to mention that the existence of a "motivic spectral sequence" is not strictly speaking a part of the theory discussed here, but rather of a yet to be constructed stable homotopy theory of schemes.

1. a theory of sheaves of relative cycles.

2. a theory of sheaves and presheaves with transfers.

3. the Nisnevich and cdh-topologies on the category of schemes.

4. the Friedlander-Lawson moving lemma for families of algebraic cycles.

The theory of sheaves of relative cycles is developed in the first paper of this volume "Relative cycles and Chow sheaves". The basic idea of this theory comes from two independent sources - one is Lawson homology theory and the classical theory of Chow varieties and another is the sheaf theoretic approach to finite relative cycles used in [22] and [21].

An important aspect of the theory of relative cycles is that it is "elementary" in the sense that it only uses very basic properties of schemes. We develop this theory in the very general context of schemes of finite type over an arbitrary Noetherian scheme, although applications considered in this volume concern varieties over a field.

Another of the main tools appearing in this paper is the cdh-topology. This plays an important role in the theory, because different kinds of "localization" sequences for sheaves of relative cycles become exact in the cdh-topology.

The theory of presheaves with transfers and more specifically homotopy invariant presheaves with transfers is the main theme of the second paper "Cohomological theory of presheaves with transfers". The idea that transfers should play an important role in motivic theory can be traced to two sources. One is the use of transfers in the proof of a rigidity theorem by A. Suslin in [20]; another is the theory of qfh-sheaves considered in [22]. These were first combined in [21], leading to a proof of A. Suslin's conjecture relating the algebraic singular homology with finite coefficients and the etale cohomology. The results on presheaves with transfers obtained in the second paper together with the Nisnevich (or "completely decomposable") topology [19] enable us to prove the Mayer-Vietoris exact sequence for algebraic singular homology; this result is an example of a solution to a "moving" probem inaccessible to either classical methods or S. Bloch's "moving by blow-ups" techniques.

The Friedlander-Lawson moving lemma for families of algebraic cycles ([11]) appears in the third paper in which we construct a bivariant theory called bivariant cycle cohomology[4]. In this paper, duality between cohomology and homology is studied, following duality studied by Eric M. Fried-

---

[4]We should mention that the word "bivariant" is used by us in its "naive" sense and not in the sense of [12].

lander and H.B. Lawson in the context of Lawson homology [10]. While the definition of our bivariant theory (at least in the case of a smooth first argument) is very elementary, to prove basic properties such as Mayer-Vietoris and localization requires all the machinery of the preceding papers together with the moving lemma for families of algebraic cycles. As before, we have had two main sources of inspiration. One is the theory of morphic cohomology which is a bivariant analog of Lawson homology developed by Eric M. Friedlander and H.B. Lawson. Another is the approach to localization problems based on the theory of presheaves with transfers and the cdh-topology which was developed by V. Voevodsky.

The fifth and last paper in the volume gives a proof of the fact that the groups of the Borel-Moore homology component of the bivariant cycle cohomology are canonically isomorphic (in appropriate cases) to Bloch's higher Chow groups, thereby providing a link between our theory and Bloch's original approach to motivic (co-)homology.

The reader will find separate introductions at the beginning of each paper in this volume. To obtain a broader view of motivic cohomology and its relationship to Grothendieck's original goal of a good category of motives, the reader may wish to consult [1].

V. Voevodsky is pleased to acknowledge encouragement and useful conversations with A. Beilinson, A. Goncharov, and D. Kazhdan.

References

1. *Motives*, volume 54-55 of *Proc. of Symp. in Pure Math.* AMS, 1994.
2. A. Beilinson. Height pairing between algebraic cycles. In *K-theory, Arithmetic and Geometry.*, volume 1289 of *Lecture Notes in Math.*, pages 1–26. Springer-Verlag, 1987.
3. A. Beilinson, R. MacPherson, and V. Schechtman. Notes on motivic cohomology. *Duke Math. J.*, pages 679–710, 1987.
4. A. A. Beilinson, A. B. Goncharov, V. V. Schechtman, and A. N. Varchenko. Aomoto dilogarithms, mixed Hodge structures and motivic cohomology of pairs of triangles on the plane. In *The Grothendieck festchrift*, volume 1, pages 135–172. Birkhauser, Boston, 1990.
5. S. Bloch. Algebraic cycles and higher K-theory. *Adv. in Math.*, 61:267–304, 1986.
6. S. Bloch. Algebraic cycles and the Lie algebra of mixed Tate motives. *JAMS*, 4:771–791, 1991.
7. S. Bloch. The moving lemma for higher Chow groups. *J. Algebr. Geom.*, 3(3):537–568, Feb. 1994.
8. S. Bloch and I. Kriz. Mixed Tate motives. *Ann. of Math.*, 140:557–605, 1994.
9. S. Bloch and S. Lichtenbaum. A spectral sequence for motivic cohomology. *Preprint*, 1994.

10. Eric M. Friedlander and H. Blaine Lawson. Duality relating spaces of algebraic cocycles and cycles. *Topology*, 36(2):533–565, 1997.
11. Eric M. Friedlander and H. Blaine Lawson. Moving algebraic cycles of bounded degree. *Invent. Math.*, 132(1):91–119, 1998.
12. W. Fulton and R. MacPherson. *Categorical framework for the study of singular spaces*. Number 243 in Mem. AMS. AMS, 1981.
13. A. B. Goncharov. Geometry of configurations, polylogarithms, and motivic cohomology. *Adv. Math.*, 114(2):197–318, 1995.
14. Daniel R. Grayson. Weight filtrations via commuting automorphisms. *K-Theory*, 9(2):139–172, 1995.
15. Masaki Hanamura. Mixed motives and algebraic cycles. I. *Math. Res. Lett.*, 2(6):811–821, 1995.
16. Marc Levine. *Mixed motives*. American Mathematical Society, Providence, RI, 1998.
17. S. Lichtenbaum. Values of zeta-functions at non-negative integers. In *Number theory*, volume 1068 of *Lecture Notes in Math.*, pages 127–138. Springer-Verlag, 1983.
18. S. Lichtenbaum. New results on weight-two motivic cohomology. In *The Grothendieck festchrift*, volume 3, pages 35–55. Birkhauser, Boston, 1990.
19. Y. Nisnevich. The completely decomposed topology on schemes and associated descent spectral sequences in algebraic K-theory. In *Algebraic K-theory: connections with geometry and topology*, pages 241–342. Kluwer Acad. Publ., Dordrecht, 1989.
20. A. Suslin. On the K-theory of algebraically closed fields. *Invent. Math*, 73:241–245, 1983.
21. Andrei Suslin and Vladimir Voevodsky. Singular homology of abstract algebraic varieties. *Invent. Math.*, 123(1):61–94, 1996.
22. V. Voevodsky. Homology of schemes. *Selecta Mathematica, New Series*, 2(1):111–153, 1996.

2

Relative Cycles and Chow Sheaves

Andrei Suslin and Vladimir Voevodsky

Contents

1 Introduction   **11**

2 Generalities   **14**
  2.1 Universally equidimensional morphisms 14
  2.2 Universally equidimensional closed subschemes 17
  2.3 Cycles on Noetherian schemes 20

3 Relative cycles   **23**
  3.1 Relative cycles 23
  3.2 Cycles associated with flat subschemes 27
  3.3 Chow presheaves 30
  3.4 Relative cycles over geometrically unibranch schemes 39
  3.5 Multiplicities of components of inverse images of equidimensional cycles over geometrically unibranch schemes 41
  3.6 Functoriality of Chow presheaves 48
  3.7 Correspondence homomorphisms 53

4 Chow sheaves in the h-topologies   **59**
  4.1 The h-topologies 59
  4.2 Sheaves in the h-topologies associated with Chow presheaves 61
  4.3 Fundamental exact sequences for Chow sheaves 70
  4.4 Representability of Chow sheaves 75

1. Introduction

Let X be a scheme. A cycle on X is a formal linear combination of points of the Zariski topological space of X. A cycle is called an effective cycle if all points appear in it with non negative coefficients. Suppose that X is a projective scheme over a field k of characteristic zero. Then for any projective embedding $i : X \to \mathbf{P}^n$ the classical construction produces a projective variety $C_r(X, i)$ called the Chow variety of effective cycles of dimension r on X such that k-valued points of $C_r(X, i)$ are in natural bijection with effective cycles of dimesnion r on X. Moreover, for any field extension E/k an E-valued point of $C_r(X, i)$ defines a cycle on $X \times_{Spec(k)} Spec(E)$. In particular for any Noetherian scheme S over k and any morphism $\phi : S \to C_r(X)$ we get a cycle $\mathcal{Z}_\phi$ on $S \times_{Spec(k)} X$ which lies over generic points of S. For any such cycle $\mathcal{Z}_\phi$ and any morphism $f : S' \to S$ of Noetherian schemes over k the composition $f \circ \phi$ defines a new cycle $cycl(f)(\mathcal{Z})$ on $X \times_{Spec(k)} S'$. Thus existence of Chow verieties implies that for any Noetherian scheme S over k there is a natural class of cycles on $X \times_{Spec(k)} S$ which are contravariantly functorial with respect to all morphisms $S' \to S$ of Noetherian schemes over k.

Let now $X \to S$ be any scheme of finite type over a Noethcrian scheme S. In this paper we introduce a class of cycles on X which are called *relative cycles* on X over S. Their most important property is the existence of well defined base change homomorphisms for arbitrary morphisms $S' \to S$ of Noetherian schemes. In the case when $X = X_k \times_{Spec(k)} S$ where X_k is a projective scheme over a field of characteristic zero and S is a semi-normal Noetherian scheme over k the class of effective relative cycles on X over S coincides with the class of cycles of the form $\mathcal{Z}_\phi$ considered above.

Informally speaking a *relative cycle* on X over S is a cycle on X which lies over generic points of S and has a well defined specialization to any fiber of the projection $X \to S$. We denote the group of relative cycles of relative dimension r on X over S by $Cycl(X/S, r)$. Unfortunately if S is not a scheme of characterstic zero the specializations $\mathcal{Z}_s$ of a relative cycle $\mathcal{Z}$ on X over S to points s of S do not have in general integral coefficients - the characteristic of the residue field of s may appear in denominators. We denote by $z(X/S, r)$ the subgroup in $Cycl(X/S, r)$ which consists of relative cycles $\mathcal{Z}$ such that for any point s of S the specialization $\mathcal{Z}_s$ of $\mathcal{Z}$ to s has integral coefficients. The groups $z(X/S, r)$ are contravariantly functorial with respect to S, i.e. for any morphism of Noetherian schemes $f : S' \to S$ there is a homomorphism $cycl(f) : z(X/S, r) \to z(X \times_S S'/S', r)$ and for any composable pair of morphisms f, g one has $cycl(f \circ g) = cycl(g) \circ cycl(f)$.

It gives us a presheaf of abelian groups on the category of Noetherian schemes over S which we also denote by $z(X/S, r)$. These *Chow presheaves* are the main objects of our study.

The paper is organized as follows. We start in sections 2.1, 2.2 with some elementary properties of equidimensional morphisms and equidimensional closed subschemes. Besides the standard theorems of Chevalley (Theorem 2.1.1, Proposition 2.1.7(3)) our main technical tool here as well as in the rest of the paper is the "platification theorem" (Theorem 2.2.2).

In section 2.3 we prove some basic results about cycles on Noetherian schemes. All material here is well known and is included only for reader's convenience.

In section 3.1 we introduce the groups $Cycl(X/S, r)$ of relative cycles on a scheme X of finite type over a Noetherian scheme S. Relative cycles over general Noetherian schemes demonstrate all kinds of "pathological" behavior. For instance the group $Cycl(X/S, r)$ is not necessarily generated by the corresponding submonoid $Cycl^{eff}(X/S, r)$ of effective relative cycles (see example 3.4.7) and supports of noneffective relative cycles of relative dimension r over S do not have to be equidimensional of relative dimension r over S (example 3.1.9).

We also define in this section different versions of our main object - the groups of relative cycles with proper support and the groups of relative cycles with equidimensional support.

The main theorem 3.3.1 of section 3.3 says that for any morphism $S' \to S$ of Noetherian schemes there is a well defined base change homomorphism $Cycl(X/S, r) \to Cycl(X \times_S S'/S', r) \otimes \mathbf{Q}$. It gives us a construction of presheaves $Cycl(X/S, r)_{\mathbf{Q}}$ on the category of Noetherian schemes over S such that for any such scheme S' the group $Cycl(X/S, r)_{\mathbf{Q}}(S')$ is the group of relative cycles of dimension r on $X \times_S S'$ over S' with *rational* coefficients.

An example of a relative cycle with integral coefficients whose specialization does not have integral coefficients which is due to A.Merkurjev is given in 3.5.10(1).

In order to obtain a good definition of Chow presheaves with integral coefficients we consider a formal condition on a cycle to be *universally* with integral coefficients. It turns out that the corresponding groups $z(X/S, r)$ are subgroups in the groups $Cycl(X/S, r)$ with torsion quotient for all Noetherian schemes S (Proposition 3.3.14). We call the corresponding presheaves of abelian grous $z(X/S, r)$ the *Chow presheaves* of relative cycles on X over S.

We also define in the similar way the presheaves $c(X/S, d)$ which correspond to proper relative cycles and presheaves $z_{equi}(X/S, r)$, $c_{equi}(X/S, r)$

of relative cycles (resp. proper relative cycles) with equidimensional support.

In section 3.4 we consider relative cycles over geometrically unibranch schemes (in particular over normal schemes). It turns out that over such a scheme S the group of relative cycles with equidimensional support is generated by the cycles of integral closed subschemes of X which are equidimensional over S (Theorem 3.4.2) which implies that in this case our definition coincide with the naive one. On the other hand the corresponding statement does not hold in generally for the groups $z(X/S, r)$ as shown in example 3.5.10(2).

One of the important properties of presheaves $z(X/S, r)$ is that for any regular base scheme S the groups $z(X/S, r)$ coincide with the groups $Cycl(X/S, r)$, i.e. in this case any relative cycle has universally integral coefficients. Our base change homomorphisms over a regular base scheme coincide with base change homomorphisms which one can define by means of the Tor-formula (see the end of the section 3.5). We also show in section 3.5 that our base change homomorphisms coincide in the case of finite cycles over normal schemes with base change homomorphisms defined in [16].

In section 3.6 we study functoriality of Chow presheaves $z(X/S, r)$ and $c(X/S, r)$ with respect to X. We show in particular that for any morphism (resp. any proper morphism) $p : X \to Y$ of schemes of finite type over S there is a push-forward homomorphism of presheaves $p_* : c(X/S, r) \to c(Y/S, r)$ (resp. $p_* : z(X/S, r) \to z(Y/S, r)$) and for any flat (resp. flat and proper) equidimensional morphism $f : X \to Y$ of relative dimension n there is a pull-back homomorphism of presheaves $f^* : z(Y/S, r) \to z(X/S, r+n)$ (resp. $f^* : c(Y/S, r) \to c(X/S, r+n)$).

In the next section we define for Chow presheaves the *correspondence homomorphisms* which were considered in context of Chow varieties by Eric M. Friedlander. In particular we show that there is a well defined homomorphism of external product

$$z(X/S, r_1) \otimes z(Y/S, r_2) \to z(X \times_S Y/S, r_1 + r_2)$$

for any schemes of finite type X, Y over S.

In the last section we consider Chow presheaves as sheaves in the h-topologies. In particular we construct in section 4.3 some exact sequences of Chow sheaves which are important for localization-type theorems in algebraic cycle homology and Suslin homology (see [17]). Finally in section 4.4 we study representability of Chow sheaves. One can easily show that except for some trivial cases the Chow presheaves are not representable as presheaves. To avoid this difficulty we introduce a notion of the h-representability. We construct then the Chow scheme $C_{r,d}$ of cycles of degree d and dimension r on $\mathbf{P}^n$ over $Spec(\mathbf{Z})$ using essentially the classical construction due to Chow (see also [14]) and show that it h-represents the Chow sheaf $z_d^{eff}(\mathbf{P}^n/Spec(\mathbf{Z}), r)$ of effective relative cycles of relative dimension r and degree d on $\mathbf{P}^n$. A rather formal reasoning shows then that for any quasi-projective scheme X over a Noetherian scheme S and any $r \geq 0$ the Chow sheaf $c(X/S, r)$ is h-representable by disjoint union of quasi-projective schemes over S. As an application of this representability result we show that for a quasi-projective scheme X over a Noetherian scheme S the group $z(X/S, r)_h$ of sections of the h-sheaf associated with the Chow presheaf $z(X/S, r)$ can be described using the notion of " continuous algebraic maps" introduced by Eric M. Friedlander and O. Gabber ([3]) which generalizes a similar result obtained in [2] for quasi-projective schemes over a field.

Everywhere in this text a scheme means a separated scheme.

2. Generalities

2.1. Universally equidimensional morphisms

For a scheme X we denote by $dim(X)$ the dimension of the Zariski topological space of X. By definition $dim(X)$ is either a positive integer of infinity. If $x \in X$ is a point of a locally Noetherian scheme X we denote by $dim_x(X)$ the limit $\lim dim(U)$ taken over the partially ordered set of open neighborhoods of x in X. One can easily see that it is well defined and equals $dim(U)$ for a sufficiently small U (see [8, Ch.0,14.4.1]).

For a morphism $p : X \to S$ denote by $dim(X/S)$ the function on the set of points of X of the form $dim(X/S)(x) = dim_x(p^{-1}(p(x)))$. The most important property of the dimension functions is given by the following well known theorem.

Theorem 2.1.1 (Chevalley) *Let $p : X \to S$ be a morphism of finite type. Then for any $n \geq 0$ the subset $\{x \in X : dim_{X/S}(x) \geq n\}$ is closed in X.*

Proof: See [8, Th. 13.1.3].

Definition 2.1.2 *A morphism of schemes $p : X \to S$ is called an equidimensional morphism of dimension r if the following conditions hold:*

1. p is a morphism of finite type.

2. The function $dim(X/S)$ is constant and equals r.

3. Any irreducible component of X dominates an irreducible component of S.

A morphism of schemes $p : X \to S$ is called universally equidimensional of dimension r if for any morphism $S' \to S$ the projection $X \times_S S' \to S'$ is equidimensional of dimension r.

Finally, we say that $p : X \to S$ is a morphism of dimension $\leq r$ if $dim(X/S)(x) \leq r$ for all points x of X.

One can easily see that in the definition of equidimensional morphism given above one can replace the condition $dim(X/S) = r$ by the condition that for any point y of S the dimension of all irreducible components of the topological space $p^{-1}(p(x))$ equals r.

Proposition 2.1.3 Let $p : X \to S$ be a morphism of finite type of Noetherian schemes. Then p is equidimensional of dimension r if and only if for any point x of X there is an open neighborhood U in X and a factorization of the morphism $p_U : U \to S$ of the form $U \xrightarrow{p_0} \mathbf{A}_S^r \to S$ such that p_0 is a quasi-finite morphism and any irreducible component of U dominates an irreducible component of $\mathbf{A}_S^r$.

Proof: See [8, 13.3.1(b)].

Definition 2.1.4 A morphism of schemes $p : X \to S$ is called an open morphism if for any open subset U of X the subset $p(U)$ is open in S. It is called universally open if for any morphism $S' \to S$ the projection $X \times_S S' \to S'$ is an open morphism.

Definition 2.1.5 Let S be a Noetherian scheme. It is called unibranch (resp. geometrically unibranch) if for any point s of S the scheme $Spec(\mathcal{O}_{s,S}^h)$ where $\mathcal{O}_{s,S}^h$ is the henselization of the local ring of s in S (resp. the scheme $Spec(\mathcal{O}_{s,S}^{sh})$ where $\mathcal{O}_{s,S}^{sh}$ is the strict henselization of the local ring of s in S) is irreducible.

Remark: Our definition is consistent with the one given in [8, 6.15.1] in view of [8, 18.8.15].

Proposition 2.1.6 Let S be a Noetherian geometrically unibranch scheme and $f : S' \to S$ be a proper birational morphism. Then for any point s of S the fiber S'_s of f over s is geometrically connected.

Proof: It follows from [7, 4.3.5] and [8, 18.8.15(c)].

Proposition 2.1.7 *Let* $p : X \to S$ *be a morphism of finite type of Noetherian schemes. Then the following implications hold:*

1. *If* p *is a universally equidimensional morphism then* p *is universally open.*

2. *If* $dim(X/S) = r$ *and* p *is open (resp. universally open) then* p *is equidimensional (resp. universally equidimensional) of dimension* r.

3. *If* S *is geometrically unibranch and* p *is equidimensional then* p *is universally equidimensional (and hence universally open).*

Proof: (1) It follows from [8, 14.4.8.1 and 14.4.4].
(2) Obvious.
(3) It is known that any equidimensional morphism over a geometrically unibranch scheme is universally open (see [8, 14.4.4]). It implies immediately that for any morphism $S' \to S$ the projection $p' : X \times_S S' \to S'$ satisfies the condition (3) of Defenition 2.1.2. Since the conditions (1) and (2) are obviously stable under a base change the morphism p' is equidimensional.

Remarks:

1. Note that an equidimensional morphism does not have to be open.

2. One can easily see that the inverse statement to the third part of this proposition holds. Namely a Noetherian scheme X is geometrically unibranch if any equidimensional morphism over X is universally equidimensional.

3. Let $p : X \to S$ be a morphism of finite type such that $S = Spec(k)$ where k is a field. Proposition 2.1.7(3) implies that p is universally equidimensional of dimension r if and only if all irreducible components of X have dimension r.

Proposition 2.1.8 *Let* $p : X \to S$ *be a flat morphism of finite type. Then* p *is universally equidimensional of dimension* r *if and only if for any generic point* $y : Spec(K) \to S$ *of* S *the projection* $X \times_S Spec(K) \to Spec(K)$ *is equidimensional of dimension* r.

Proof: Since any flat morphism of finite type is universally open ([11, I.2.12]) it is sufficient to verify that under the conditions of the proposition we have $dim(X/S) = r$. It follows immediately from [8, 12.1.1(iv)] and Theorem 2.1.1.

Proposition 2.1.9 *Let* $p : X \to S$ *be an equidimensional morphism of relative dimension* r *such that* X *is irreducible. Suppose that* p *admits a decomposition of the form* $X \xrightarrow{p_0} W \xrightarrow{p_1} S$ *such that* p_0 *is surjective and proper and* p_1 *has at least one fiber of dimension* r. *Then* p_1 *is equidimensional of dimension* r *and* p_0 *is finite in the generic point of* W.

Proof: It follows easily from Theorem 2.1.1.

Lemma 2.1.10 *Let* $p : X \to S$ *be a morphism such that any irreducible component of* X *dominates an irreducible component of* S *and* $i : X_0 \to X$ *be a closed embedding which is an isomorphism over the generic points of* S. *Then* i *is defined by a nilpotent sheaf of ideals. In particular,* p *is a universally equidimensional morphism of dimension* r *if and only if* $p_0 : X_0 \to S$ *is a universally equidimensional morphism of dimension* r.

Lemma 2.1.11 *Let* $X \to S$ *be a scheme of finite type over a Noetherian scheme* S, $Z \subset X$ *be a closed subscheme universally equidimensional of relative dimension* r *over* S *and* $S' \to S$ *be a blow-up of* S. *Let* $\tilde{Z}$ *be the proper transform of* Z *in* $X \times_S S'$. *Then* $\tilde{Z}$ *is a closed subscheme of* $Z \times_S S'$ *defined by a nilpotent sheaf of ideals.*

Proof: Since $Z \times_S S'$ is equidimensional over S' and hence its generic points lie over generic points of S' our statement follows from Lemma 2.1.10.

2.2. Universally equidimensional closed subschemes

Let $p : X \to S$ be a morphism of finite type. We denote by $Z_i(X/S)$ (resp. by $Z_{\leq i}(X/S)$) the set of closed reduced subschemes Z in X such that the morphism $p_{|Z} : Z \to S$ is universally equidimensional of relative dimension i (resp. of dimension $\leq i$).

For any morphism $f : S' \to S$ and any reduced closed subscheme Z of X we denote by $f^{-1}(Z)$ the maximal reduced subscheme of the closed subscheme $Z' = Z \times_S S'$ of $X' = X \times_S S'$. One can easily see that if Z is an element of $Z_{\leq i}(X/S)$ (resp. an element of $Z_i(X/S)$) then $f^{-1}(Z)$ is an element of $Z_{\leq i}(X'/S')$ (resp. an element of $Z_i(X'/S')$).

Lemma 2.2.1 *Let* $p : X \to S$ *be a morphism of schemes and* Z_1, Z_2 *be a pair of elements in* $Z_i(X/S)$ *(resp. in* $Z_{\leq i}(X/S)$*). Then* $Z_1 \cup Z_2$ *is an element of* $Z_i(X/S)$ *(resp. of* $Z_{\leq i}(X/S)$*).*

Proof: Obvious.

Theorem 2.2.2 *Let* $p : S' \to S$ *be a morphism of Noetherian schemes and* U *be an open subscheme in* S *such that* p *is flat over* U. *Then there exists a closed subscheme* Z *in* S *such that* $U \cap Z = \emptyset$ *and the proper transform of* S' *with respect to the blow-up* $S_Z \to S$ *of* S *with center in* Z *is flat over* S_Z.

Proof: See [13, 5.2].

Theorem 2.2.3 *Let* $p : X \to S$ *be a morphism of finite type,* U *be an open subscheme of* X, Z *be an element of* $Z_i(U/S)$ *and* V *be an open subscheme of* S *such that the closure* $\bar{Z}$ *of* Z *in* X *is flat over* V. *Then there exists a blow-up* $f : S' \to S$ *of* S *such that the closure of* $f^{-1}(Z)$ *in* $X' = X \times_S S'$ *belongs to* $Z_i(X'/S')$ *and the morphism* $f^{-1}(V) \to V$ *is an isomorphism.*

Proof: By Theorem 2.2.2 there is a blow-up $f : S' \to S$ such that the proper transform $\tilde{Z}$ of $\bar{Z}$ with respect to f is flat over S'. One can easily see now that the closure $\bar{Z}'$ of $Z' = Z \times_S S'$ in $X' = X \times_S S'$ is a closed subscheme in $\tilde{Z}$ and the corresponding closed embedding is an isomorphism over the generic points of S'. Therefore, $\bar{Z}'$ is universally equidimensional over S' by Lemma 2.1.10.

Definition 2.2.4 *Let* $f : S' \to S$ *be a morphism of Noetherian schemes. We say that* f *is an* abstract blow-up *of* S *if the morphism* f *is proper, any irreducible component of* S' *dominates an irreducible component of* S *and there exists a dense open subscheme* U *in* S *such that the morphism* $(f^{-1}(U)_{red}) \to U_{red}$ *is an isomorphism (here* X_{red} *denote the maximal reduced subscheme of* X).

Note that any abstract blow-up in the sense of Defenition 2.2.4 is surjective.

The following lemma lists some trivial properties of abstract blow-ups.

Lemma 2.2.5   1. *A composition of abstract blow-ups is an abstract blow-up.*

2. *If* S *is a Noetherian scheme and* $S' \to S$, $S'' \to S$ *is a pair of abstract blow-ups of* S *then the morphism* $S' \times_S S'' \to S$ *is an abstract blow-up of* S.

Corollary 2.2.6 *Let* $p : X \to S$ *be a morphism of finite type and* U *be an open subscheme of* X, Z *be an element of* $Z_i(U/S)$. *Then there exists an abstract blow-up* $f : S' \to S$ *of* S *such that the closure of* $f^{-1}(Z)$ *in* $X' = X \times_S S'$ *belongs to* $Z_i(X'/S')$

Proof: It is sufficient to note that we may replace S by the disjoint union of its reduced irreducible components and that in the case of integral S there is an open nonempty subset V in S such that the closure $\bar{Z}$ is flat over V.

Corollary 2.2.7 *Let* $p : X \to S$ *be a universally eqidimensional quasi-projective morphism of dimension* r. *Then there is an abstract blow-up* $S' \to S$, *a universally equidimensional projective morphism* $\bar{X}' \to S'$ *of dimension* r *and an open embedding* $i : X \times_S S' \to \bar{X}'$ *over* S'.

Proof: Obvious.

Theorem 2.2.8 *Let* $p : X \to S$ *be a universally equidimensional morphism and* Z *be an element in* $Z_{\leq i}(X/S)$. *Then there exists an abstract blow-up* $f : S' \to S$ *of* S *and an element* W *in* $Z_i(X \times_S S'/S')$ *such that* $f^{-1}(Z)$ *is contained in* W.

Proof: We start with the following lemma.

Lemma 2.2.9 *Let* S *be a local Noetherian scheme and* $p : X \to S$ *be an affine flat equidimensional morphism of dimension* r. *Let further* Z *be a closed subscheme of the closed fiber of* X *over* S *which does not contain any of the generic points of this fiber.*

Then there exists a closed subscheme W *of* X *which is flat and equidimensional of dimension* $r - 1$ *over* S *such that* Z *lies in* W.

Proof: There exists a finite set $\{x_1, \ldots, x_k\}$ of closed points of X such that the following conditions hold:

1. $\{x_1, \ldots, x_k\} \cap Z = \emptyset$.

2. Any irreducible component of X contains at least one of the points $x_1, \ldots, x_k$.

3. Any irreducible component of the closed fiber X_s of X contains at least one of the points $x_1, \ldots, x_k$.

Since X is affine there is a regular function f on X such that $f = 0$ on Z and $f = 1$ on the set $\{x_1, \ldots, x_k\}$. The third property of this set implies that the divisor $W = (f)$ of f is flat over S (see [11, I.2.5]) and the second one that this divisor is equidimensional of dimension $r - 1$ over generic points of S. By Proposition 2.1.8 we conclude that W is equidimensional of dimension $r - 1$ over S. Lemma is proven.

To prove our theorem it is obviously sufficient to show that if Z is an integral closed subscheme of X of dimension $\leq i$ over S where $i < r$ then there exist an abstract blow-up $S' \to S$ and an element $W \in Z_{r-1}(X'/S')$ such that Z' is contained in W.

Assume first that X is affine over S. By Theorem 2.2.2 it is sufficient to consider the case of a flat morphism $p : X \to S$. Let z be the generic point of Z and $s = p(z)$. Consider the fiber X_s of X over s. Since Z_s is of codimension at least one in X_s by Lemma 2.2.9 there is an open subscheme U of S which contains s and a closed subscheme W_U in $p^{-1}(U)$ which is flat and equidimensional of dimension $r - 1$ over U and which contains $Z \cap p^{-1}(U)$. By Theorem 2.2.3 there is an abstract blow-up $f : S' \to S$ such that $f^{-1}(U) \to U$ is an isomorphism and the closure $\bar{W}'$ of $W' = W_U \times_S S'$ in $X' = X \times_S S'$ belongs to $Z_{r-1}(X'/S')$.

In particular the closure of the pre-image in X' of the generic point of Z is contained in $\bar{W}'$. Let Q be the complement to U in S and Z' be the intesection $Z \cap p^{-1}(Q)$. By our construction we have $dim(Z') < dim(Z)$. Using induction on $dim(Z)$ we may assume that there is an abstract blow-up $S'' \to S$ such that the $Z'' = Z \times_S S''$ is contained in an element W_1 of $Z_{d-1}(X''/S'')$. Considering an abstract blow-up $S''' \to S$ which dominates both S' and S'' (Lemma 2.2.5) we conclude that the statement of our theorem is correct for affine morphisms $p : X \to S$.

Let now $p : X \to S$ be an arbitrary universally equidimensional morphism of dimension r. Then there is a finite open covering $X = \cup U_i$ such that the morphisms $p_i : U_i \to S$ are affine. Since our statement holds for affine morphisms there is an abstract blow-up $S' \to S$ and a family of elements $W_i \in Z_{d-1}(U_i'/S')$ such that $Z' \cap U_i'$ is contained in W_i. By Theorem 2.2.3 we may choose S' such that the closure $\bar{W}_i$ of each W_i in X' is universally equidimensional of dimension $r - 1$. It implies the result we need since one obviously has $Z' \subset \cap \bar{W}_i$. Theorem is proven.

2.3. Cycles on Noetherian schemes

Let X be a Noetherian scheme. We denote by $Cycl(X)$ (resp. by $Cycl^{eff}(X)$) the free abelian group (resp. the free abelian monoid) generated by points of the Zariski topological space of X.

For any element $\mathcal{Z}$ of $Cycl(X)$ we denote by $supp(\mathcal{Z})$ the closure of the set of points on X which appear in $\mathcal{Z}$ with nonzero coefficients. We consider $supp(\mathcal{Z})$ as a reduced closed subscheme of X.

Let Z be a closed subscheme of X and ζ_i, $i = 1, \ldots, k$ be the generic points of the irreducible components of Z. We define an element $Cycl_X(Z)$

of the abelian monoid $Cycl^{eff}(X)$ as the formal linear combination of the form

$$cycl_X(Z) = \sum_{i=1}^{k} m_i \zeta_i$$

where $m_i = length(\mathcal{O}_{Z,\zeta_i})$. Each number m_i is a positive integer which is called the multiplicity of Z in the point ζ_i.

This construction gives us a map from the set of closed subschemes of X to the abelian monoid $Cycl^{eff}(X)$ which can be canonically extended to a homomorphism from the free abelian monoid generated by this set to $Cycl^{eff}(X)$. We denote this homomorphism by $cycl_X$.

Let $p : X \to S$ be a flat morphism of Noetherian schemes and let $\mathcal{Z} = \sum n_i z_i$ be a cycle on S. Denote by Z_i the closure of the point z_i which we consider as a closed integral subscheme in S and set $p^*(\mathcal{Z}) = \sum n_i cycl_X(Z_i \times_S X)$. In this way we get a homomorphism (flat pull-back) $p^* : Cycl(S) \to Cycl(X)$. The following lemma is straightforward (cf. [4, Lemma 1.7.1]).

Lemma 2.3.1   1. *If Z is any closed subscheme of S then*
$$p^*(cycl_S(Z)) = cycl_X(Z \times_S X).$$

2. *$supp(p^*(\mathcal{Z})) = (p^{-1}(supp(\mathcal{Z})))_{red}$. In particular the homomorphism $p^* : Cycl(S) \to Cycl(X)$ is injective provided that p is surjective.*

Let $X \to Spec(k)$ be a scheme of finite type over a field k and let L/k be any field extension. The corresponding morphism $p : X_L \to X$ is flat and hence defines a homomorphism $p^* : Cycl(X) \to Cycl(X_L)$. The image of a cycle $\mathcal{Z} \in Cycl(X)$ under this homomorphism will be usually denoted by $\mathcal{Z} \otimes_k L$ of $\mathcal{Z}_L$.

Lemma 2.3.2 *Let $X \to Spec(k)$ be a scheme of finite type over a field k and let k'/k be a finite normal field extension with the Galois group G. If $\mathcal{Z}' \in Cycl(X_{k'})^G$ then there is a unique cycle $\mathcal{Z} \in Cycl(X)$ such that $[k':k]_{insep}\mathcal{Z}' = \mathcal{Z}_{k'}$.*

Proof: The uniqueness of $\mathcal{Z}$ follows immediately from Lemma 2.3.1(2). To prove the existence note that the group $Cycl(X_{k'})^G$ is generated by cycles of the form $\mathcal{Z}' = \sum_{\tau \in G/H} \tau(z')$ where z' is a point of $X_{k'}$ and $H = Stab_G(z')$. Let z be the image of z' in X and let Z be the closure of z which we consider as a closed integral subscheme of X. The points $\tau(z')$ are precisely the generic points of the scheme $Z' = Z \times_{Spec(k)} Spec(k')$ and the multiplicities with which they appear in the cycle $\mathcal{Z}_{k'}$ are all equal to the length of the local Artinian ring $\mathcal{O}_{Z',z'}$. The elementary Galois theory shows that this length is a factor of $[k' : k]_{insep}$. Thus the cycle $\mathcal{Z} = [k' : k]_{insep}z/length(\mathcal{O}_{Z',z'})$ has the required property.

Corollary 2.3.3 *In the assumptions and notations of the previous lemma denote by p the exponential characteristic of the field k. Then the homomorphism*

$$Cycl(X)[1/p] \to (Cycl(X_{k'})[1/p])^G$$

is an isomorphism.

Let $X \to Spec(k)$ be a scheme of finite type over a field k. Then we have a direct sum decomposition $Cycl(X) = \coprod Cycl(X, r)$ where $Cycl(X, r)$ is a subgroup of $Cycl(X)$ generated by points of dimension r. Furthermore one sees easily that for a field extension k'/k the homomorphism

$$Cycl(X) \to Cycl(X_{k'})$$

preserves this decomposition.

Let S be a Noetherian scheme and $p : X \to S$ be a proper morphism of finite type. For any cycle $\mathcal{Z} = \sum n_i z_i \in Cycl(X)$ set

$$p_*(\mathcal{Z}) = \sum n_i m_i p(z_i)$$

where m_i is the degree of the field extension $k_{z_i}/k_{p(z_i)}$ if this extension is finite and zero otherwise. The proof of the following statement is similar to that of Proposition 1.7 of [4] and we omit it.

Proposition 2.3.4 *Consider a pull-back square of morphisms of finite type of Noetherian schemes*

$$\begin{array}{ccc} \tilde{X} & \xrightarrow{f_X} & X \\ \tilde{p}\downarrow & & \downarrow p \\ \tilde{S} & \xrightarrow{f} & S \end{array}$$

in which f is flat and p is proper. Then for any cycle $\mathcal{Z} \in Cycl(X)$ we have

$$f^*(p_*(\mathcal{Z})) = \tilde{p}_*(f_X^*(\mathcal{Z})).$$

The following lemma is straighforward.

Lemma 2.3.5 *Let $f : X \to Y$ be a finite flat morphism of connected Noetherian schemes. Denote by $deg(f)$ the degree of f. Then $f_* f^* = deg(f)Id_{Cycl(Y)}$.*

3. Relative cycles

3.1. Relative cycles

Definition 3.1.1 *Let S be a Noetherian scheme, k be a field and $x : Spec(k) \to S$ be a k-point of S. A fat point over x is a triple (x_0, x_1, R), where R is a discrete valuation ring and $x_0 : Spec(k) \to Spec(R)$, $x_1 : Spec(R) \to S$ are morphisms such that*

1. $x = x_1 \circ x_0$

2. *The image of x_0 is the closed point of $Spec(R)$.*

3. x_1 *takes the generic point of $Spec(R)$ to a generic point of S.*

Usually we will abbreviate the notation (x_0, x_1, R) to (x_0, x_1).

Lemma 3.1.2 *Let S be a Noetherian scheme, $X \to S$ be a scheme over S and Z be a closed subscheme in X. Let further R be a discrete valuation ring and $f : Spec(R) \to S$ be a morphism. Then there exists a unique closed subscheme $\phi_f(Z)$ in $Z \times_S Spec(R)$ such that:*

1. *The closed embedding $\phi_f(Z) \to Z \times_S Spec(R)$ is an isomorphism over the generic point of $Spec(R)$.*

2. $\psi_f(Z)$ *is flat over $Spec(R)$.*

Proof: See [8, 2.8.5].

Let S be a Noetheiran scheme, $X \to S$ be a scheme of finite type over S and Z be a closed subscheme of X. For any fat point (x_0, x_1) over a k-point x of S we denote by $(x_0, x_1)^*(Z/S)$ the cycle on $X \times_{Spec(k)} S$ associated with the closed subscheme $\phi_{x_1}(Z) \times_{Spec(R)} Spec(k)$.

If $\mathcal{Z} = \sum m_i z_i$ is any cycle on X we denote by $(x_0, x_1)^*(\mathcal{Z})$ the cycle $\sum m_i (x_0, x_1)^*(Z_i)$ where Z_i is the closure of the point z_i (considered as a reduced closed subscheme of X).

Definition 3.1.3 *Let S be a Noetherian scheme and $X \to S$ be a scheme of finite type over S. A relative cycle on X over S is a cycle $\mathcal{Z} = \sum m_i z_i$ on X satisfying the following requirements:*

1. *The points z_i lie over generic points of S.*

2. *For any field k, k-point x of S and a pair of fat points (x_0, x_1), (y_0, y_1) of S over x one has:*

$$(x_0, x_1)^*(\mathcal{Z}) = (y_0, y_1)^*(\mathcal{Z}).$$

We say that $\mathcal{Z} = \sum n_i z_i$ is a relative cycle of dimension r if each point z_i has dimension r in its fiber over S. We denote the corresponding abelian groups by $Cycl(X/S, r)$.

We say that $\mathcal{Z}$ is an equidimensional relative cycle of dimension r if $supp(\mathcal{Z})$ is equidimensional of dimesnion r over S. We denote the corresponding abelian groups by $Cycl_{equi}(X/S, r)$.

We say that $\mathcal{Z}$ is a proper relative cycle if $supp(\mathcal{Z})$ is proper over S. We denote the corresponding abelian groups by $PropCycl(X/S, r)$ and $PropCycl_{equi}(X/S, r)$.

We will also use the notations $Cycl^{eff}(X/S, r)$, $PropCycl^{eff}(X/S, r)$ etc. for the corresponding abelian monoids of effective relative cycles.

The following lemma gives us means to construct fat points.

Lemma 3.1.4 *Let S be a Noetherian scheme, η be a generic point of S and s be a point in the closure of η. Let further L be an extension of finite type of the field of functions on S in η. Then there is a discrete valuation ring R and a morphism $f : Spec(R) \to S$ such that the following conditions hold:*

1. *f maps the generic point of $Spec(R)$ to η and the field of functions of R is isomorphic to L over k_η,*

2. *f maps the closed point of $Spec(R)$ to s.*

Proof: See [6, 7.1.7].

Let S be a Noetherian scheme, $X \to S$ be a scheme of finite type over S and $\mathcal{Z} = \sum n_i z_i$ be a cycle on X such that the points z_i lie over generic points of S and are of dimension r in the corresponding fibers. Let Z_i denote the closure of z_i considered as a closed integral subscheme of X. It is clear from the above definition that $\mathcal{Z} \in Cycl(X/S, r)$ if and only if $\mathcal{Z} \in Cycl(X_{red}/S_{red}, r)$. Furthemore the schemes Z_i are flat over generic points of S_{red} and according to Theorem 2.2.2 one can find a blow-up $S' \to S_{red}$ such that the proper transforms $\tilde{Z}_i$ of Z_i's are flat over S'. Now we can formulate the following usefull criterion.

Proposition 3.1.5 *Under the above assumptions the following conditions are equivalent:*

1. *$\mathcal{Z} \in Cycl(X/S, r)$.*

2. *If $x : Spec(k) \to S$ is any geometric point of S and $x_1', x_2' : Spec(k) \to S'$ is a pair of its liftings to S' then the cycles $\mathcal{W}_1$,*

$\mathcal{W}_2$ on $X_x = X \times_S Spec(k)$ given by the formulae

$$\mathcal{W}_1 = \sum_{i=1}^{k} n_i cycl_{X_s}(\tilde{Z}_i \times_{x'_1} Spec(k))$$

$$\mathcal{W}_2 = \sum_{i=1}^{k} n_i cycl_{X_s}(\tilde{Z}_i \times_{x'_2} Spec(k))$$

coincide.

Proof: **(1=>2)** the geometric points x, x'_1, x'_2 give us set-theoretical points $s \in S$, $s'_1, s'_2 \in S'$ such that s'_1, s'_2 lie over s. We may assume that s (and hence also s'_1, s'_2) is not generic. Using Lemma 3.1.4 we construct discrete valuation rings R'_i and morphisms $Spec(R'_i) \to S'$ which map the closed point of $Spec(R'_i)$ to s'_i and the generic point of $Spec(R'_i)$ to a generic point of S'.

Denote the residue fields of R'_i by k'_i. One checks easily that the scheme $(Spec(k'_1) \times_S Spec(k'_2)) \times_{S' \times_S S'} Spec(k)$ is not empty. Choosing any geometric L-point of this scheme for a field L we get a commutative diagram

$$
\begin{array}{ccccc}
 & Spec(k'_1) & \to & Spec(R'_1) & \to & S' \\
 & \nearrow & & & \scriptstyle{x'_1}\nearrow & & \searrow \\
Spec(L) & & \to & & Spec(k) & \xrightarrow{x} & S \\
 & \searrow & & & \scriptstyle{x'_2}\searrow & & \nearrow \\
 & Spec(k'_2) & \to & Spec(R'_2) & \to & S'
\end{array}
$$

Thus we get a geometric point $Spec(L) \to S$ and two fat points $Spec(L) \to Spec(R'_i) \to S$ over it. The inverse images of the cycle $\mathcal{Z}$ with respect to these fat points are equal to $\sum n_i cycl[(\tilde{Z}_i \times_{x'_1} Spec(k)) \times_{Spec(k)} Spec(L)]$ and $\sum n_i cycl[(\tilde{Z}_i \times_{x'_2} Spec(k)) \times_{Spec(k)} Spec(L)]$ respectively. These cycles coincide according to the condition $\mathcal{Z} \in Cycl(X/S, r)$. Lemma 2.3.1 shows now that $\mathcal{W}_1 = \mathcal{W}_2$.

(2=>1) Let $x : Spec(k) \to S$ be a geometric point of S and let (x_0, x_1), (y_0, y_1) be a pair of fat points over x. According to the valuative criterion of properness (see [6, 7.3]) these fat points have canonical liftings to fat points $(x'_0, x'_1), (y'_0, y'_1)$ of S'. This gives us two geometric points

$$x' = x'_1 \circ x'_0 : Spec(k) \to S'$$

$$y' = y'_1 \circ y'_0 : Spec(k) \to S'$$

of S' over x. Our statement follows now from obvious equalities:

$$(x_0, x_1)^*(\mathcal{Z}) = (x_0, x_1')^*(\sum n_i \tilde{Z}_i) = \sum n_i cycl(\tilde{Z}_i \times_{x'} Spec(k))$$

$$(y_0, y_1)^*(\mathcal{Z}) = (y_0, y_1')^*(\sum n_i \tilde{Z}_i) = \sum n_i cycl(\tilde{Z}_i \times_{y'} Spec(k)).$$

Corollary 3.1.6 *Let k be a field and $X \to Spec(k)$ be a scheme of finite type over k. Then the group $Cycl(X/Spec(k), r)$ is the free abelian group generated by points of dimension r on X, i.e., one has*

$$Cycl(X/Spec(k), r) = Cycl_{equi}(X/Spec(k), r) = Cycl(X, r).$$

Proposition 3.1.7 *Let S be a Noetherian scheme, $X \to S$ be a scheme of finite type over S and $\mathcal{Z} = \sum_{i=1}^{k} n_i z_i$ be an effective cycle on X which belongs to $Cycl(X/S, r)$ for some $r \geq 0$. Denote by Z_i the closure of the point z_i in X which we consider as an integral closed subscheme in X. Then Z_i is equidimensional of dimension r over S.*

Proof: According to the Chevalley theorem 2.1.1 all components of all fibers of the projection $Z_i \to S$ are of dimension $\geq r$. Assume that there exists a point s of S such that the fiber $(Z_0)_s$ of Z_0 over s has a component of dimension $> r$. Let η be the generic point of this component.

By Theorem 2.2.2 there is a blow-up $f : S' \to S_{red}$ of S_{red} such that the proper transforms $\tilde{Z}_i$ of the subschemes Z_i with respect to f are flat (and hence equidimensional of dimension r over S'). The morphism $\tilde{Z}_0 \to Z_0$ is proper and dominant and hence surjective. Let τ be any point of $\tilde{Z}_0$ over η and let s_1' be its image in S'. On the other hand let s_2' be any closed point of the fiber $(S')_s$. Choosing an algebraically closed field k which contains a composite of the fields $k_{s_1'}$ and $k_{s_2'}$ over k_s we get two geometric points $x_1', x_2' : Spec(k) \to S'$ over the same geometric point $x : Spec(k) \to S$. Consider the following cycles on $X_k = X \times_S Spec(k)$:

$$\mathcal{W}_1 = \sum n_i cycl_{X_k}(\tilde{Z}_i \times_{s_1'} Spec(k))$$

$$\mathcal{W}_2 = \sum n_i cycl_{X_k}(\tilde{Z}_i \times_{s_2'} Spec(k)).$$

In view of Proposition 3.1.5 it is sufficient to show that these cycles are different. We will do so by showing that the images of $supp(\mathcal{W}_1)$ and $supp(\mathcal{W}_2)$ in X_s are different. The image of $supp(\mathcal{W}_2)$ coincides with the image of $\cup(\tilde{Z}_i)_{s_2'} \subset X_{k_{s_2'}}$ in X_s and hence is of dimension $\leq r$ (since the morphism $X_{s_2'} \to X_s$ is finite). On the other hand the image of $supp(\mathcal{W}_1)$ contains η and hence is of dimension $> r$.

Corollary 3.1.8 *Let S be a Noetherian scheme and $X \to S$ be a scheme of finite type over S. Then one has:*

$$Cycl^{eff}_{equi}(X/S, r) = Cycl^{eff}(X/S, r)$$

$$PropCycl^{eff}_{equi}(X/S, r) = PropCycl^{eff}(X/S, r).$$

Example 3.1.9 Proposition 3.1.7 fails for cycles which are not effective. Consider the scheme $X = \mathbf{P}^1_k \times \mathbf{A}^2$ over $S = \mathbf{A}^2_k$ where k is a field. Consider the following two rational functions on S:

$$f(x, y) = y/x$$

$$g(x, y) = y/(x + y^2).$$

Let $\Gamma_f, \Gamma_g \in X$ be the graphs of these functions and let $\mathcal{Z}$ be the cycle on X of the form $\Gamma_f - \Gamma_g$. Obviously $supp(\mathcal{Z})$ is not equidimensional over S. We claim that $\mathcal{Z}$ nevertheless belongs to $Cycl(X/S, 0)$.

Let $S' \to S$ be the blow-up of the point $(0, 0)$ in $\mathbf{A}^2$. Denote by f', g' the rational functions on S' which correspond to f and g. One can easily see that f', g' are in fact regular on S' and moreover if $S'_0 \subset S'$ denotes the exceptional divisor of S' we have $f'_{S'_0} = g'_{S'_0}$. Since the proper transforms of the closed subschemes Γ_f, Γ_g are the graphs of f' and g' in $S' \times_S X = S' \times \mathbf{P}^1_k$ Proposition 3.1.5 shows that $\mathcal{Z}$ indeed belongs to $Cycl(X/S, 0)$.

3.2. Cycles associated with flat subschemes

Let $p : X \to S$ be a morphism of finite type of Noetherian schemes. We denote by $Hilb(X/S, r)$ (resp. by $PropHilb(X/S, r)$) the set of closed subschemes Z of $X \times_S S$ which are flat (resp. flat and proper) and equidimensional of dimension r over S.

Let $\mathbf{N}(Hilb(X/S, r))$, $\mathbf{N}(PropHilb(X/S, r))$ (resp. $\mathbf{Z}(Hilb(X/S, r))$, $\mathbf{Z}(PropHilb(X/S, r)))$ be the corresponding freely generated abelian monoids (resp. abelian groups).

The assigment $S'/S \to \mathbf{N}(Hilb(X \times_S S'/S', r))$ etc. defines a presheaf of abelian monoids (groups) on the category of Noetherian schemes over S. If $\mathcal{Z} = \sum n_i Z_i$ is an element of $\mathbf{Z}(\text{Hilb}(X/S, r))$ and S' is a Noetherian scheme over S we denote by $\mathcal{Z} \times_S S'$ the corresponding element $\sum n_i(Z_i \times_S S')$ of $\mathbf{Z}(\text{Hilb}(X \times_S S'/S', r))$.

Lemma 3.2.1 *Let $p : X \to S$ be a finite flat morphism of Noetherian schemes of constant degree and S'/S be any Noetherian scheme over S.*

Let further τ' be a generic point of S' and $\eta'_1, \ldots, \eta'_k$ be all points of $X' = X \times_S S'$ lying over τ'. Then one has

$$\sum_{j=1}^{k} length(\mathcal{O}_{X', \eta'_j})[k_{\eta'_j} : k_{\tau'}] = deg(p)length(\mathcal{O}_{S', \tau'})$$

Proof: The module $\coprod \mathcal{O}_{X', \eta'_j} = [(p')_*(\mathcal{O}_{X'})]_{\tau'}$ is a free $\mathcal{O}_{S', \tau'}$-module of rank $deg(p)$. Computing its length over $\mathcal{O}_{S', \tau'}$ in two different ways we get the desired equality.

Proposition 3.2.2 *Let $X \to S$ be a scheme of finite type over a Noetherian scheme S and $S' \to S$ be any Noetherian scheme over S. Let further $\mathcal{Z} = \sum n_i Z_i$ be an element of $\mathbf{Z}(\text{Hilb}(X/S, \mathbf{r}))$. If $cycl_X(\mathcal{Z}) = 0$ then $cycl_{X \times_S S'}(\mathcal{Z} \times_S S') = 0$.*

Proof: Replacing X by $\cup Z_i$ we may assume that X is equidimensional of relative dimension r over S. Generic points of $Z_i \times_S S'$ lie over generic points of S' and are generic in their fibers. Let $\eta' \in X' = X \times_S S'$ be any of these generic points. Computing the multiplicity of η' in $cycl(\mathcal{Z} \times_S S')$ we may replace X by any open neighborhood of the point $\eta = pr_1(\eta') \in X$.

Lemma 3.2.3 *Let $p : X \to S$ be a flat equidimensional morphism of dimension r and x be a generic point of a fiber of p. Then for any decomposition of p of the form $X \overset{p_0}{\to} \mathbf{A}^r_S \to S$ such that p_0 is an equidimensional quasi-finite morphism there exists an open neighborhood U of x in X such that $(p_0)_{|U}$ is a flat quasi-finite morphism.*

Proof: It follows immediately from the fact that p is flat and [5, Ex. IV, Cor. 5.9,p.99].

Using Lemma 3.2.3 and Proposition 2.1.3 and replacing S by $\mathbf{A}^r_S$ we see that it is sufficient to treat the case $r = 0$. Furthemore replacing S' by $Spec(\mathcal{O}_{S', \tau'})$ and S by $Spec(\mathcal{O}_{s, \tau})$ (where $\tau = p(\eta)$ and $\tau' = p'(\eta')$) we may assume that S and S' are local schemes, S' is Artinian and $f : S' \to S$ takes the closed point of S' to the closed point of S.

Let S^{sh} (resp. $(S')^{sh}$) denote the strict henselization of the local scheme S (resp. S') in the closed point. Lemma 2.3.1 shows that $cycl(\mathcal{Z} \times_S S^{sh}) = 0$ and moreover that it is sufficient to check that

$$cycl((\mathcal{Z} \times_S S') \times_{S'} (S')^{sh}) = cycl(\mathcal{Z} \times_S (S')^{sh}) = 0.$$

Since there exists a morphism $(S')^{sh} \to S^{sh}$ over S we see that we may replace S and S' by S^{sh} and $(S')^{sh}$ and assume the schemes S and S' to be strictly henselian.

Since η lies over the closed point of S and S is henselian we conclude that $Spec(\mathcal{O}_{X,\eta})$ is an open neighborhood of η in X and is finite over S (see [11, I.4.2(c)]). Thus replacing X by $Spec(\mathcal{O}_{X,\eta})$ we may additionally assume that X is local and finite over S. In these assumptions η is the only point over τ (and hence η' is the only point over τ') and the schemes Z_i are finite and flat over S of constant degree $deg(Z_i/S)$. Lemma 3.2.1 shows now that the multiplicity of η' in $cycl_{X'}(\mathcal{Z} \times_S S')$ is equal to

$$\sum_i n_i length(\mathcal{O}_{Z'_i,\eta'}) = (\sum_i n_i deg(Z_i/S)) length(\mathcal{O}_{S',\tau'})/[k_{\eta'} : k_{\tau'}].$$

We only have to show that $\sum n_i deg(Z_i/S) = 0$. To do so let τ^0 be a generic point of S and let $\eta_1^0, \ldots, \eta_k^0$ be all points of X over τ^0. The multiplicity of η_j^0 in $cycl(\mathcal{Z})$ is equal to $\sum n_i length(\mathcal{O}_{Z_i,\eta_j^0})$. Using once again Lemma 3.2.1 we get

$$0 = \sum_j [k_{\eta_j^0} : k_{\tau^0}] \sum_i n_i length(\mathcal{O}_{Z_i,\eta_j^0}) = (\sum_i n_i deg(Z_i/S)) length(\mathcal{O}_{S,\tau^0}).$$

Proposition is proven.

Corollary 3.2.4 *Let $X \to S$ be a scheme of finite type over a Noetherian scheme S, Z be an element of $Hilb(X/S,r)$ and (x_0, x_1, R) be a fat point over a k-point $x : Spec(k) \to S$ of S. Then*

$$(x_0, x_1)^*(cycl_X(Z)) = cycl_{X \times_S Spec(k)}(Z \times_S Spec(k)).$$

Proof: Let Z_i be the irreducible components of Z, z_i be their generic points and n_i be their multiplicities such that $cycl_X(Z) = \sum n_i z_i$. One checks easily that

$$cycl_{X \times_S Spec(R)}(Z \times_S Spec(R)) = \sum n_i cycl_{X \times_S Spec(R)}(\phi_{x_1}(Z_i)).$$

Proposition 3.2.2 shows now that

$$(x_0, x_1)^*(cycl_X(Z)) = \sum n_i cycl_{X \times_S Spec(k)}(\phi_{x_1}(Z_i) \times_{Spec(R)} Spec(k)).$$

Corollary 3.2.5 *Let $X \to S$ be a scheme of finite type over a Noetherian scheme S. Then the image of the cycle map $\mathbf{Z}(Hilb(X/S,r)) \to Cycl(X)$ lies in $Cycl_{equi}(X/S,r)$.*

Corollary 3.2.6 *Let R be a discrete valuation ring and $X \to Spec(R)$ be a scheme of finite type over $Spec(R)$. Then a cycle $\mathcal{Z} = \sum n_i Z_i$ belongs to $Cycl(X/Spec(R), r)$ if and only if the points z_i belong to the generic fiber of X over $Spec(R)$ and are of dimension r in this fiber.*

3.3. Chow presheaves

Theorem 3.3.1 *Let $X \to S$ be a scheme of finite type over a Noetherian scheme S, $\mathcal{Z}$ be an element of the group $Cycl(X/S, r)$ and $f : T \to S$ be a Noetherian scheme over S.*

Then there is a unique element $\mathcal{Z}_T$ in $Cycl(X \times_S T/T, r) \otimes_{\mathbf{Z}} \mathbf{Q}$ such that for any commutative diagram of the form

$$
\begin{array}{ccc}
 & Spec(A) & \overset{y_1}{\to} \; T \\
{\scriptstyle y_0} \nearrow & & \\
Spec(k) & & \Big\downarrow {\scriptstyle f} \\
{\scriptstyle x_0} \searrow & & \\
 & Spec(R) & \overset{x_1}{\to} \; S
\end{array}
$$

where (x_0, x_1) and (y_0, y_1) are fat k-points of S and T respectively one has:

$$(y_0, y_1)^*(\mathcal{Z}_T) = (x_0, x_1)^*(\mathcal{Z}).$$

Proof: The uniqueness of $\mathcal{Z}_T$ follows easily form 3.1.4 and 2.3.1(2).

In the proof of existence we start with the special case $T = Spec(k_s)$ where s is a set-theoretic point of S. In this case we have the following slightly more precise statement.

Lemma 3.3.2 *Denote by p the exponential characteristic of the field k_s. Then there exists a unique cycle $\mathcal{Z}_s$ in $Cycl(X_s, r)[1/p]$ such that for any field extension k/k_s and any fat point (x_0, x_1) over the k-point $Spec(k) \to Spec(k_s) \to S$ one has $(x_0, x_1)^*(\mathcal{Z}) = \mathcal{Z}_s \otimes_{k_s} k$.*

Proof: Denote by Z_i the closure of z_i which we consider as an integral closed subscheme in X and choose a blow-up $S' \to S_{red}$ such that the proper transforms $\tilde{Z}_i$ of Z_i are flat over S'. If k/k_s is a field extension such that the k-point $Spec(k) \to Spec(k_s) \to S$ admits a lifting to S' we get a cycle $\mathcal{Z}_k = \sum n_i cycl(\tilde{Z}_i \times_{S'} Spec(k)) \in Cycl(X \times_S Spec(k), r)$ which is independent of the choice of the lifting according to Proposition 3.1.5. Moreover if L/k is a field extension then the cycle $\mathcal{Z}_L$ is also defined and we have

$$\mathcal{Z}_L = \mathcal{Z}_k \otimes_k L$$

The morphism $S' \to S$ being a surjective morphism of finite type we can find a finite normal extension k_0/k_s such that the point $Spec(k_0) \to Spec(k_s) \to S$ admits a lifting to S'. The formula above shows that the cycle $\mathcal{Z}_{k_0}$ is $Gal(k_0/k_s)$-invariant and hence descends to a cycle $\mathcal{Z}_s \in Cycl(X_s, r)[1/p]$ by Lemma 2.3.2.

Let now k be any extension of k_s such that the point $Spec(k) \to Spec(k_s) \to S$ admits a lifting to S' and let L be a composite of k and k_0 over k_s. Then

$$\mathcal{Z}_k \otimes_k L = \mathcal{Z}_L = \mathcal{Z}_{k_0} \otimes_{k_0} L = \mathcal{Z}_s \otimes_{k_s} L = (\mathcal{Z}_s \otimes_{k_s} k) \otimes_k L$$

and hence $\mathcal{Z}_k = \mathcal{Z}_s \otimes_{k_s} k$.

Finally let k/k_s be a field extension and (x_0, x_1, R) be a fat point over a k-point $Spec(k) \to Spec(k_s) \to S$. The morphism $x_1 : Spec(R) \to S$ has a canonical lifting to S' (according to the valuative criterion of properness). This gives us a lifting to S' of our k-point $Spec(k) \to S$ and it follows immediately from the construction of $\mathcal{Z}_s$ that one has $(x_0, x_1)^*(\mathcal{Z}) = \mathcal{Z}_k = \mathcal{Z}_s \otimes_{k_s} k$. Lemma is proven.

In the course of the proof of Lemma 3.3.2 we have established that the cycle $\mathcal{Z}_s$ has the following somewhat more general property.

Lemma 3.3.3 *Let $S' \to S$ be a blow up such that the proper transforms $\tilde{Z}_i$ of Z_i are flat over S' and let k/k_s be a field extension such that the k-point $Spec(k) \to Spec(k_s) \to S$ admits a lifting to S'. Then $\mathcal{Z}_0 \otimes_{k_s} k = \sum n_i cycl(\tilde{Z}_i \times_{S'} Spec(k))$.*

Let $\tau_1, \ldots, \tau_n$ be the generic points of T and $\sigma_1, \ldots, \sigma_n$ be their images in S. Consider the cycles

$$\mathcal{Z}_{\sigma_j} \otimes_{k_{\sigma_j}} k_{\tau_j} = \sum_l n_{jl} z_{jl} \in Cycl(X \times_S Spec(k_{\tau_j}), r) \otimes_{\mathbf{Z}} \mathbf{Q}.$$

Here n_{jl} are rational numbers and z_{jl} are points of $X \times_S T$ lying over τ_j and having dimension r in their fibers. Set $\mathcal{Z}_T = \sum_{j,l} n_{jl} z_{jl}$. We are going to show that $\mathcal{Z}_T$ belongs to $Cycl(X \times_S T/T, r) \otimes_{\mathbf{Z}} \mathbf{Q}$ and has the desired property. To do so we need the following lemma.

Lemma 3.3.4 *Let $f : S' \to S$ be a proper surjective morphism of finite type of Noetherian schemes, A be a discrete valuation ring and $Spec(A) \to S$ be a morphism of schemes. Then there exists a commutative diagram of the form:*

$$
\begin{array}{ccc}
Spec(A') & \longrightarrow & S' \\
\downarrow{\scriptstyle g} & & \downarrow{\scriptstyle f} \\
Spec(A) & \longrightarrow & S
\end{array}
$$

In which A' is a discrete valuation ring and the morphism g is surjective.

Proof: Let K be the quotient field of A. Since f is a surjective morphism of finite type there exists a finite extension K' of K such that the K'-point $Spec(K') \to Spec(K) \to S$ admits a lifting to S'. Let ν be the discrete valuation of K which corresponds to A. It can be extended to a discrete valuation ν' of K' and we take A' to be the corresponding discrete valuation ring. The valuative criterion of properness ([6, 7.3]) shows that the morphism $Spec(A') \to Spec(A) \to S$ admits a lifting to S'.

Consider a commutative diagram of the form

$$
\begin{array}{ccc}
 & Spec(A) & \overset{y_1}{\to} \ T \\
{\scriptstyle y_0}\nearrow & & \\
Spec(k) & & \Big\downarrow f \\
{\scriptstyle x_0}\searrow & & \\
 & Spec(R) & \overset{x_1}{\to} \ S
\end{array}
$$

in which (x_0, x_1) (resp. (y_0, y_1)) is a fat k-point of S (resp. of T). Let as before $S' \to S_{red}$ denote a blow-up of S_{red} such that the proper transforms $\tilde{Z}_i$ of Z_i are flat over S'. Lemma 3.3.4 shows that there is a discrete valuation ring A' and a surjective morphism $Spec(A') \to Spec(A)$ such that the morphism $Spec(A') \to S$ admits a lifting to S'. Denote the residue field of A (resp. of A') by k_A (resp. by $k_{A'}$) and let k' be a composite of k and $k_{A'}$ over k_A so that we have the following commutative diagram:

$$
\begin{array}{ccccc}
Spec(k') & \to & Spec(A') & \to & S' \\
\downarrow & & \downarrow & & \downarrow \\
Spec(k) & \overset{y_0}{\to} & Spec(A) & \overset{y_1}{\to} T \to & S
\end{array}
$$

Assume that y_1 maps the generic point of $Spec(A)$ to τ_1 and consider the following two elements in $\mathbf{Q}(\mathrm{Hilb}(X \times_S Spec(A')/Spec(A'), r))$:

$$
\mathcal{W} = \sum_i n_i(\tilde{Z}_i \times_{S'} Spec(A'))
$$

$$
\mathcal{W}_1 = \sum_{j,l} n_{jl}(\phi_{y_1}(Z_{jl}) \times_{Spec(A)} Spec(A'))
$$

$$
= \sum_l n_{1l}(\phi_{y_1}(Z_{1l}) \times_{Spec(A)} Spec(A'))
$$

where Z_{jl} denotes the closure of the points z_{jl} considered as a closed integral subscheme of $X \times_S T$.

Lemma 3.3.5

$$
cycl(\mathcal{W}) = cycl(\mathcal{W}_1)
$$

Proof: Let K (resp. K') denote the quotient field of A (resp. of A'). Since the map

$$Cycl(X \times_S Spec(A')/Spec(A'), r) \otimes \mathbf{Q}$$
$$\to Cycl(X \times_S Spec(K')/Spec(K'), r) \otimes \mathbf{Q}$$

is clearly injective we may replace A' by K' everywhere. Furthermore

$$cycl(\mathcal{W} \times_{Spec(A')} Spec(K)) = \sum n_i cycl(\tilde{Z}_i \times_{S'} Spec(K'))$$

and according to Lemma 3.3.3 this cycle is equal to $\mathcal{Z}_{\sigma_1} \otimes_{k_{\sigma_1}} K'$.

On the other hand we have:

$$cycl(\mathcal{W}_1 \times_{Spec(A')} Spec(K')) =$$

$$= \sum_l n_{1l} cycl([\phi_{y_1}(Z_{1l}) \times_{Spec(A)} Spec(K)] \times_{Spec(K)} Spec(K')) =$$

$$= \sum_l n_{1l} cycl((Z_{1l} \times_T Spec(K)) \times_{Spec(K)} Spoo(K')) =$$

$$= [\sum_l n_{1l} cycl(Z_{1l} \times_T Spec(k_{\tau_1}))] \otimes_{k_{\tau_1}} K' =$$

$$= (\sum_l n_{1l} z_{1l}) \otimes_{k_{\tau_1}} K' = (\mathcal{Z}_{\sigma_1} \otimes_{k_{\sigma_1}} k_{\tau_1}) \otimes_{k_{\tau_1}} K'$$

$$= \mathcal{Z}_{\sigma_1} \otimes_{k_{\sigma_1}} K'.$$

Lemma is proven.

Proposition 3.2.2 implies now that

$$cycl(\mathcal{W} \times_{Spec(A')} Spec(k')) = cycl(\mathcal{W}_1 \times_{Spec(A')} Spec(k'))$$

i.e.

$$\sum n_i cycl(\tilde{Z}_i \times_{S'} Spec(k')) = \sum_l n_{1l} cycl(\phi_{y_1}(Z_{1l}) \times_{Spec(A')} Spec(k')) =$$

$$= (y_0, y_1)^*(\mathcal{Z}_T) \otimes_k k'.$$

On the other hand the cycle $(x_0, x_1)^*(\mathcal{Z}) \otimes_k k'$ is equal to $\sum n_i cycl(\tilde{Z}_i \times_{S'} Spec(k'))$ where this time the morphism $Spec(k') \to S'$ is a lifting of the same point $Spec(k') \to Spec(k) \to Spec(R) \to S$ obtained using the canonical lifting of the morphism $Spec(R) \to S$. Proposition 3.1.5 shows that $(x_0, x_1)^*(\mathcal{Z}) \otimes_k k' = (y_0, y_1)^*(\mathcal{Z}_T) \otimes_k k'$ and hence $(x_0, x_1)^*(\mathcal{Z}) = (y_0, y_1)^*(\mathcal{Z}_T)$. Theorem 3.3.1 is proven.

Remark: In general cycles of the form $cycl(f)(\mathcal{Z})$ do not have integral coefficients. See example 3.5.10(1) below.

Lemmas 3.3.6 and 3.3.8 below describe the behavior of supports of cycles with respect to the base change homomorphisms.

Lemma 3.3.6 *In the notations and assumptions of Theorem 3.3.1 we have*

$$supp(\mathcal{Z}_T) \subset (supp(\mathcal{Z}))_T = supp(\mathcal{Z}) \times_S T.$$

Proof: Since $supp(\mathcal{Z}_T) = \cup supp(\mathcal{Z}_{k_{\tau_j}})$ where τ_j are the generic points of T it is sufficient to consider the case $T = Spec(k)$ where k is a field. According to Lemma 3.1.4 there exists an extension k'/k and a fat point (x_0, x_1, R) over the k'-point $Spec(k') \to Spec(k) \to S$. The defining property of the cycle $\mathcal{Z}_k$ shows that

$$supp(\mathcal{Z}_k) \times_{Spec(k)} Spec(k') = supp(\mathcal{Z}_{k'}) = supp((x_0, x_1)^*(\mathcal{Z})) \subset$$

$$\subset \cup_i \phi_{x_1}(Z_i) \times_{Spec(R)} Spec(k') \subset \cup_i (Z_i \times_S Spec(R)) \times_{Spec(R)} Spec(k') =$$

$$= supp(\mathcal{Z}) \times_S Spec(k') = (supp(\mathcal{Z}) \times_S Spec(k)) \times_{Spec(k)} Spec(k').$$

Since the morphism $X_{k'} \to X_k$ is surjective the above inclusion implies the desired one $supp(\mathcal{Z}_k) \subset supp(\mathcal{Z}) \times_S Spec(k)$.

Lemma 3.3.7 *Consider a pull-back square of morphisms of finite type of Noetherian schemes of the form*

$$\begin{array}{ccc} X' & \to & X \\ \downarrow & & \downarrow \\ S' & \xrightarrow{f} & S \end{array}$$

and assume that the morphism f is universally open and any generic point of X lies over a generic point of S. Then any generic point of X' lies over a generic point of S'.

Proof: Any generic point of X' obviously lies over a generic point of S. Replacing S by this point we may assume that $S = Spec(k)$ where k is a field. Then the morphism $X \to S$ is universally open (being flat) and hence the morphism $X' \to S'$ is open, which implies that it takes generic points to generic points.

Lemma 3.3.8 *Let $X \to S$ be a scheme of finite type over a Noetherian scheme S, $\mathcal{Z} = \sum n_i z_i$ be an element of $Cycl(X/S, r) \otimes \mathbf{Q}$, $f : S' \to S$ be a Noetherian scheme over S and $\mathcal{Z}' = cycl(f)(\mathcal{Z})$ be the corresponding element of $Cycl(X \times_S S'/S', r) \otimes \mathbf{Q}$.*

1. *If f is a universally open morphism then*

$$supp(\mathcal{Z}') = (supp(\mathcal{Z}) \times_S S')_{red}.$$

2. *If f is dominant then $Supp(\mathcal{Z})$ is the closure of $(X \times_S f)(Supp(\mathcal{Z}'))$.*

Proof: 1. The inclusion

$$supp(\mathcal{Z}') \subset (supp(\mathcal{Z}) \times_S S')_{red}$$

follows from Lemma 3.3.6. Lemma 3.3.7 implies immediately that generic points of $supp(\mathcal{Z}) \times_S S'$ lie over generic points of S'. We may assume therefore that $S' = Spec(k)$ and the image of S' in S is a generic point η of S. Then k is an extension of k_η and according to Lemma 2.3.1(2) we have

$$Supp(\mathcal{Z}') = Supp(\mathcal{Z}_\eta \otimes_{k_\eta} k) = (Supp\mathcal{Z}_\eta \times_{Spec(k_\eta)} Spec(k))_{red}.$$

Now it suffices to note that

$$\mathcal{Z}_\eta = \sum_{z_i/\eta} n_i z_i$$

(the sum being taken over those points z_i which lie over η) and hence $Supp(\mathcal{Z}_\eta) = Supp(\mathcal{Z}) \times_S Spec(k_\eta)$. 3. It suffices to show that $z_i \in (X \times_S f)(Supp(\mathcal{Z}'))$. Denote by η_i the image of z_i in S and let η_i' be a generic point of S' over η_i. Using Lemma 3.3.6 and the part (1) of the present lemma we get

$$Supp(\mathcal{Z} \times_{Spec(k_{\eta_i})} Spec(k_{\eta_i'}))_{red} = Supp(\mathcal{Z}_{k_{\eta_i'}}) =$$

$$= Supp(\mathcal{Z}'_{k_{\eta_i'}}) \subset Supp(\mathcal{Z}')$$

It suffices to note now that the morphism

$$Supp(\mathcal{Z}) \times_{Spec(k_{\eta_i})} Spec(k_{\eta_i'}) \to Supp(\mathcal{Z})$$

is surjective.

Let $f : T \to S$ be a morphism of Noetherian schemes and $X \to S$ be a scheme of finite type over S. We denote by

$$cycl(f) : Cycl(X/S, r) \otimes_{\mathbf{Z}} \mathbf{Q} \to Cycl(X \times_S T/T, r) \otimes_{\mathbf{Z}} \mathbf{Q}$$

the homomorphism $\mathcal{Z} \to \mathcal{Z}_T$ constructed in Theorem 3.3.1. Lemma 3.3.6 shows that this homomorphism takes $Cycl_{equi}(X/S,r) \otimes_{\mathbf{Z}} \mathbf{Q}$ to $Cycl_{equi}(X \times_S T/T,r) \otimes_{\mathbf{Z}} \mathbf{Q}$, $PropCycl(X/S,r) \otimes_{\mathbf{Z}} \mathbf{Q}$ to $PropCycl(X \times_S T/T,r) \otimes_{\mathbf{Z}} \mathbf{Q}$ etc. We use the same notation $cycl(f)$ for the homomorphisms induced on the corresponding groups and monoids.

It follows immediately from definitions and Lemma 3.1.4 that

$$cycl(g \circ f) = cycl(f) \circ cycl(g)$$

for any composable pair of morphisms of Noetherian schemes.

This shows that for any scheme of finite type $X \to S$ over a Noetherian scheme S there is a presheaf of $\mathbf{Q}$-vector spaces $Cycl(X/S,r)_{\mathbf{Q}}$ on the category of Noetherian schemes over S such that for any Noetherian scheme S' over S one has $Cycl(X/S,r)_{\mathbf{Q}}(S') = Cycl(X \times_S S'/S',r) \otimes \mathbf{Q}$.

Similarly we have presheaves of $\mathbf{Q}$-vector spaces $PropCycl(X/S,r)_{\mathbf{Q}}$, $Cycl_{equi}(X/S,r)_{\mathbf{Q}}$, $PropCycl_{equi}(X/S,r)_{\mathbf{Q}}$ and presheaves of uniquely divisible abelian monoids $Cycl^{eff}(X/S,r)_{\mathbf{Q}+}$, $PropCycl^{eff}(X/S,r)_{\mathbf{Q}+}$.

For any Noetherian scheme S' over S we have a canonical lattice $Cycl(X \times_S S',r)$ in the $\mathbf{Q}$-vector space $Cycl(X/S,r)_{\mathbf{Q}}(S')$. Unfortunately these lattices do not form in general a subpresheaf in $Cycl(X/S,r)_{\mathbf{Q}}$ (see example 3.5.10(1)).

Lemma 3.3.9 *Let S be a Noetherian scheme, $X \to S$ be a scheme of finite type over S and $\mathcal{Z}$ be an element of $Cycl(X/S,r)$. Then the following conditions are equivalent:*

1. *For any Noetherian scheme T over S the cycle $\mathcal{Z}_T$ belongs to $Cycl(X \times_S T/T,r)$.*

2. *For any point $s \in S$ the cycle $\mathcal{Z}_s$ belongs to $Cycl(X_s,r)$.*

3. *For any point $s \in S$ there exists a separable field extension k/k_s such that the cycle $\mathcal{Z}_k = \mathcal{Z}_s \otimes_{k_s} k$ belongs to $Cycl(X \times_S Spec(k),r)$.*

Proof: The implication (3=>2) follows from Lemma 2.3.2. The other implications are obvious.

We denote by $z(X/S,r)$ (resp. $c(X/S,r)$, $z_{equi}(X/S,r)$, $c_{equi}(X/S,r)$) the subgroup of $Cycl(X/S,r)$ (resp. of $PropCycl(X/S,r)$, $Cycl_{equi}(X/S,r)$, $PropCycl_{equi}(X/S,r)$) consisting of cycles satisfying the equivalent conditions of Lemma 3.3.9. It is clear that $z(X/S,r)$ (resp. ...) is a subpresheaf in the presheaf $Cycl(X/S,r)_{\mathbf{Q}}$. Moreover

$$c(X/S,r) = z(X/S,r) \cap PropCycl(X/S,r)_{\mathbf{Q}}$$

etc.

Lemma 3.3.10 *Let $X \to S$ be a scheme of finite type over a Noetherian scheme S, $T \to S$ be a Noetherian scheme over S and $\mathcal{W}$ be an element of $\mathbf{Z}(\mathrm{Hilb}(X/S, r))$. Then*

$$[cycl_X(\mathcal{W})]_T = cycl_{X \times_S T}(\mathcal{W}_T).$$

Proof: It is sufficient to treat the case $\mathcal{W} = Z$ where Z is a closed subscheme of X flat and equidimensional of relative dimension r over S. Consider any commutative diagram of the form

$$
\begin{array}{ccc}
& Spec(A) & \overset{y_1}{\to} \ T \\
{\scriptstyle y_0} \nearrow & & \downarrow {\scriptstyle f} \\
Spec(k) & & \\
{\scriptstyle x_0} \searrow & & \\
& Spec(R) & \overset{x_1}{\to} \ S
\end{array}
$$

in which (x_0, x_1) (resp. (y_0, y_1)) is a fat k-point of S (resp. of T). Corollary 3.2.4 shows that

$$(y_0, y_1)^*(cycl_{X \times_S T}(Z \times_S T)) = cycl_{X \times_S Spec(k)}(Z \times_S Spec(k)) =$$

$$= (x_0, x_1)^*(cycl_X(Z))$$

Thus the cycle $cycl_{X \times_S T}(Z \times_S T)$ satisfies the requirements defining the cycle $[cycl_X(Z)]_T$.

Corollary 3.3.11 *The homomorphisms $cycl_X$ define a homomorphism of presheaves*

$$cycl : \mathbf{Z}(\mathrm{Hilb}(X/S, r)) \dashrightarrow z_{\mathrm{equi}}(X/S, r).$$

Lemma 3.3.12 *Let $X \to S$ be a scheme of finite type over a Noetherian scheme S and $f : S' \to S$ be a flat morphism of Noetherian schemes. Assume further that the schemes S, S' are reduced. Then for any element $\mathcal{Z}$ in $Cycl(X/S, r)$ one has*

$$cycl(f)(\mathcal{Z}) = f_X^*(\mathcal{Z})$$

where $f_X = pr_1 : X \times_S S' \to X$ and f_X^* is the flat pull-back defined in section 2.3.

Proposition 3.3.13 *Let S be a Noetherian scheme of exponential characteristic n and $X \to S$ be a scheme of finite type over S. Then the subgroups $Cycl(X \times_S T/T, r)[1/n]$ in $Cycl(X \times_S T/T, r) \otimes \mathbf{Q}$ for Noetherian schemes T over S form a subpresheaf $Cycl(X/S, r)[1/n]$ in the presheaf $Cycl(X/S, r)_{\mathbf{Q}}$.*

Proof: It follows easily from Lemmas 3.3.2 and 3.3.9.

Proposition 3.3.14 *Let $X \to S$ be a scheme of finite type over a Noetherian scheme S. Then the quotient presheaf $Cycl(X/S, r)_{\mathbf{Q}}/z(X/S, r)$ is a presheaf of torsion abelian groups.*

Proof: We have to show that for any $\mathcal{Z} = \sum n_i z_i \in Cycl(X/S, r)$ there exists an integer $N > 0$ such that $N\mathcal{Z} \in z(X/S, r)$. Using Noetherian induction we may assume that for any closed subscheme $T \subset S$ such that $T \neq S$ there exists an integer $N(T)$ such that $N(T)\mathcal{Z}_T \in z(X \times_S T/T, r)$. Denote by Z_i the closure of the point z_i considered as an integral closed subscheme of X. Let $S' \to S_{red}$ be a blow-up such that the proper transforms $\tilde{Z}_i$ of Z_i are flat over S'. Let U be a dense open subset of S over which the morphism $S' \to S_{red}$ is an isomorphism and let T be the closed reduced subscheme $S - U$. We claim that $N(T)\mathcal{Z} \in z(X/S, r)$. By Lemma 3.3.9 it is sufficient to verify that $N(T)\mathcal{Z}_s \in Cycl(X_s, r)$ for any point $s \in S$. If $s \in T$ it follows from our choice of $N(T)$. If $s \in U$ it follows from Lemma 3.3.3.

Proposition 3.3.15 *Let S be a regular Noetherian scheme. Then for any scheme of finite type X over S and any $r \geq 0$ one has:*

$$Cycl(X/S, r) = z(X/S, r)$$

$$Cycl_{equi}(X/S, r) = z_{equi}(X/S, r)$$

etc.

Proof: Let $\mathcal{Z}$ be an element of $Cycl(X/S, r)$ and let $s \in S$ be a point of S. We have to show that $\mathcal{Z}_s \in Cycl(X_s, r)$. We proceed by induction on $n = dim(\mathcal{O}_{S,s})$.

If $n = 1$ then $\mathcal{O}_{S,s}$ is a discrete valuation ring so that we have a canonical fat point $Spec(k_s) \overset{x_0}{\to} Spec(\mathcal{O}_{S,s}) \overset{x_1}{\to} S$ over the point $Spec(k_s) \to S$. Thus $\mathcal{Z}_s = (x_0, x_1)^*(\mathcal{Z}) \in Cycl(X_s, r)$.

If $n > 1$ choose a regular system of parameters $t_1, \ldots, t_n$ in $\mathcal{O}_{S,s}$. Replacing S by an appropriate open neighborhood of s we may assume that S is affine , $t_1, \ldots, t_n \in \mathcal{O}(S)$ and the closed subscheme T of S defined by the equation $t_1 = 0$ is regular and irreducible. Let τ be the generic point of T. Since $dim(\mathcal{O}_{S,\tau}) = 1$ we conclude that $\mathcal{Z}_\tau \in Cycl(X_\tau, r)$ and hence $\mathcal{Z}_T \in Cycl(X \times_S T/T, r)$. The induction hypothesis shows now that $\mathcal{Z}_s \in Cycl(X_s, r)$.

3.4. Relative cycles over geometrically unibranch schemes

Lemma 3.4.1 *Let k be a field, $X \to Spec(k)$, $S \to Spec(k)$ be two schemes of finite type over k and Z be a closed subscheme in $X \times_{Spec(k)} S$ defined by nilpotent sheaf of ideals which is flat over S. Let further E be an extension of k and s_1, s_2 be two E-points of S over k. If S is geometrically connected then the cycles associated with the closed subschemes $Z \times_{s_1} Spec(E)$ and $Z \times_{s_2} Spec(E)$ in $X \times_{Spec(k)} Spec(E)$ coincide.*

Proof: We may replace E by its algebraic closure and thus assume that E is algebraically closed. Next we may replace X by $X_E = X \times_{Spec(k)} Spec(E)$, S be $S_E = S \times_{Spec(k)} Spec(E)$ and Z by $Z_E = Z \times_{Spec(k)} Spec(E)$ and thus assume that $E = k$. The scheme S being connected we may find a chain of rational points of S starting with s_1 and ending with s_2 and such that any pair of consecutive points belongs to the same irreducible component of S. Thus we may assume that S is an integral scheme. Let $X_1, \ldots, X_n$ denote the irreducible components of X which are considered as closed integral subschemes of X. Since S is integral and the base field is algebraically closed we conclude that the schemes $X_i \times S$ are integral and coincide with the irreducible components of $X \times S$.

This shows that the cycle $cycl_{X \times S}(Z)$ may be written (uniquely) in the form $\sum n_i cycl_{X \times S}(X_i \times S)$. Proposition 3.2.2 shows now that the cycles $cycl_X(Z \times_{s_j} Spec(k))$ both coincide with $\sum n_i cycl_X(X_i)$.

Theorem 3.4.2 *Let S be a Noetherian geometrically unibranch scheme and $X \to S$ be a scheme of finite type over S. Let further $Z \subset X$ be a closed subscheme which is equidimensional of relative dimension r over S. Then $cycl_X(Z) \in Cycl_{equi}(X/S, r)$.*

Proof: Replacing Z by its irreducible components (which are also equidimensional over S) we may assume that the scheme Z is integral. Choose a blow-up $S' \to S_{red}$ such that the proper transform $\tilde{Z}$ of Z is flat over S'. Let further k be a field, $s : Spec(k) \to S$ be a k-point of S and $s_1, s_2 : Spec(k) \to S'$ be two liftings of s to S'. According to Proposition 3.1.5 we have to show that the cycles $cycl(\tilde{Z} \times_{s_1} Spec(k))$, $cycl(\tilde{Z} \times_{s_2} Spec(k))$ coincide. Note that according to Proposition 2.1.7 and Lemma 2.1.11 the closed subscheme $\tilde{Z}$ in $Z \times_S S'$ is defined by a nilpotent sheaf of ideals and hence $\tilde{Z} \times_S Spec(k)$ is a closed subscheme of $(Z \times_S S') \times_S Spec(k) = (Z \times_S Spec(k)) \times_{Spec(k)} (S' \times_S Spec(k))$ defined by a nilpotent sheaf of ideals. The scheme $S' \times_S Spec(k)$ is geometrically connected according to Proposition 2.1.6. Thus our statement follows from Lemma 3.4.1.

Corollary 3.4.3 *Let S be a Noetherian geometrically unibranch scheme and $X \to S$ be a scheme of finite type over S. Then the abelian group $Cycl_{equi}(X/S,r)$ (resp. $PropCycl_{equi}(X/S,r)$) is freely generated by cycles of integral closed subschemes Z in X which are equidimensional (resp. proper and equidimensional) of dimension r over S.*

Corollary 3.4.4 *Let S be a Noetherian geometrically unibranch scheme and $X \to S$ be a scheme of finite type over S. Then the abelian group $Cycl_{equi}(X/S,r)$ (resp. the abelian group $PropCycl_{equi}(X/S,r)$) is generated by the abelian monoid $Cycl^{eff}(X/S,r)$ (resp. by the abelian monoid $PropCycl^{eff}(X/S,r)$).*

Corollary 3.4.5 *Let S be a Noetherian regular scheme and X be a scheme of finite type over S. Then abelian group $z_{equi}(X/S,r)(S)$ (resp. the abelian monoid $z^{eff}(X/S,r)(S)$) is the free abelian group (resp. the free abelian monoid) generated by closed integral subschemes of X which are equidimensional of dimension r over S.*

Corollary 3.4.6 *Let S be a Noetherian regular scheme and X be a scheme of finite type over S. Then the abelian group $c_{equi}(X/S,r)(S)$ (resp. the abelian monoid $c^{eff}(X/S,r)(S)$) is the free abelian group (resp. the free abelian monoid) generated by closed integral subschemes of X which are proper and equidimensional of dimension r over S.*

Example 3.4.7 The statement of Corollary 3.4.4 is false for schemes S which are not geometrically unibranch. Let us consider the following situation.

Let $S = S_1 \cup S_2$ be a union of two copies of affine line (i.e. $S_1 \cong S_2 \cong \mathbf{A}^1$) such that the point $\{0\}$ (resp. the point $\{1\}$) of S_1 is identified with the point $\{0\}$ (resp. the point $\{1\}$) of S_2.

We take X to be abstractly isomorphic to S, i.e. $X = X_1 \cup X_2$ also is a union of two copies of affine line glued together in the same way. Consider the morphism $X \to S$ which maps X_1, X_2 identically on S_1. Using Proposition 3.1.5 one can easily see that:

$$Cycl^{eff}(X/S,0) = 0$$

$$Cycl(X/S,0) = \mathbf{Z}$$

and hence the abelian group of relative cycles is not generated by abelian monoid of effective relative cycles in this case.

Remarks:

1. The statement of Corollary 3.4.3 is false for the abelian groups $Cycl(X/S, r)$ and $PropCycl(X/S, r)$ since if $dim(S) > 1$ there exist elements $\mathcal{Z}$ in these groups such that $supp(\mathcal{Z})$ is not equidimensional over S (see exmple 3.1.9).

2. It is not true in general that for a geometrically unibranch scheme S the groups $z_{equi}(X/S, r)$, $c_{equi}(X/S, r)$ (or abelian monoids $z^{eff}(X/S, r)$, $c^{eff}(X/S, r)$ are generated by elements which correspond to integral closed subschemes of X (see example 3.5.10(2)) but in view of Proposition 3.3.15 it is true for regular schemes S.

Proposition 3.4.8 *Let S be a normal Noetherian scheme and $X \to S$ be a smooth scheme of finite type of dimension r over S. Then one has:*

$$z_{equi}(X/S, r - 1) = Cycl_{equi}(X/S, r - 1)$$

$$c_{equi}(X/S, r - 1) = PropCycl_{equi}(X/S, r - 1)$$

and the first group is isomorphic to the group of relative Cartier divisors on X over S.

Proof: Note first that any normal scheme is geometrically unibranch by [8]. By Corollary 3.4.3 the group $Cycl_{equi}(X/S, r - 1)$ is generated by integral closed subschemes Z in X which are equidimensional of relative dimension $r - 1$ over S. By [8, 21.14.3] any such Z is flat over S and our result follows from the definition of relative Cartier divisor (see [9]) and Corollary 3.3.11.

3.5. Multiplicities of components of inverse images of equidimensional cycles over geometrically unibranch schemes

Let S be a Noetherian geometrically unibranch scheme and $X \to S$ be a scheme of finite type over S. Corollary 3.4.3 shows that the abelian group $Cycl_{equi}(X/S, r)$ is the free abelian group generated by generic points of closed integral subschemes $Z \subset X$ of X which are equidimensional of relative dimension r over S. Let now $T \to S$ be a morphism of Noetherian schemes. Lemma 3.3.6 shows that the cycle $[Cycl_X(Z)]_T$ is a formal linear combination of generic points of $Z \times_S T$ with certain multiplicities. The aim of this section is to give an explicit formula for these multiplicities. It is sufficient to consider the case $T = Spec(k_s)$ where s is a certain point of S. Moreover since the groups $Cycl_{equi}(X/S, r)$ and $Cycl_{equi}(X_{red}/S_{red}, r)$

coincide we may assume that the scheme S is reduced and since irreducible components of S do not intersect we may further assume that S is integral. The formula we are going to provide involves multiplicities of certain ideals so we start by recalling briefly the necessary definitions and results from commutative algebra (see [10]).

Let $\mathcal{O}$ be a local Noetherian ring of dimension r and M be a finitely generated $\mathcal{O}$ module. An ideal $I \subset \mathcal{O}$ is called an ideal of definition if I contains a certain power of the maximal ideal. For any ideal of definition I of $\mathcal{O}$ one may consider the so called Samuel function:

$$\chi_M^I(n) = length(M/I^{n+1}M).$$

It is known that for n big enough $\chi_M^I(n)$ is a polynomial in n of degree at most r. Furthermore it may be written in the form

$$\chi_M^I(n) = (e/r!)n^r + \text{(terms of lower degree)}$$

where e is a nonnegative integer called the multiplicity of I with respect to M. We will denote it by $e(I, M)$. The integer $e(I, \mathcal{O})$ is called the multiplicity of I and is denoted $e(I)$.

Proposition 3.5.1 *Let* $\mu_1, \ldots, \mu_k$ *be all minimal prime ideals of* $\mathcal{O}$ *such that* $dim(\mathcal{O}/\mu) = r$, *then*

$$e(I, M) = \sum_{i=1}^{k} e(I, \mathcal{O}/\mu_i) length_{\mathcal{O}_{\mu_i}}(M_{\mu_i}) = \sum_{i=1}^{k} e(I(\mathcal{O}/\mu_i)) length_{\mathcal{O}_{\mu_i}} M_{\mu_i}.$$

Proof: See [10] Theorem 14.7.

Lemma 3.5.2 *Let* $\mathcal{O} \to \mathcal{O}'$ *be a local homomorphism of Noetherian local rings. Let* $\mathcal{M}_{\mathcal{O}}$ *be the maximal ideal of* $\mathcal{O}$ *and suppose that* $\mathcal{O}'$ *is a flat* $\mathcal{O}$-*algebra and* $\mathcal{M}_{\mathcal{O}}\mathcal{O}'$ *is an ideal of definition of* $\mathcal{O}'$. *Then for any ideal of definition* I *of* $\mathcal{O}$ *the following formula holds*

$$e(I\mathcal{O}') = e(I) length_{\mathcal{O}'}(\mathcal{O}'/\mathcal{M}_{\mathcal{O}}\mathcal{O}').$$

Proof: Flatness of $\mathcal{O}'$ over $\mathcal{O}$ implies that for any n we have

$$\chi_{\mathcal{O}'}^{I\mathcal{O}'}(n) = length_{\mathcal{O}'}(\mathcal{O}' \otimes_{\mathcal{O}} \mathcal{O}/I^n) = length_{\mathcal{O}}(\mathcal{O}/I^n) length_{\mathcal{O}'}(\mathcal{O}'/\mathcal{M}_{\mathcal{O}}\mathcal{O}') =$$

$$= \chi_{\mathcal{O}}^I(n) length_{\mathcal{O}'}(\mathcal{O}'/\mathcal{M}_{\mathcal{O}}\mathcal{O}').$$

The following property of multiplicities is obvious.

Lemma 3.5.3 *Assume that $\mathcal{O}'$ is a finite local $\mathcal{O}$ algebra such that $dim(\mathcal{O}') = dim(\mathcal{O})$ and M is a finitely generated $\mathcal{O}'$-module. Let k and k' be the residue fields of $\mathcal{O}$ and $\mathcal{O}'$. Then for any ideal of definition I of $\mathcal{O}$ the following formula holds:*

$$e(I\mathcal{O}', M) = e(I, M)/[k' : k].$$

Let $Z \to S$ be a scheme equidimensional of relative dimension r over an integral Noetherian geometrically unibranch scheme S. Let further s be a point of S, I be an ideal of definition of the local ring $\mathcal{O}_{S,s}$ and z be a generic point of the fiber $Z_s = Z \times_{Spec(k_s)} S$. We set $n_I(z) = e(I\mathcal{O}_{Z,z})/e(I)$.

Lemma 3.5.4   1. *$n_I(z)$ is independent of the choice of I.*

     2. *If Z is flat over S then $n_I(z)$ coincides with the multiplicity of z in the cycle $(cycl_Z(Z))_s = cycl_{Z_s}(Z_s)$.*

Proof: The second statement follows immediately from Lemma 3.5.2. Replacing Z by an appropriate neighborhood of z we may assume that the morphism $Z \to S$ admits a decomposition of the form $Z \xrightarrow{p_0} \mathbf{A}_S^r \to S$ where p_0 is an equidimensional quasi-finite morphism (see Proposition 2.1.3). Denote $p_0(z)$ by x. Then x is the (unique) generic point of the fiber of $\mathbf{A}_S^r$ over s and according to the part (2) of our proposition we have $e(I\mathcal{O}_{\mathbf{A}_S^r,x}) = e(I)$. This shows that $n_I(z) = e(I\mathcal{O}_{Z,z})/e(I\mathcal{O}_{\mathbf{A}_S^r,x})$ so that we may replace S by $\mathbf{A}_S^r$ and assume that $r = 0$.

Denote by S' the henselization of S at s and by s' the closed point of S'. Set also $Z' = Z \times_S S'$. The fiber of Z' over s' is isomorphic to the fiber of Z over s and thus there exists exactly one point z' lying over z. Since $\mathcal{O}_{S',s'} = \mathcal{O}_{S,s}^h$ is ind-etale over $\mathcal{O}_{S,s}$ and $\mathcal{O}_{Z',z'}$ is ind-etale over $\mathcal{O}_{Z,z}$ we conclude from Lemma 3.5.2 that $e(I) = e(I\mathcal{O}_{S',s})$, $e(I\mathcal{O}_{Z,z}) = e(I\mathcal{O}_{Z',z'})$ and hence $n_I(z) = n_{I\mathcal{O}_{S',s'}}(z')$. Thus replacing S by S' and Z by Z' we may assume that S is a henselian local scheme (note that S' is integral since S is geometrically unibranch). In this case $\mathcal{O}_{Z,z}$ is a finite $\mathcal{O}_{S,s}$-algebra and according to Lemma 3.5.3 and Proposition 3.5.1 we have

$$n_I(z) = \frac{e(I\mathcal{O}_{Z,z})}{e(I)} = \frac{e(I, \mathcal{O}_{Z,z})}{[k_z : k_s]e(I)} = \frac{e(I)dim_{F(S)}(\mathcal{O}_{X,x} \otimes_{\mathcal{O}_{S,s}} F(S))}{e(I)[k_z : k_s]} =$$

$$= \frac{dim_{F(S)}(\mathcal{O}_{X,x} \otimes_{\mathcal{O}_{S,s}} F(S))}{[k_z : k_s]}$$

where $F(S)$ is the quotient field of the integral domain $\mathcal{O}_S$. Thus $n_I(z)$ does not depend on the choice of I.

We will use the notation $n(z)$ or $n_{Z/S}(z)$ for the common value of multiplicities $n_I(z)$. We will denote by $[Z/S]_s$ the element of $Cycl(Z_s, r)$ of the form $\sum n(z)z$ where the sum is taken over all generic points of $Z_s = Z \times_S Spec(k_s)$. In the course of the proof of Lemma 3.5.4 we have established some useful properties of multiplicities $n(z)$ which we would like to list now for future use.

Corollary 3.5.5   1. *Assume that the morphism $Z \to S$ is factorized in the form $Z \to \mathbf{A}_S^r \to S$ where the first morphism is quasi-finite and equidimensional. Then for any point $z \in Z$ generic in its fiber over S one has $n_{Z/S}(z) = n_{Z/\mathbf{A}_S^r}(z)$.*

2. *Let S' be the henselization of S at s, s' be the closed point of S' and z' be the only point of $Z' = Z \times_S S'$ lying over z. Then $n_{Z'/S'}(z') = n_{Z/S}(z)$.*

3. *Let $Z \to S$ be a finite equidimensional morphism. Assume that the fiber Z_s consists of only one point z. Then*

$$n(z) = \frac{dim_{F(S)}(\mathcal{O}_{Z,z} \otimes_{\mathcal{O}_{S,s}} F(S))}{[k_z : k_s]}$$

where $F(S)$ is the function field of the integral scheme S.

Proposition 3.5.6 *Let $p : Z \to S$ be an equidimensional scheme of relative dimension r over an integral Noetherian geometrically unibranch scheme S. Let S' be another integral Noetherian geometrically unibranch scheme and let $f : S' \to S$ be a dominant morphism. Set $Z' = Z \times_S S'$. Let finally s' be any point of S' and s be its image in S. Then $[Z'/S']_{s'} = [Z/S]_s \otimes_{k_s} k_{s'}$.*

In other words if z' is a generic point of the fiber $Z'_{s'}$ and z is its image in Z then $n(z') = n(z)length((k_{s'} \otimes_{k_s} k_z)_{\mu_{z'}})$ where $\mu_{z'}$ is the ideal $ker(k_{s'} \otimes_{k_s} k_z \to k_{z'})$.

Proof: Replacing Z by an appropriate neighborhood of z we may assume by Proposition 2.1.3 that the morphism $Z \to S$ admits a decomposition of the form $Z \overset{p_0}{\to} \mathbf{A}_S^r \to S$ where p_0 is an equidimensional quasi-finite morphism. Corollary 3.5.5(1) shows that $n_{Z/S}(z) = n_{Z/\mathbf{A}_S^r}(z)$ and $n_{Z'/S'}(z') = n_{Z'/\mathbf{A}_{S'}^r}(z')$. Furthermore denoting by x (resp. x') the image of z (resp. z') in $\mathbf{A}_S^r$ (resp. $\mathbf{A}_{S'}^r$) one checks easily that the local rings $(k_{s'} \otimes_{k_s} k_z)_{\mu_{z'}}$ and $(k_{x'} \otimes_{k_x} k_z)_{\mu_{z'}}$ coincide. Thus we may replace S by $\mathbf{A}_S^r$, S' by $\mathbf{A}_{S'}^r$ and assume that $r = 0$. We certainly may also assume S, S' to be local schemes and s, s' to be their closed points.

Consider first the special case $S' = Spec(\mathcal{O}_{S,s}^{sh})$ where $\mathcal{O}_{S,s}^{sh}$ is the strict henselization of $\mathcal{O}_{S,s}$. Since $\mathcal{O}_{S',s'}$ (resp. $\mathcal{O}_{Z',z'}$) is ind-etale over $\mathcal{O}_{S,s}$ (resp. $\mathcal{O}_{Z,z}$) we conclude from Lemma 3.5.2 that $e(I\mathcal{O}_{Z',z'}) = e(I\mathcal{O}_{Z,z})$, $e(I\mathcal{O}_{S',s'}) = e(I\mathcal{O}_{S,s})$ for any ideal of definition I of $\mathcal{O}_{S,s}$ and hence $n(z) = n(z')$. On the other hand $k_{s'}$ is a separable algebraic extension of k_s. Thus $(k_{s'} \otimes_{k_s} k_z)_{\mu_{z'}}$ is a reduced local Artinian ring and hence a field, i.e.

$$length((k_{s'} \otimes_{k_s} k_z)_{\mu_{z'}}) = 1.$$

In the general case there exists a morphism $f' : Spec(\mathcal{O}_{S',s'}^{sh}) \to Spec(\mathcal{O}_{S,s}^{sh})$ such that the diagram

$$
\begin{array}{ccc}
Spec(\mathcal{O}_{S',s'}^{sh}) & \xrightarrow{f'} & Spec(\mathcal{O}_{S,s}^{sh}) \\
\downarrow & & \downarrow \\
S' & \xrightarrow{f} & S
\end{array}
$$

commutes. In view of the special case considered above we may replace f by f' and assume that S and S' are strictly henselian local schemes. Finally replacing Z by $Spec(\mathcal{O}_{Z,z})$ we may assume that Z is finite over S and z is its only point over s. In this situation k_z is a purely inseparable extension of k_s and hence the Artinian ring $k_{s'} \otimes_{k_s} k_z$ is local (i.e. z' is the only point of Z' over s') and its residue field coincides with $k_{z'}$. Using Corollary 3.5.5(3) we conclude that

$$n(z) = \frac{dim_{F(S)}(\mathcal{O}_{Z,z} \otimes_{\mathcal{O}_{S,s}} F(S))}{[k_z : k_s]}$$

$$n(z') = \frac{dim_{F(S')}(\mathcal{O}_{Z',z'} \otimes_{\mathcal{O}_{S',s'}} F(S'))}{[k_{z'} : k_{s'}]}$$

$$length((k_{s'} \otimes_{k_s} k_z)_{\mu_{z'}}) = length(k_{s'} \otimes_{k_s} k_z) = \frac{[k_z : k_s]}{[k_{z'} : k_{s'}]}$$

Now it is sufficient to note that

$$\mathcal{O}_{Z',z'} \otimes_{\mathcal{O}_{S',s'}} F(S') = (\mathcal{O}_{Z,z} \otimes_{\mathcal{O}_{S,s}} \mathcal{O}_{S',s'}) \otimes_{\mathcal{O}_{S',s'}} F(S') =$$

$$= (\mathcal{O}_{Z,z} \otimes_{\mathcal{O}_{S,s}} F(S)) \otimes_{F(S)} F(S')$$

and hence

$$dim_{F(S)}(\mathcal{O}_{Z,z} \otimes_{\mathcal{O}_{S,s}} F(S)) = dim_{F(S')}(\mathcal{O}_{Z',z'} \otimes_{\mathcal{O}_{S',s'}} F(S')).$$

Lemma 3.5.7 *Under the assumptions of Proposition 3.5.6 let Z' be a closed subscheme of Z defined by a nilpotent sheaf of ideals and such that the closed embedding $Z' \to Z$ is an isomorphism over the generic point of S. Then for any point s of S the cycles $[Z'/S]_s$ and $[Z/S]_s$ coincide.*

Proof: Let z be a generic point of the fiber Z_s. We have to show that $n_{Z/S}(z) = n_{Z'/S}(z)$. Let I be an ideal of definition of the local ring $\mathcal{O}_{S,s}$, let $\mu_1, \ldots, \mu_n$ be all minimal prime ideals of $\mathcal{O}_{Z,z}$ and let ν be the kernel of the natural surjective homomorphism $\mathcal{O}_{Z,z} \to \mathcal{O}_{Z',z}$. The ideal ν being nilpotent is contained in $\mu_1 \cap \ldots \cap \mu_n$ and the ideals $\mu_1/\nu, \ldots, \mu_n/\nu$ are the minimal prime ideals of $\mathcal{O}_{Z',z}$. Furthermore our assumptions imply that $(\mathcal{O}_{Z,z})_{\mu_i} = (\mathcal{O}_{Z',z})_{\mu_i/\nu}$. Proposition 3.5.1 shows that

$$e(I\mathcal{O}_{Z,z}) = \sum_i e(I(\mathcal{O}_{Z,z}/\mu_i))length((\mathcal{O}_{Z,z})_{\mu_i}) =$$

$$= \sum_i e(I(\mathcal{O}_{Z',z}/(\mu_i/\nu)))length((\mathcal{O}_{Z',z})_{\mu_i/\nu}) = e(I\mathcal{O}_{Z',z}).$$

Theorem 3.5.8 *Let $X \to S$ be a scheme of finite type over an integral Noetherian geometrically unibranch scheme S. Let further Z be a closed integral subscheme of X equidimensional of relative dimension r over S and let $\mathcal{Z}$ be the corresponding element of $Cycl(X/S,r)$. Then the cycles $\mathcal{Z}_s$ and $[Z/S]_s$ coincide for any point $s \in S$ of S.*

Proof: Let k be an extension of k_s and let (x_0, x_1, R) be a fat point of S over the k-point $Spec(k) \to Spec(k_s) \to S$. To prove the theorem we have to show that $(x_0, x_1)^*(\mathcal{Z}) = [Z/S]_s \otimes_{k_s} k$. Since $Z \times_S Spec(R)$ is equidimensional over $Spec(R)$ and the closed embedding $\phi_{x_1}(Z) \to Z \times_S Spec(R)$ is an isomorphism over the generic point of $Spec(R)$ we conclude by Lemma 2.1.10 that $\phi_{x_1}(Z)$ is defined by a nilpotent sheaf of ideals. Let s' be the closed point of $Spec(R)$. Since $\phi_{x_1}(Z)$ is flat over $Spec(R)$ we see from 3.5.4(2),3.5.7 and 3.5.6 that

$$(x_0, x_1)^*(\mathcal{Z}) = cycl(\phi_{x_1}(Z) \times_{Spec(R)} Spec(k)) = [\phi_{x_1}(Z)/Spec(R)]_{s'} =$$

$$= [Z \times_S Spec(R)/Spec(R)]_{s'} = [Z/S]_s \otimes_{k_s} k.$$

For cycles equidimensional over regular Noetherian schemes the above formula for multiplicities of components of the inverse image reduces to the usual Tor-formula as one sees from the following lemma.

Lemma 3.5.9 *Let $Z \to S$ be an equidimensional scheme of relative dimension r over a regular scheme S and $s \in S$ be a point of S. If z is a generic point of the fiber Z_s of Z over s then one has*

$$n_{Z/S}(z) = \sum_{i=0}^{dim(\mathcal{O}_{S,s})} (-1)^i length_{\mathcal{O}_{Z,z}}(Tor_i^{\mathcal{O}_{S,s}}(\mathcal{O}_{Z,z}, k_s)).$$

Proof: Let $t_1, \ldots, t_k$ be a regular system of parameters of the regular local ring $\mathcal{O}_{S,s}$. Take $I = \mathcal{M} = t_1 \mathcal{O}_{S,s} + \ldots + t_n \mathcal{O}_{S,s}$ to be the maximal ideal of this ring. Since $\mathcal{O}_{S,s}$ is regular the multiplicity $e(I)$ equals one (see [10]). Thus $n(z) = e(I\mathcal{O}_{Z,z})$. Theorem of Serre ([15]) shows that

$$e(I\mathcal{O}_{Z,z}) = \sum_i (-1)^i length_{\mathcal{O}_{Z,z}}(H_i(K(\underline{t}, \mathcal{O}_{Z,z})))$$

where $K(\underline{t}, \mathcal{O}_{Z,z})$ is the Koszul complex corresponding to the sequence $\underline{t} = (t_1, \ldots, t_n)$. On the other hand $K(\underline{t}, \mathcal{O}_{S,s})$ is a projective resolution of k_s over $\mathcal{O}_{S,s}$ and hence

$$H_i(K(l, \mathcal{O}_{Z,z})) = H_i(K(\underline{t}, \mathcal{O}_{S,s}) \otimes_{\mathcal{O}_{S,s}} \mathcal{O}_{Z,z}) = Tor_i^{\mathcal{O}_{S,s}}(\mathcal{O}_{Z,z}, k_s).$$

Example 3.5.10   1. Let F be a field of characteristic $p > 0$ and let $a, b \in F^*$ be elements independent modulo $(F^*)^p$. Set

$$A = F[T_0, T_1, T_2]/(aT_0^p + bT_1^p - T_2^p)$$

and let $S = Spec(A)$. One verifies easily that A is an integrally closed domain and hence S is a normal integral scheme. Let X be the normalization of S in the field $F(S)(\gamma)$ where $\gamma^p = b/a$. It is easy to check that X is isomorphic to $Spec(F(\alpha, \beta)[T_1, T_2])$ where $\alpha^p = a$, $\beta^p = b$ and the homomorphism $A = F[S] \to F[X] = F(\alpha, \beta)[T_1, T_2]$ maps T_0 to $\alpha^{-1} T_2 - \gamma^{-1} T_1$.

Take s to be the only singular point of S (i.e. $T_0(s) = T_1(s) = T_2(s) = 0$) and x to be the only point of X over s. Take I to be the maximal ideal of $\mathcal{O}_{S,s}$. Then $I\mathcal{O}_{X,x}$ is the maximal ideal of the regular local ring $\mathcal{O}_{X,x}$ and hence $e(I\mathcal{O}_{X,x}) = 1$. On the other hand one checks easily that $e(I) = p$ (see [10, 14.5]). Thus $n(x) = 1/p$. Therefore cycles of the form $cycl(f)(\mathcal{Z})$ do not have in general integral coefficients[1].

2. In the notations of previous example let Y be the scheme obtained by gluing of p copies of X in the singular point. One can easily see that the fundamental cycle on Y is an element of $c(Y/S, 0)$ which can

---

[1] This example is due to A.S.Merkurjev.

not be represented as a sum of cycles which correspond to integral subschemes of Y.

3.6. Functoriality of Chow presheaves

Let $f : S_1 \to S_2$ be a morphism of Noetherian schemes. We say that a closed subscheme Z of S_1 is proper with respect to f if the restriction of f to Z is a proper morphism. We say that a point s of S_1 is proper with respect to f if the closure of s in S_1 which we consider as a reduced closed subscheme is proper with respect to f.

Let S be a Noetherian scheme and $f : X \to Y$ be a morphism of schemes of finite type over S. Let further $\mathcal{Z} = \sum n_i z_i$ be a cycle on X which lies over generic points of S. We say that $\mathcal{Z}$ is proper with respect to f if all the points z_i are proper with respect to f. We define then a cycle $f_*(\mathcal{Z})$ on Y as the sum $\sum n_i m_i f(z_i)$ where m_i is the degree of the field extension $k_{f(z_i)}/k_{z_i}$ if this extension is finite and zero otherwise.

Theorem 3.6.1 *Let S be a Noetherian scheme, $p : X_1 \to X_2$ be a morphism of schemes of finite type over S, and $f : S' \to S$ be a Noetherian scheme over S. Set $X_i' = X_i \times_S S'$ $(i = 1, 2)$ and denote by $p' : X_1' \to X_2'$ be the corresponding morphism over S'. Let further $\mathcal{Z} = \sum n_i Z_i$ (resp. $\mathcal{W} = \sum m_j W_j$) be an element of $\mathbf{Z}(Hilb(X_1/S, r))$ (resp. of $\mathbf{Z}(Hilb(X_2/S, r)))$. Assume that the closed subschemes Z_i are proper with respect to p and*

$$p_*(cycl_{X_1}(\mathcal{Z})) = cycl_{X_2}(\mathcal{W}).$$

Then the cycle $cycl_{X_1'}(\mathcal{Z} \times_S S')$ is proper with respect to p' and we have

$$p_*'(cycl_{X_1'}(\mathcal{Z} \times_S S')) = cycl_{X_2'}(\mathcal{W} \times_S S').$$

Proof: Replacing X_1 by $\cup Z_i$ and X_2 by $(\cup W_j) \cup (\cup p(Z_i))$ we may assume that p is proper, X_1 is equidimensional of relative dimension r over S and all fibers of X_2 over S are of dimension $\leq r$.

Both cycles in question are linear combinations of points of X_2' which lie over generic points of S', are generic in their fibers over S' and are of dimension r in these fibers. Let $\eta' \in X_2'$ be any of such points and let η be its image in X_2. Computing the multiplicities of η' in our cycles we may replace X_2 by any open neighborhood of η. Let V be an irreducible component of X_2 which is not equidimensional of dimension r over S. Then $V \subset p(Z_i)$ for a certain i and hence V is dominated by a component of Z_i. Proposition 2.1.9 shows that all fibers of V over S are of dimension $< r$ and hence η does not belong to V. Thus throwing away bad components we may assume that X_2 is equidimensional of relative dimension r over S. Proposition 2.1.9 shows also that we may assume that p is a finite morphism. According to Proposition 2.1.3 we may assume further that the morphism $X_2 \to S$ has a factorization of the form $X_2 \to \mathbf{A}_S^r \to S$ where the first arrow is an equidimensional quasi-finite morphism. Since the morphism p maps components of X_1 onto components of X_2 we conclude that the composition $X_1 \to X_2 \to \mathbf{A}_S^r$ is also equidimensional (and quasifinite). Let Z_i^0 (resp. W_j^0) be the closed subset of Z_i (resp. of W_j) consisting of points where Z_i (resp. W_j) is not flat over $\mathbf{A}_S^r$. Lemma 3.2.3 shows that Z_i^0 and W_j^0 contain no points generic in their fibers over S. Thus η does not belong to $(\cup W_j^0) \cap (\cup p(Z_i^0))$ and shrinking X_2 further around η we may assume that Z_i and W_j are flat over $\mathbf{A}_S^r$. This shows that we may replace S by $\mathbf{A}_S^r$ and assume that $r = 0$.

Let τ' (resp. τ) be the image of η' (resp. η) in S' (resp. S). We may replace S' by $Spec(\mathcal{O}_{S',\tau'})$ and assume that S' is a local Artinian scheme. Let $\mathcal{O}_{S',\tau'}^{sh}$ and $\mathcal{O}_{S,\tau}^{sh}$ be the strict henselizations of the local rings $\mathcal{O}_{S',\tau'}$ and $\mathcal{O}_{S,\tau}$ respectively. Find a morphism $f_0 : Spec(\mathcal{O}_{S',\tau'}^{sh}) \to Spec(\mathcal{O}_{S,\tau}^{sh})$ making the following diagram commute

$$
\begin{array}{ccc}
Spec(\mathcal{O}_{S',\tau'}^{sh}) & \xrightarrow{f_0} & Spec(\mathcal{O}_{S,\tau}^{sh}) \\
\downarrow & & \downarrow \\
Spec(\mathcal{O}_{S',\tau'}) = S' & \to & S.
\end{array}
$$

In view of Lemma 2.3.1 it is sufficient to check that the flat pull-backs of cycles in question to the scheme $X_2' \times_{S'} Spec(\mathcal{O}_{S',\tau'}^{sh})$ coincide. Proposition 2.3.4 shows that these pull-backs are equal to $p_*'(cycl(\mathcal{Z} \times_S Spec(\mathcal{O}_{S',\tau'}^{sh})))$ and $cycl(\mathcal{W} \times_S Spec(\mathcal{O}_{S',\tau'}^{sh}))$ respectively. Thus we may replace S' by $Spec(\mathcal{O}_{S',\tau'}^{sh})$ in the same way we may replace S by $Spec(\mathcal{O}_{S,\tau}^{sh})$ and thus assume that S and S' are strictly henselian local schemes and τ, τ' are their closed points. Finally replacing X_2 by $Spec(\mathcal{O}_{X_2,\eta})$ we may assume that X_2 is a local scheme finite over S. Since S is strictly local it implies that X_2' is also local i.e. η' is the only point of X_2'. Let $\alpha_1', \ldots, \alpha_k'$ be all the points of X_1'. Using Lemma 3.2.1 we see that the multiplicity of η' in the cycle $p_*'(cycl_{X_1'}(\mathcal{Z} \times_S S'))$ is equal to

$$
\sum_{l=1}^{k} [k_{\alpha_l'} : k_{\eta'}] \sum_i length\mathcal{O}_{Z_i',\alpha_l'} =
$$
$$
= \sum_i \frac{n_i}{[k_{\eta'} : k_{\tau'}]} \sum_l [k_{\alpha_l'} : k_{\tau'}] length\mathcal{O}_{Z_i',\alpha_l'} =
$$
$$
= \frac{length\mathcal{O}_{S',\tau'}}{[k_{\eta'} : k_{\tau'}]} \sum_i n_i deg(Z_i/S).
$$

On the other hand the multiplicity of η' in the cycle $cycl_{X_2'}(\mathcal{W} \times_S S')$ is equal to

$$\sum_j m_j length\mathcal{O}_{W_j',\eta'} = \frac{length\mathcal{O}_{S',\tau'}}{[k_{\eta'} : k_{\tau'}]} \sum_j m_j deg(W_j/S).$$

Thus we have to show that $\sum_i n_i deg(Z_i/S) = \sum_j m_j deg(W_j/S)$. To do so choose a generic point τ^0 of S, let $\alpha_1^0, \ldots, \alpha_n^0$ be all points of X_2 over τ^0 and for each $s = 1, \ldots, n$ let $\alpha_{st}^0 (t = 1, \ldots, n_s)$ be the points of X_1 over α_s^0. Our assumptions imply that for every $s = 1, \ldots, n$ we have the following equality

$$\sum_t [k_{\alpha_{st}^0} : k_{\alpha_s^0}] \sum_i length\mathcal{O}_{Z_i,\alpha_{st}^0} = \sum_j m_j length\mathcal{O}_{W_j,\alpha_s^0}.$$

Taking the sum of this equalities with coefficients $[k_{\alpha_s^0} : k_{\tau^0}]$ and using once again Lemma 3.2.1 we get what we wanted.

Proposition 3.6.2 *Let $p : X \to Y$ be a morphism of schemes of finite type over a Noetherian scheme S and $\mathcal{Z} = \sum n_i z_i$ be an element of $Cycl(X/S, r)$ such that the points z_i are proper with respect to p. Then the following statements hold:*

1. *The cycle $p_*(\mathcal{Z})$ on Y belongs to $Cycl(Y/S, r)$.*

2. *For any morphism $f : S' \to S$ of Noetherian schemes the cycle $cycl(f)(\mathcal{Z})$ has the form $\sum m_j z_j'$ where the points z_j' are proper with respect to $p' = p \times_S S'$ and moreover*

$$p_*'(cycl(f)(\mathcal{Z})) = cycl(f)(p_*(\mathcal{Z})).$$

Proof: Denote by Z_i (resp. W_i) the closure of z_i (resp. $p(z_i)$) considered as an integral closed subscheme of X (resp. of Y). Replacing X by $\cup Z_i$ we may assume that the morphism p is proper. Let k be a field, $x : Spec(k) \to S$ be a k-point of S and (x_0, x_1, R) be a fat point of S over x. Set

$$\mathcal{Z}_0 = \sum n_i \phi_{x_1}(Z_i) \in \mathbf{Z}(\text{Hilb}(X \times_S Spec(R)/Spec(R), r))$$

$$\mathcal{W}_0 = \sum n_i m_i \phi_{x_1}(W_i) \in \mathbf{Z}(\text{Hilb}(Y \times_S Spec(R)/Spec(R), r))$$

where $m_i = [k_{z_i} : k_{p(z_i)}]$ if this field extension is finite and zero otherwise.

It is clear from the definitions that $cycl(\mathcal{W}_0) = (p \times_S Spec(R))_*(cycl(\mathcal{Z}_0))$. Theorem 3.6.1 implies now that

$$(x_0, x_1)^*(p_*(\mathcal{Z})) = cycl(\mathcal{W}_0 \times_{Spec(R)} Spec(k)) =$$

$$= (p \times_S Spec(k))_* cycl(\mathcal{Z}_0 \times_{Spec(R)} Spec(k)) = (p \times_S Spec(k))_*((x_0, x_1)^*(\mathcal{Z})).$$

Thus the cycle $(x_0, x_1)^*(p_*(\mathcal{Z}))$ is independent of the choice of fat point (x_0, x_1, R) over x. The same argument shows now that for any morphism $f : S' \to S$ of Noetherian schemes the cycle $p'_*(cycl(f)(\mathcal{Z}))$ meets the property defining the cycle $cycl(f)(p_*(\mathcal{Z}))$ and hence is equal to this cycle.

Corollary 3.6.3 *Let S be a Noetherian scheme and $f : X \to Y$ be a morphism (resp. a proper morphism) of schemes of finite type over S. Then there are homomorphisms:*

$$f_* : c(X/S, r) \to c(Y/S, r)$$

$$f_* : c_{equi}(X/S, r) \to c_{equi}(Y/S, r)$$
$$f_* : c^{eff}(X/S, r) \to c^{eff}(Y/S, r)$$

(resp. homomorphisms

$$f_* : z(X/S, r) \to z(Y/S, r)$$

$$f_* : z_{equi}(X/S, r) \to z_{equi}(Y/S, r)$$
$$f_* : z^{eff}(X/S, r) \to z^{eff}(Y/S, r))$$

such that for any composable pair of morphisms $X \xrightarrow{f} Y \xrightarrow{g} Z$ of schemes of finite type over S one has $(gf)_* = g_* f_*$.

Let us consider now the contravariant functoriality of Chow presheaves.

Lemma 3.6.4 *Let S be a Noetherian scheme and $f : X \to Y$ be a flat equidimensional morphism of dimension n of schemes of finite type over S. Then for any element $\mathcal{Z}$ in $Cycl(Y/S, r) \otimes \mathbf{Q}$ one has $f^*(\mathcal{Z}) \in Cycl(X/S, r+n)$ and for any Noetherian scheme $g : S' \to S$ we have*

$$cycl(g)(f^*(\mathcal{Z})) = (f \times_S S')^*(cycl(g)(\mathcal{Z})).$$

Proof: Easy.

Let S be a Noetherian scheme, $f : X \to Y$ be a flat (resp. flat and proper) equidimensional morphism of relative dimension n of schemes of finite type over S and $F(-, -)$ be one of the presheaves $z(-, -)$, $z^{eff}(-, -)$, $z_{equi}(-, -)$ (resp. $c(-, -)$, $c^{eff}(-, -)$, $c_{equi}(-, -)$). If $\mathcal{Z}$ is a cycle on Y which belongs to $F(Y/S, r)$ then by Lemma 3.6.4 the cycle $f^*(\mathcal{Z})$ belongs to $F(X/S, r+n)$ and this construction gives us homomorphisms of presheaves

$$f^* : F(Y/S, r) \to F(X/S, r+n).$$

For any composable pair $X \xrightarrow{f} Y \xrightarrow{g} Z$ of flat (resp. flat and proper) equidimensional morphisms of schemes of finite type over S we obviously have $(gf)^* = f^*g^*$.

Proposition 3.6.5 *Let S be a Noetherian scheme. Consider a pull-back square of schemes of finite type over S of the form:*

$$
\begin{array}{ccc}
Y' & \xrightarrow{g} & Y \\
{\scriptstyle p'}\downarrow & & \downarrow{\scriptstyle p} \\
X' & \xrightarrow{f} & X
\end{array}
$$

such that the morphism f is flat and equidimensional of dimension r. Assume further that f is proper and $F(-,-)$ is one of the presheaves $c(-,-)$, $c^{eff}(-,-)$, $c_{equi}(-,-)$ or that p is proper and $F(-,-)$ is one of the presheaves $z(-,-)$, $z^{eff}(-,-)$, $z_{equi}(-,-)$.

Then the following diagram of presheaves commutes:

$$
\begin{array}{ccc}
F(Y/S,n) & \xrightarrow{g^*} & F(Y'/S,n+d) \\
{\scriptstyle p_*}\downarrow & & \downarrow{\scriptstyle p'_*} \\
F(X/S,n) & \xrightarrow{f^*} & F(X'/S,n+d)
\end{array}
$$

Proof: It follows immediately from our definitions and 2.3.4.

As an application of our construction of push-forward homomorphisms we will show that for finite cycles over normal schemes our construction of base change homomorphisms gives the same answer as the one used in [16]. To start we will recall briefly the latter construction. For an integral scheme X we denote by $F(X)$ its field of functions.

Definition 3.6.6 *A finite surjective morphism of integral Noetherian schemes $f : Y \to S$ is called a pseudo-Galois covering if the field extension $F(Y)/F(S)$ is normal and the canonical homomorphism*

$$
Aut_S(Y) \to Aut_{F(S)}(F(Y)) = Gal(F(Y)/F(S))
$$

is an isomorphism.

Let S be a normal integral Noetherian scheme, X be an integral scheme and $p : X \to S$ be a finite surjective morphism. Let $g : S' \to S$ be any Noetherian integral scheme over S. Denote by X'_i the irreducible components of $X' = X \times_S S'$ and by x_i (resp. x) denote the generic point of X'_i (resp. of X). Since any normal scheme is geometrically unibranch Theorem 3.4.2 implies that $x \in Cycl_{equi}(X/S,0)$. Consider the cycle $cycl(g)(x) = \sum n_i x'_i$.

Assume that there exists[2] a pseudo-Galois covering $f : Y \to S$ and an S-morphism $q : Y \to X$. Let G be the Galois group $Gal(F(Y)/F(S)) = Aut_S(Y)$. Denote by Y'_j the irreducible components of $Y' = Y \times_S S'$. It is easy to check that the action of G permutes the components Y'_j transitively so that in particular the field extensions $F(Y'_j)/F(S')$ are all isomorphic. Denote by $l(i)$ the number of components Y'_j lying over X'_i and by l the total number of components Y'_j.

Proposition 3.6.7 *In the above notations one has:*

$$n_i = \frac{[F(X) : F(S)]l(i)}{[F(X'_i) : F(S')]l}.$$

Proof: Denote the generic point of Y (resp. of Y'_j) by y (resp. by y'_j). The cycle y is in $Cycl(Y/S, 0)$ by Theorem 3.4.2 and has the following obvious properties:

1. $f_*(y) = [F(Y) : F(S)]s$ where s is the generic point of S.
2. $q_*(y) = [F(Y) : F(X)]x$
3. $\sigma_*(y) = y$ for any $\sigma \in G$.

Consider the cycle $cycl(g)(y) = \sum m_j y'_j$. Proposition 3.6.2 shows that $(f \times_S S')_*(cycl(g)(y)) = [F(Y) : F(S)]s$ i.e. $\sum m_j[F(Y'_j) : F(S')] = [F(Y) : F(S)]$. Moreover for any $\sigma \in G$ we have

$$(\sigma \times_S S')_*(cycl(g)(y)) = cycl(g)(\sigma_*(y)) = cycl(g)(y).$$

Since the action of G on the set $y'_1, \ldots, y'_l$ is transitive we conclude that all multiplicities m_j are the same and equal to $\frac{[F(Y):F(S)]}{l[F(Y'_j):F(S')]}$. Finally $cycl(g)(x) = \frac{1}{[F(Y):F(X)]}(q \times_S S')_*(cycl(g)(y))$ and hence

$$n_i = \frac{1}{[F(Y) : F(X)]} \sum_{y'_j/x'_i} m_j[F(Y'_j) : F(X'_i)] = \frac{[F(X) : F(S)]l(i)}{[F(X'_i) : F(S')]l}.$$

3.7. Correspondence homomorphisms

Let $Y \to X$ be a scheme of finite type over a Noetherian scheme X. For any cycle $\mathcal{Y} \in Cycl(Y/X, r) \otimes \mathbf{Q}$ define a homomorphism

$$Cor(\mathcal{Y}, -) : Cycl(X) \otimes \mathbf{Q} \to Cycl(Y) \otimes \mathbf{Q}$$

---

[2]Such a covering always exists. For an excellent scheme S it follows trivially from the finiteness of normalizations of S in finite extensions of its field of functions. The proof in general case is a little more complicated.

as follows. Let $\mathcal{X} = \sum n_i x_i$ be an element of $Cycl(X) \otimes \mathbf{Q}$. Denote by X_i the closure of the point x_i which we consider as an integral closed subscheme of X. Let $i_{X_i} : X_i \to X$ be the corresponding embedding. We set

$$Cor(\mathcal{Y}, \mathcal{X}) = \sum n_i (X_i \times_X Y \to Y)_* (cycl(i_{X_i})(\mathcal{Y})).$$

Lemma 3.7.1 *Consider a pull-back square of Noetherian schemes of the form*

$$\begin{array}{ccc} Y' & \xrightarrow{q} & Y \\ \downarrow & & \downarrow \\ X' & \xrightarrow{p} & X. \end{array}$$

Assume that the morphism $Y \to X$ is of finite type and let $\mathcal{Y}$ be an element of $Cycl(Y/X, r) \otimes \mathbf{Q}$ and $\mathcal{X}'$ be a cycle on X' which is proper with respect to p. Then one has:

$$Cor(\mathcal{Y}, p_*(\mathcal{X}')) = q_*(Cor(cycl(p)(\mathcal{Y}), \mathcal{X}')).$$

Proof: We may assume that $\mathcal{X}' = cycl_{X'}(Z')$ where Z' is an integral closed subscheme of X' and $p_{Z'} : Z' \to X$ is a proper morphism. We have

$$q_*(Cor(cycl(p)(\mathcal{Y}), \mathcal{X}')) = (Z' \times_X Y \to Y)_* cycl(Z' \to X)(\mathcal{Y})$$

and

$$Cor(\mathcal{Y}, p_*(\mathcal{X}')) = n(p(Z') \times_X Y \to Y)_* cycl(p(Z') \to X)(\mathcal{Y})$$

where $n = [F(Z') : F(p(Z'))]$ if this extension is finite and zero otherwise. We have further

$$(Z' \times_X Y \to Y)_* cycl(Z' \to X)(\mathcal{Y}) =$$

$$= (p(Z') \times_X Y \to Y)_* (q \times_X Y)_* cycl(q) cycl(p(Z') \to X)(\mathcal{Y}).$$

where q is the morphism $Z' \to p(Z')$. It is sufficient to show that

$$(Z' \times_X Y \to p(Z') \times_X Y)_* cycl(Z' \to p(Z'))(\mathcal{W}) = n(\mathcal{W})$$

for any element $\mathcal{W}$ in $Cycl(p(Z') \times_X Y/p(Z'))$. To prove the last statement we may replace $p(Z')$ by its generic point $Spec(F(Z'))$ and assume that $\mathcal{W} = cycl(W)$ where W is an integral closed subscheme of $Y \times_X Spec(F(Z'))$. For an infinite field extension $F(Z')/F(p(Z'))$ we immediately conclude that the left hand side of our equality equals zero (as well as the right hand side). For finite field extension our equality follows from Lemmas 3.3.12 and 2.3.5.

The following lemma is straightforward.

Lemma 3.7.2 *Consider a pull-back square of Noetherian schemes of the form*

$$
\begin{array}{ccc}
Y' & \xrightarrow{q} & Y \\
\downarrow & & \downarrow \\
X' & \xrightarrow{p} & X.
\end{array}
$$

Assume that the morphism $Y \to X$ is of finite type, schemes X' and X are reduced and the morphism p is flat. Then for any cycle $\mathcal{X}$ in $Cycl(X) \otimes \mathbf{Q}$ and any cycle $\mathcal{Y}$ in $Cycl(Y/X, r) \otimes \mathbf{Q}$ one has:

$$
q^* Cor(\mathcal{Y}, \mathcal{X}) = Cor(q^*(\mathcal{Y}), p^*(\mathcal{X})).
$$

Theorem 3.7.3 *Let S be a Noetherain scheme and $f : Y \to X$ be a morphism of schemes of finite type over S. Let further $\mathcal{Y} = \sum n_i y_i$ be an element of $Cycl(Y/X, n) \otimes \mathbf{Q}$ and $\mathcal{X}$ be an element of $Cycl(X/S, m) \otimes \mathbf{Q}$. Then the element $Cor(\mathcal{Y}, \mathcal{X})$ belongs to the group $Cycl(Y/S, n + m)$. Moreover for any Noetherian scheme $g : S' \to S$ over S one has*

$$
cycl(g)(Cor(\mathcal{Y}, \mathcal{X})) = Cor(cycl(g \times_S X)(\mathcal{Y}), cycl(g)(\mathcal{X})).
$$

Proof: We will first consider the following particular case of our theorem.

Lemma 3.7.4 *The statement of the theorem holds if $\mathcal{Y} = cycl_Y(Y_0)$, $\mathcal{X} = cycl_X(X_0)$ where Y_0 (resp. X_0) is a closed subscheme of Y (resp. of X) which is flat over X (resp. over S).*

Proof: Set $X' = X \times_S S'$, $Y' = Y \times_S S'$. By Lemma 3.3.10 $Cor(\mathcal{Y}, \mathcal{X})$ coincides with the cycle associated with the closed subscheme $Y_0 \times_X X_0$ in Y which is clearly flat over S. Therefore $Cor(\mathcal{Y}, \mathcal{X})$ is a relative cycle over S by 3.2.5. We have further

$$
cycl(g)(Cor(\mathcal{Y}, \mathcal{X})) = cycl(g)(cycl_Y(Y_0 \times_X X_0)) = cycl_{Y'}((Y_0 \times_X X_0) \times_S S')
$$

and

$$
Cor(cycl(g \times_S X)(\mathcal{Y}), cycl(g)(\mathcal{X}))
$$
$$
= Cor(cycl_{Y'}(Y_0 \times_S S'), cycl_{X'}(X_0 \times_S S'))
$$
$$
= cycl_{Y'}((Y_0 \times_S S') \times_{X'} (X_0 \times_S S'))
$$

Since

$$
(Y_0 \times_S S') \times_{X'} (X_0 \times_S S') \cong (Y_0 \times_X X_0) \times_S S'
$$

we conclude that our equality holds.

Let k be a field, $x : Spec(k) \to S$ be a k-valued point of S and (x_0, x_1, R) be a fat point of S over x. It is clearly sufficient to show that

$$(x_0, x_1)^*(Cor(\mathcal{Y}, \mathcal{X})) = Cor(cycl(x \times_S X)(\mathcal{Y}), cycl(x)(\mathcal{X})).$$

Let (x_0, Id) be the obvious fat point of $Spec(R)$. By Lemma 3.7.2 (and Lemma 3.3.12) we have

$$(x_0, x_1)^*(Cor(\mathcal{Y}, \mathcal{X})) = (x_0, Id)^*(Cor(cycl(x_1 \times_S X)(\mathcal{Y}), cycl(x_1)(\mathcal{X}))).$$

We may assume now that $S = Spec(R)$ where R is a discrete valuation ring. In this case all cycles which are formal linear combinations of points over the generic point of S are relative cycles over S (by Corollary 3.2.6). Thus we may assume that $\mathcal{X} = cycl_X(X_0)$ where X_0 is a closed integral subscheme of X which is equidimensional of relative dimension m over S and we have to show that

$$cycl(x)(Cor(\mathcal{Y}, \mathcal{X})) = (Cor(cycl(x \times_S X)(\mathcal{Y}), cycl(x)(\mathcal{X}))).$$

Denote by i the closed embedding $X_0 \to X$. We set

$$
\begin{aligned}
&Y_0 = Y \times_Y X_0 \\
&X_x = X \times_S Spec(k) & &Y_x = Y \times_S Spec(k) \\
&(X_0)_x = X_0 \times_S Spec(k) & &(Y_0)_x = Y_0 \times_S Spec(k) \\
&u = pr_1 : Y_0 \to Y & &u_x = u \times_S Spec(k) : (Y_0)_x \to Y_x \\
&v = pr_1 : X_x \to X & &v_0 = pr_1 : (X_0)_x \to X_0
\end{aligned}
$$

We have

$$x^*(Cor(\mathcal{Y}, \mathcal{X})) = x^*(u_*(cycl(i)(\mathcal{Y}))) = ((u_x)_*(cycl(x)(cycl(i)(\mathcal{Y})))) =$$

$$= (u_x)_*(cycl(x)(Cor(cycl(i)(\mathcal{Y}), cycl_{X_0}(X_0))))$$

and

$$Cor(cycl(v)(\mathcal{Y}), cycl(x)(\mathcal{X})) =$$

$$= (u_x)_*(Cor(cycl(v_0)cycl(i)(\mathcal{Y}), cycl(x)(cycl_{X_0}(X_0)))).$$

We may replace now X by X_0 and assume further that X is integral and equidimensional of relative dimension m over S and $\mathcal{X}$ is the fundamental cycle of X.

Denote by Y_i the closures of the points y_i which we consider as integral closed subschemes of Y. Let $f_X : \tilde{X} \to X$ be a blow-up of X_{red} such that the proper transforms of Y_i with respect to f_X are flat and equidimensional of relative dimension n over $\tilde{X}$. Consider the pull-back square

$$
\begin{array}{ccc}
\tilde{Y} & \overset{f_{\tilde{X}}}{\to} & Y \\
\downarrow & & \downarrow \\
\tilde{X} & \overset{f_X}{\to} & X.
\end{array}
$$

Let $\tilde{\mathcal{X}}$ be the fundamental cycle of $\tilde{X}$. Since S is the spectrum of a discrete valuation ring we have $\tilde{\mathcal{X}} \in Cycl(\tilde{X}/S, m)$. Consider the following diagram of abelian groups (we write $Cycl(-)$ instead of $Cycl(-,-) \otimes \mathbf{Q}$):

$$
\begin{array}{ccccccc}
Cycl(Y/X) & \overset{cycl(f_X)}{\longrightarrow} & Cycl(\tilde{Y}/\tilde{X}) & \overset{Cor(-,\tilde{\mathcal{X}})}{\longrightarrow} & Cycl(\tilde{Y}/S) & \overset{(f_Y)_*}{\longrightarrow} & Cycl(Y/S) \\
\downarrow cycl(j) & & \downarrow cycl(\tilde{j}) & & \downarrow cycl(x) & & \downarrow cycl(x) \\
Cycl(Y_x/X_x) & \overset{cycl(f_{X_x})}{\longrightarrow} & Cycl(\tilde{Y}_x/\tilde{X}_x) & \overset{Cor(-,\tilde{\mathcal{X}}_x)}{\longrightarrow} & Cycl(\tilde{Y}_x/S_x) & \overset{(f_{Y_x})_*}{\longrightarrow} & Cycl(Y_x/S_x)
\end{array}
$$

where
$$
\begin{array}{l}
S_s = Spec(k) \\
X_x = X \times_S Spec(k) \quad Y_x = Y \times_S Spec(k) \\
\tilde{X}_x = \tilde{X} \times_S Spec(k) \quad \tilde{Y}_x = \tilde{Y} \times_S Spec(k)
\end{array}
$$

and
$$
j : X_x \to X \quad \tilde{j} : \tilde{X}_x \to \tilde{X}
$$

are the obvious morphisms and $\tilde{\mathcal{X}}_x = cycl(x)(\tilde{\mathcal{X}})$.

The composition of the upper horizontal arrows equals $Cor(-, \mathcal{X})$ by Lemma 3.7.1. By Proposition 3.6.2

$$
(\tilde{X}_x \to \tilde{X})_*(\tilde{\mathcal{X}}_x) = cycl(x)(\mathcal{X})
$$

and thus the composition of lower horizontal arrows equals $Cor(-, cycl(x)(\mathcal{X}))$ by Lemma 3.7.1. We have only to show now that our diagram is commutative on $\mathcal{Y} \in Cycl(Y/X)$. The first square is commutative since $f_X \circ \tilde{j} = j \circ f_{X_x}$. The last square is commutative by Proposition 3.6.2. Finally the middle square is commutative on $cycl(f_X)(\mathcal{Y})$ by our choice of the blow-up $\tilde{X} \to X$ and Lemma 3.7.4.

Corollary 3.7.5 *Let S be a Noetherian scheme and $X_1 \to X_2 \overset{p}{\to} S$ be a morphism of schemes of finite type over S. Let $F(-,-)$ be one of the presheaves $z(-,-), z^{eff}(-,-), z_{equi}(-,-), c(-,-), c_{equi}(-,-), c^{eff}(-,-)$. Then for any $n, m \geq 0$ there is a canonical morphism of presheaves of the form*

$$
Cor_{X_1/X_2} : p_*(F(X_1/X_2, n)) \otimes F(X_2/S, m) \to F(X_1/S, m+n)
$$

Lemma 3.7.1 implies immediately the following proposition.

Proposition 3.7.6 *Let S be a Noetherian scheme. Consider a pull-back square of schemes of finite type over S of the form:*

$$
\begin{array}{ccc}
Y' & \overset{g}{\to} & Y \\
\downarrow & & \downarrow \\
X' & \overset{f}{\to} & X
\end{array}
$$

Denote by $p : X \to S$, $p' : X' \to S$ the structural morphisms of X and X' respectively. Assume further that $F(-)$ is one of the presheaves $c(-,-)$, $c^{eff}(-,-)$, $c_{equi}(-,-)$ or that f is proper and $F(-,-)$ is one of the presheaves $z(-,-)$, $z^{eff}(-,-)$, $z_{equi}(-,-)$.

Then the following diagram of morphisms of presheaves commutes:

$$
\begin{array}{ccc}
p_*F(Y/X) \otimes F(X'/S) & \overset{Id \otimes f_*}{\Longrightarrow} & p_*F(Y/X) \otimes F(X/S) \\
\downarrow & & \searrow \\
 & & \qquad F(Y/S) \\
p'_*F(Y'/X') \otimes F(X'/S) & \overset{Cor_{Y'/X'}}{\Longrightarrow} \quad F(Y'/S) & \nearrow
\end{array}
$$

Proposition 3.7.7 *Let $Z \to Y \to X$ be a composable pair of morphisms of finite type of Noetherian schemes. Let further $\mathcal{X}$ be an element of $Cycl(X) \otimes \mathbf{Q}$ and $\mathcal{Z}, \mathcal{Y}$ be elements of $Cycl(Z/Y, n) \otimes \mathbf{Q}$ and $Cycl(Y/X, m) \otimes \mathbf{Q}$ respectively. Then one has*

$$
Cor(Cor(\mathcal{Z}, \mathcal{Y}), \mathcal{X}) = Cor(\mathcal{Z}, Cor(\mathcal{Y}, \mathcal{X})).
$$

Proof: It follows easily from definitions and Theorem 3.7.3.

Corollary 3.7.8 *Let S be a Noetherian scheme and $Z \overset{f}{\to} Y \overset{g}{\to} X$ be a composable pair of morphisms of schemes of finite type over S. Denote by $h : Z \to S$ the structural morphism and let $F(-)$ be one of the presheaves $z(-,-)$, $z^{eff}(-,-)$, $z_{equi}(-,-)$, $c(-,-)$, $c_{equi}(-,-)$, $c^{eff}(-,-)$.*

Then the following diagram of morphisms commutes:

$$
\begin{array}{ccc}
h_*g_*F(Z/Y) \otimes h_*F(Y/X) \otimes F(X/S) & \to & h_*F(Z/X) \otimes F(X/S) \\
\downarrow & & \downarrow \\
h_*g_*F(Z/Y, n) \otimes F(Y/S) & \to & F(Z/S).
\end{array}
$$

Existence of correspondence homomorphisms for relative cycles allows us to construct the homomorphism of external product (see [4]). Let $p_X : X \to S$, $p_Y : Y \to S$ be two schemes of finite type over a Noetherian scheme S. Let further $F(-,-)$ be one of the groups $z(-,-)$, $z_{equi}(-,-)$, $z^{eff}(-,-)$, $c(-,-)$, $c_{equi}(-,-)$, $c^{eff}(-,-)$. We define the external product homomorphism

$$F(X/S, n) \otimes F(Y/S, m) \to F(X \times_S Y/S, n+m)$$

as the following composition

$$F(X/S, n) \otimes F(Y/S, m) \overset{cycl(p_Y) \otimes Id}{\longrightarrow} F(X \times_S Y/Y, n) \otimes F(Y/S, m)$$

$$\overset{Cor_{X \times_S Y/Y}}{\longrightarrow} F(X \times_S Y/S, n+m).$$

One can verify easily using results of this section that it satisfies all the standard properties of external products of cycles and defines a homomorphism of the corresponding Chow presheaves.

4. Chow sheaves in the h-topologies

4.1. The h-topologies

In this section we will remind briefly the definitions and some basic properties of three Grothendieck topologies (the h-topology, the qfh-topology and the cdh-topology) on the categories of schemes. For more information on the h- and the qfh-topologies see [18] or [16].

A morphism of schemes $p : X \to Y$ is called a topological epimorphism if the underlying Zariski topological space of Y is a quotient space of the underlying Zariski topological space of X (i.e. p is surjective and a subset A of Y is open if and only if $p^{-1}(A)$ is open in X), p is called a universal topological epimorphism if for any Z/Y the morphism $p_Z : X \times_Y Z \to Z$ is a topological epimorphism.

An h-covering of a scheme X is a finite family of morphisms of finite type $\{p_i : X_i \to X\}$ such that $\coprod p_i : \coprod X_i \to X$ is a universal topological epimorphism.

A qfh-covering of a scheme X is an h-covering $\{p_i\}$ such that all the morphisms p_i are quasi-finite.

h-coverings (resp. qfh-coverings) define a pretopology on the category of schemes, the h-topology (resp. the qfh-topology) is the associated topology.

The definition of the cdh-topology is a little less natural. Namely the cdh-topology on the category of schemes is the minimal Grothendieck topology such that the following two types of coverings are cdh-coverings.

1. Nisnevich coverings, i.e. etale coverings $\{U_i \xrightarrow{p_i} X\}$ such that for any point x of X there is a point x_i on one of the U_i such that $p_i(x_i) = x$ and the morphism $Spec(k_{x_i}) \to Spec(k_x)$ is an isomorphism.

2. Coverings of the form $X' \coprod Z \xrightarrow{p_{X'} \coprod p_Z} X$ such that $p_{X'}$ is a proper morphism, p_Z is a closed embedding and the morphism $p_{X'}^{-1}(X - p_Z(Z)) \to X - p_Z(Z)$ is an isomorphism.

Obviously the h-topology is stronger than both the qfh- and the cdh-topology, the cdh-topology is stronger than the Zariski topology and standard results on flat morphisms show that qfh-topology is stronger than the flat topology.

Lemma 4.1.1 *Let $\{U_i \xrightarrow{p_i} X\}$ be an h-covering of a Noetherian scheme X. Denote by $\coprod V_j$ the disjoint union of all irreducible components of $\coprod U_i$ which dominate an irreducible component of X. Then the morphism $q : \coprod V_j \to X$ is surjective.*

Proof: See [18].

Remark: In fact the property of h-coverings considered in Lemma 4.1.1 is characteristic. Namely one can show that a morphism of finite type $X \to S$ of Noetherian schemes is an h-covering if and only if for any Noetherian scheme T over S the union of irreducible components of $X \times_S T$ which dominate irreducible components of T is surjective over T.

Proposition 4.1.2 *Let X be a Noetherian scheme and $\mathbf{U} = \{U_i \to X\}$ be an h-covering of X. Then there is a refinement $\{V_j \to X\}$ of $\mathbf{U}$ such that each morphism $q_j : V_j \to X$ admits a decomposition of the form $V_j \xrightarrow{q_j^f} W_j \xrightarrow{q_j^p} X$ such that q_j^f is a faithfully flat morphism, W_j is irreducible and q_j^p is an abstract blow-up (see Defenition 2.2.4) of an irreducible component of X.*

Proof: To prove our proposition we may assume that X is integral and our covering is of the form $U \to X$. By Theorem 2.2.2 there is a blow-up $p : W \to X$ such that the proper transform $\tilde{U}$ of U is flat over W. It is sufficient to show that the morphism $\tilde{U} \to W$ is surjective. Let Z be a closed subscheme in X such that $Z \neq X$ and the morphism $p : W \to X$ is an isomorphism outside Z. Since $W \times_X U \to W$ is an h-covering and the closure of the complement $W \times_X U - \tilde{U}$ lies over $p^{-1}(Z)$ and, therefore is not dominant over any irreducible component of W, the surjectivity of the morphism $\tilde{U} \to W$ follows from Lemma 4.1.1.

Lemma 4.1.3 *Let S be a Noetherian scheme and $p : X \to S$ be a scheme of finite type over S. Suppose that there is an h-covering $f : S' \to S$ such that the scheme $X' = X \times_S S'$ is proper over S'. Then X is proper over S.*

Proof: Denote the projections $X \times_S S' \to X$, $X \times_S S' \to S'$ by f' and p' respectively. Let Z be a closed subset in X. It is sufficient to show that $p(Z)$ is closed in S. Since $f : S' \to S$ is an h-covering $p(Z)$ is closed in S if and only if $f^{-1}(p(Z))$ is closed in S'. We obviously have $f^{-1}(p(Z)) = p'((f')^{-1}(Z))$. Since p' is proper we conclude that this set is closed.

Lemma 4.1.4 *Let S be a Noetherian scheme and $\phi : F \to G$ be a morphism of presheaves on the category of Noetherian schemes over S. Assume further that for any integral Noetherian scheme T over S and any section $a \in G(T)$ of G over T there is an abstract blow-up $f : T' \to T$ such that $f^*(a)$ belongs to the image of $\phi_{T'} : F(T') \to G(T')$. Then $\phi_{cdh} : F_{cdh} \to G_{cdh}$ is an epimorphism of the associated cdh-sheaves.*

Proof: It is sufficient to show that for any section $a \in G(S)$ of G over S there is a cdh-covering $\{X_i \xrightarrow{p_i} S\}$ such that $p_i^*(a)$ belongs to the image of $\phi_{X_i} : F(X_i) \to G(X_i)$. Our condition implies that there is a closed subscheme S_1 in S such that $S_1 \neq S$ and a proper surjective morphism $q : X_1 \to S$ such that $X_1 - q^{-1}(S_1) \to S - S_1$ is an isomorphism and $q^*(a)$ belongs to $Im(\phi_{X_1})$. Since $X_1 \coprod S_1 \to S$ is a cdh-covering of S we reduced our problem to S_1. Repeating this construction we get a sequence of closed subschemes $\ldots S_i \subset S_{i-1} \subset \ldots \subset S_1 \subset S$ such that $S_i \neq S_{i-1}$. Since the scheme S is Noetherian this sequence must be finite, i.e. $S_i = \emptyset$ for $i > n$. The family of morphisms $\{X_i \to S_i \to S\}_{i=1,\ldots,n}$ is then a cdh-covering of S with the required property.

4.2. Sheaves in the h-topologies associated with Chow presheaves

Lemma 4.2.1 *Let $X \to S$ be a scheme of finite type over a Noetherian scheme S and $\mathcal{Z}$ be an element of $Cycl(X/S, r)$. Then there is an abstract blow-up $f : S' \to S$ such that*

$$cycl(f)(\mathcal{Z}) = \sum n_i cycl(Z_i') = \sum n_i cycl((Z_i')_{red})$$

where Z_i' are irreducible closed subschemes of $X \times_S S'$ which are flat and equidimensional of relative dimension r over S'.

Proof: It follows immediately from Theorem 2.2.2.

Theorem 4.2.2 *Let $X \to S$ be a scheme of finite type over a Noetherian scheme S. Then the presheaves $Cycl(X/S, r)_{\mathbf{Q}}$ and $Cycl^{eff}(X/S, r)_{\mathbf{Q}}$ are h-sheaves and if S is a scheme of exponential characteristic p the same holds for the presheaves $z(X/S, r) \otimes \mathbf{Z}[1/p]$ and $z^{eff}(X/S, r) \otimes \mathbf{Z}[1/p]$.*

Proof: We will only consider the case of $Cycl(X/S, r)_{\mathbf{Q}}$. The proof for $Cycl^{eff}(X/S, r)_{\mathbf{Q}}$ is similar.

Note first that the presheaves $Cycl(X/S, r)_{\mathbf{Q}}$ are separated with respect to the h-topology, i.e. the canonical morphisms of presheaves $Cycl(X/S, r)_{\mathbf{Q}} \to (Cycl(X/S, r)_{\mathbf{Q}})_h$ are monomorphisms. Therefore according to the standard construction of the sheaf associated with a presheaf (see [11],[1]) it is sufficient to show that for any cofinial class of h-coverings $\{U_i \to S\}_{i=1,...,n}$ of S the following sequence of abelian groups is exact:

$$Cycl(X/S, r)_{\mathbf{Q}}(S) \to \bigoplus_i Cycl(X/S, r)_{\mathbf{Q}}(U_i)$$

$$\to \bigoplus_{i,j} Cycl(X/S, r)_{\mathbf{Q}}(U_i \times_S U_j)$$

We may obviously replace the covering $\{U_i \to S\}$ by the covering

$$p : U = \coprod U_i \to S$$

and hence assume that $n = 1$. By Lemma 4.1.1 we may also assume that any irreducible component of U dominates an irreducible component of S.

We will use the following simple lemma.

Lemma 4.2.3 *Let $p_Y : Y \to Spec(k)$, $p_X : X \to Spec(k)$ be two schemes of finite type over a field k. Then the sequence of abelian groups*

$$Cycl(X/Spec(k), r) \otimes \mathbf{Q} \overset{cycl(p_Y)}{\longrightarrow} Cycl(X \times_{Spec(k)} Y/Y, r) \otimes \mathbf{Q} \overset{cycl(pr_1) - cycl(pr_2)}{\longrightarrow}$$

$$\to Cycl(X \times_{Spec(k)} Y \times_{Spec(k)} Y/Y \times_{Spec(k)} Y, r) \otimes \mathbf{Q}$$

(where $pr_i : Y \times_{Spec(k)} Y \to Y$ are the projections) is exact.

Let $\mathcal{Y}$ be an element of $Cycl(X/S, r)_{\mathbf{Q}}(U)$ such that $cycl(pr_1)(\mathcal{Y}) = cycl(pr_2)(\mathcal{Y})$. Since any irreducible component of U dominates an irreducible component of S Lemma 4.2.3 implies that there exists a cycle $\mathcal{Z}$ on X such that for any generic point $\eta : Spec(L) \to U$ of U one has $cycl(p \circ \eta)(\mathcal{Z}) = cycl(\eta)(\mathcal{Y})$. It is obviously sufficient to show that $\mathcal{Z}$ belongs to $Cycl(X/S, r) \otimes \mathbf{Q}$.

Let k be a field and (x_0, x_1, R), (y_0, y_1, Q) be two fat k points of S over a k-point $s : Spec(k) \to S$ of S. We have to show that $(x_0, x_1)^*(\mathcal{Z}) = (y_0, y_1)^*(\mathcal{Z})$.

Lemma 4.2.4 *Let R be a discrete valuation ring and $p : X \to Spec(R)$ be an h-covering of $Spec(R)$. Then there exist a discrete valuation ring Q, a dominant morphism $p_0 : Spec(Q) \to Spec(R)$ which takes the closed point of $Spec(Q)$ to the closed point of $Spec(R)$ and a morphism $s : spec(R) \to X$ such that $p \circ s = p_0$.*

Proof: By Lemma 4.1.1 there is an irreducible component X_0 of X which is dominant over S and whose image contains the closed point of $Spec(R)$ i.e. X_0 is irreducible and surjective over S. Then by [6] there is a discrete valuation ring Q and a dominant morphism $Spec(Q) \to X_0$ such that the image of the closed point of $Spec(Q)$ lies in the closed fiber of X_0 over $Spec(R)$.

By Lemma 4.2.4 we may construct a commutative diagram of the form

such that (x_0', x_1', R'), (y_0', y_1', Q') are fat points of U and f, g are dominant morphisms. It is obviously sufficient to show that $(x_0, x_1)^*(\mathcal{Z}) \otimes_k L = (y_0, y_1)^*(\mathcal{Z}) \otimes_k L$. We clearly have:

$$(x_0, x_1)^*(\mathcal{Z}) \otimes_k L = (x_0', x_1')^*(\mathcal{Y})$$

$$(y_0, y_1)^*(\mathcal{Z}) \otimes_k L = (y_0', y_1')^*(\mathcal{Y}).$$

It is sufficient to show that the right hand sides of these two equalities coincide. Denote the L points $x_1' \circ x_0'$ and $y_1' \circ y_0'$ of U by x' and y' respectively. Since $\mathcal{Y}$ is an element of $Cycl(X \times_S U/U, r)$ we have

$$(x_0', x_1')^*(\mathcal{Y}) = cycl(x')(\mathcal{Y}) = cycl(x' \times_S y')(cycl(pr_1)(\mathcal{Y}))$$

$$(y_0', y_1')^*(\mathcal{Y}) = cycl(y')(\mathcal{Y}) = cycl(x' \times_S y')(cycl(pr_2)(\mathcal{Y}))$$

(where $pr_i : U \times_S U \to U$ are the projections) and since

$$cycl(pr_1)(\mathcal{Y}) = cycl(pr_2)(\mathcal{Y})$$

by our condition on $\mathcal{Y}$ we conclude that $(x_0, x_1)^*(\mathcal{Z}) = (y_0, y_1)^*(\mathcal{Z})$.

Proposition 4.2.5 *Let $X \to S$ be a scheme of finite type over a scheme S. Then the presheaves $Cycl_{equi}(X/S, r)_{\mathbf{Q}}$ are qfh-sheaves and if S is a scheme of exponential characteristic p the same holds for the presheaves $z_{equi}(X/S, r) \otimes \mathbf{Z}[1/p]$.*

Proof: It is clearly sufficient to consider the case of $Cycl_{equi}(X/S, r)_{\mathbf{Q}}$. Let $\mathcal{Z}$ be an element of $Cycl(X/S, r) \otimes \mathbf{Q}$. In view of Theorem 4.2.2 it is sufficient to show that if there exists a qfh-covering $f : S' \to S$ of S such that $supp(cycl(f)(\mathcal{Z}))$ is equidimensional of dimension r over S' then $supp(\mathcal{Z})$ is equidimensional of dimension r over S. By Theorem 2.1.1 we have only to show that $dim(supp(\mathcal{Z})/S) \leq r$. By Lemma 4.1.1 we may assume that any irreducible component of S' dominates an irreducible component of S and $dim(supp(cycl(f)(\mathcal{Z}))/S') \leq r$. Then $supp(cycl(f)(\mathcal{Z}))$ is the closure in $supp(\mathcal{Z}) \times_S S'$ of the fibers of this scheme over generic points of S'. Since the projection $supp(\mathcal{Z}) \times_S S' \to supp(\mathcal{Z})$ is a qfh-covering using again Lemma 4.1.1 we conclude that the morphism $g : supp(cycl(f)(\mathcal{Z})) \to supp(\mathcal{Z})$ is surjective and hence $dim(supp(\mathcal{Z})/S) \leq r$ since the morphisms g and f are quasi-finite.

Proposition 4.2.6 *Let $X \to S$ be a scheme of finite type over a Noetherian scheme S. Then the presheaves $PropCycl(X/S, r)_{\mathbf{Q}}$ and $PropCycl^{eff}(X/S, r)_{\mathbf{Q}}$ are h-sheaves and if S is a scheme of exponential characteristic p the same holds for the presheaves $c(X/S, r) \otimes \mathbf{Z}[1/p]$ and $c^{eff}(X/S, r) \otimes \mathbf{Z}[1/p]$.*

Proof: In view of Theorem 4.2.2 it is sufficient to show that if $\mathcal{Z}$ is an element of $Cycl_{equi}(X/S, r) \otimes \mathbf{Q}$ and there exists an h-covering $f : S' \to S$ such that the support $supp(cycl(f)(\mathcal{Z}))$ is proper over S' then $supp(\mathcal{Z})$ is proper over S.

We may obviously assume that S is reduced. By 4.1.2 we may assume further that f admits a decomposition of the form $S' \xrightarrow{f_0} S'' \xrightarrow{f_1} S$ such that f_1 is an abstract blow-up and f_0 is faithfully flat. Let $cycl(f_1)(\mathcal{Z}) = \sum n_i cycl(Z_i'')$ where Z_i'' are irreducible closed subschemes of $X'' = X \times_S S''$. Since f_0 is flat the closed subsets

$$supp(cycl(f_1 f_0)(\mathcal{Z})) \quad \text{and} \quad supp(cycl(f_1)(\mathcal{Z})) \times_{S''} S'$$

coincide. Since f_0 in particular is an h-covering we conclude by Lemma 4.1.3 that $supp(cycl(f_1 f_0)(\mathcal{Z}))$ is proper over S''.

Let $\mathcal{Z} = \sum m_j cycl(Z_j)$ where Z_j are integral closed subschemes of X and $m_j \neq 0$. We have to show that the morphism $\cup Z_j \to S$ is proper.

Consider the commutative diagram:

$$\begin{array}{ccc} \cup Z_i'' & \to & \cup Z_i \\ \downarrow & & \downarrow \\ S'' & \to & S \end{array}$$

The upper horizontal arrow is proper since it is a composition of a closed embedding $Z_i'' \to Z_i \times_S S''$ with a proper morphism $Z_i \times_S S'' \to Z_i$. Being an isomorphism in generic points it is surjective. Therefore $\cup Z_i \to S$ is proper since f_1 is proper and $\cup Z_i'' \to S''$ is proper.

The following proposition follows immediately from Propositions 4.2.5, 4.2.6.

Proposition 4.2.7 *Let $X \to S$ be a scheme of finite type over a Noetherian scheme S. Then the presheaves $PropCycl_{equi}(X/S, r)_\mathbf{Q}$ are qfh-sheaves and if S is a scheme of exponential characteristic p the same holds for the presheaves $c_{equi}(X/S, r) \otimes \mathbf{Z}[1/p]$.*

Proposition 4.2.8 *Let $X \to S$ be a scheme of finite type over a Noetherian scheme S. Then for any $r \geq 0$ the h-sheaves $Cycl(X/S, r)_\mathbf{Q}$, $PropCycl(X/S, r)_\mathbf{Q}$ are the h-sheaves of abelian groups associated in the obvious way with the h-sheaves of abelian monoids $Cycl^{eff}(X/S, r)_\mathbf{Q}$, $PropCycl^{eff}(X/S, r)_\mathbf{Q}$. The same holds for the corresponding p-divisible sheaves if S is a scheme of exponential characteristic p.*

Proof: By 4.2.2, 4.2.6 the presheaves $Cycl(X/S, r)_\mathbf{Q}$, $PropCycl(X/S, r)_\mathbf{Q}$ and their effective versions are h-sheaves. By Lemma 4.1.4 it is sufficient to show that for any element $\mathcal{Z}$ of $Cycl(X/S, r)_\mathbf{Q}(S)$ there is an abstract blow-up $f : S' \to S$ such that $cycl(f)(\mathcal{Z})$ is a linear combination of elements of $Cycl^{eff}(X/S, r)(S')$. It follows immediately from Lemma 4.2.1.

Theorem 4.2.9 *Let $X \to S$ be a scheme of finite type over a Noetherian scheme S. Then one has:*

1. *For any $r \geq 0$ the presheaves $z(X/S, r)$, $c(X/S, r)$, $z^{eff}(X/S, r)$, $c^{eff}(X/S, r)$ are sheaves in the cdh-topology.*

2. *The cdh-sheaves associated with the presheaves $z_{equi}(X/S, r)$, $c_{equi}(X/S, r)$ are isomorphic to $z(X/S, r)$ and $c(X/S, r)$ respectively.*

Proof: Since we know already that $Cycl(X/S, r)_\mathbf{Q}$ etc. are h-sheaves to prove the first part of the theorem it is sufficient to show that for any element $\mathcal{Z}$ of $Cycl(X/S,r)_{\mathbf{Q}}(S)$ such that there exists a cdh-covering p : $S' \to S$ such that $cycl(p)(\mathcal{Z}) \in z(X/S,r)(S')$ one has $\mathcal{Z} \in z(X/S,r)(S)$.

It follows immediately from 3.3.9 and the fact that any cdh-covering of a spectrum of a field splits.

The second part follows trivially from the first part and Lemma 4.2.1.

Proposition 4.2.10 *Let $X \to S$ be a scheme of finite type over a Noetherian scheme S. Then for any $r \geq 0$ the cdh-sheaf $z(X/S,r)$ (resp. $c(X/S,r)$) is the cdh-sheaf of abelian groups associated in the obvious way with the cdh-sheaf of abelian monoids $z^{eff}(X/S,r)$ (resp. $c^{eff}(X/S,r)$).*

Proof: It follows from Lemma 4.1.4 and Lemma 4.2.1.

Let $p : X \to S$ be a morphism of finite type of Noetherian schemes. For any Noetherian scheme T over S consider an equivalence relation $R_T \subset \mathbf{N}(Hilb(X/S,r))(T) \times \mathbf{N}(Hilb(X/S,r))(T)$ such that $(\sum n_i Z_i, \sum m_j W_j)$ belongs to R_T if and only if $cycl_{X_T}(\sum n_i Z_i) = cycl_{X_T}(\sum m_j W_j)$ where $X_T = X \times_S T$. Proposition 3.2.2 implies that it gives us equivalence relations R on the presheaves $\mathbf{N}(Hilb(X/S,r))$, $\mathbf{N}(PropHilb(X/S,r))$, $\mathbf{Z}(Hilb(X/S,r))$, $\mathbf{Z}(PropHilb(X/S,r))$ which are obviously consistent with the additive structure of these presheaves.

Theorem 4.2.11 *Let $X \to S$ be a scheme of finite type over a Noetherian scheme S. Then one has canonical isomorphisms of cdh-sheaves:*

$$z(X/S,r) = (\mathbf{Z}(Hilb(X/S,r))/R)_{cdh}$$

$$z^{eff}(X/S,r) = (\mathbf{N}(Hilb(X/S,r))/R)_{cdh}$$

$$c(X/S,r) = (\mathbf{Z}(PropHilb(X/S,r))/R)_{cdh}$$

$$c^{eff}(X/S,r) = (\mathbf{N}(PropHilb(X/S,r))/R)_{cdh}$$

Proof: We will only consider the first isomorphism. Proof in the other cases is similar. Note first that by 3.3.11 there is a morphism of presheaves $\mathbf{Z}(Hilb(X/S,r)) \to z(X/S,r)$ and clearly the presheaf $\mathbf{Z}(Hilb(X/S,r))/R$ is the image of this morphism. The fact that the corresponding associated cdh-sheaf coincides with $z(X/S,r)$ follows immediately from Lemma 4.1.4 and Lemma 4.2.1.

Theorem 4.2.12 *Let $X \to S$ be a scheme of finite type over a Noetherian scheme S. Then one has:*

1. The sheaf $c_{equi}(X/S, 0)_{qfh}$ is canonically isomorphic to the qfh-sheaf $\mathbf{Z}_{qfh}(X/S)$ of abelian groups freely generated by the sheaf of sets representable by X.

2. The sheaf $c^{eff}(X/S, 0)_{qfh}$ is canonically isomorphic to the qfh-sheaf $\mathbf{N}_{qfh}(X/S)$ of abelian monoids freely generated by the presheaf of sets representable by X.

Proof: We will only consider the first statement. The prove of the second one is similar.

Let $\delta \in c^{eff}(X/S, 0)(X) \subset Cycl(X \times_S X/X, 0)$ be the element which corresponds to the diagonal $\Delta : X \to X \times_S X$. By the universal property of the freely generated sheaves it gives us a morphism of sheaves $\mathbf{Z}_{qfh}(X) \to c^{eff}(X/S, 0)_{qfh}$. We have to show that it is an isomorphism. Note that it is a monomorphism since the functor of associated sheaf is exact and the corresponding morphism from the presheaf of abelian groups freely generated by X to the preshcaf $c(X/S, 0)$ is obviously a monomorphism.

To show that it is an epimorphism it is sufficient to verify that for any element $\mathcal{Z}$ of $c^{eff}(X/S, 0)(S)$ there is a qfh-covering $p : S' \to S$ such that $cycl(p)(\mathcal{Z})$ is a formal linear combination of S'-points of $X \times_S S'$ over S'. We may assume that S is integral and $\mathcal{Z} = \sum n_i cycl(Z_i)$ where $Z_i \neq Z_j$ are integral closed subscheme of X which are finite and surjective over S. We will use induction on $deg(\mathcal{Z}/S) = \sum |n_i| deg(Z_i/S)$. If $N = 0$ then $\mathcal{Z} = 0$ and there is nothing to prove.

Set $Z = \coprod Z_i$. Since $p_Z : Z \to S$ is finite and surjective it is a qfh-covering. It is sufficient (by the induction hypothesis) to show that there exists a cycle $\mathcal{Z}_1$ in $c_{equi}(X \times_S Z_1/Z_1, 0)$ which is a linear combination of Z_1-points of $X \times_S Z_1$ over Z_1 and such that

$$deg(cycl(p_{Z_1})(\mathcal{Z}) - \mathcal{Z}_1) < deg(\mathcal{Z}/S).$$

The cycle $cycl(p_{Z_1})(\mathcal{Z})$ is of the form $\sum n_i(\sum m_j W_{ij})$ where W_{ij} are the irreducible components of the schemes $Z_1 \times_S Z_i$. We obviously have

$$\sum m_j deg(W_{ij}/Z_1) = deg(Z_i/S)$$

and hence

$$\sum n_i m_j deg(W_{ij}/Z_1) = deg(\mathcal{Z}/S).$$

Let W_{11} be the irreducible component of $Z_1 \times_S Z_1$ which is the image of the diagonal embedding $Z_1 \to Z_1 \times_S Z_1$. One can easily see that $cycl(W_{11}) \in c_{equi}(X \times_S Z_1/Z_1, 0)$ and hence we may set $\mathcal{Z}_1 = n_1 m_1 cycl(W_{11})$.

Lemma 4.2.13 *Let $X \to S$ be a scheme of finite type over a Noetherian scheme S. Then the h-sheaf $z_{equi}(X/S, r)_h$ (resp. the sheaf $c_{equi}(X/S, r)_h$) is isomorphic to the h-sheaf $z(X/S, r)_h$ (resp. to the sheaf $c(X/S, r)_h$).*

Proof: It follows trivially from Lemma 4.2.1

Proposition 4.2.14 *Let $X \to S$ be a scheme of finite type over a Noetherian scheme S. Then one has:*

1. *The sheaf $c(X/S, 0)_h$ is canonically isomorphic to the h-sheaf $\mathbf{Z}_h(X/S)$ of abelian groups freely generated by the sheaf of sets representable by X.*

2. *The sheaf $c^{eff}(X/S, 0)_h$ is canonically isomorphic to the h-sheaf $\mathbf{N}_h(X/S)$ of abelian monoids freely generated by the presheaf of sets representable by X.*

Proof: It follows immediately from Theorem 4.2.12 and Lemma 4.2.13.

Proposition 4.2.15 *Let $X \to S$ be a scheme of finite type over a Noetherian scheme S. Then the canonical morphisms of presheaves:*

$$z(X/S, r)_{qfh} \to z(X/S, r)_h$$

$$c(X/S, r)_{qfh} \to c(X/S, r)_h$$

etc.
are isomorphisms.

Proof: To prove this proposition we need the following two lemmas.

Lemma 4.2.16 *Let S be a Noetherian scheme and $\{x_1, \ldots, x_n\}$ be a finite set of points of S. Let $E_1, \ldots, E_n$ be finite extensions of the fields $k_{x_1}, \ldots, k_{x_n}$. Then there is a finite surjective morphism $S' \to S$ such that for any $i = 1, \ldots, n$ and any point y over x_i the field extension k_y/k_i contains E_i.*

Proof: Replacing S by the disjoint union of its irreducible components we may assume that S is integral. Using Zariski's main theorem [11] one can reduce the problem to the case when S is a semi-local scheme and x_i are the closed points of S. We may obviously assume that E_i are normal extensions of k_{x_i} and since E_i are finitely generated over k_{x_i} using induction on the minimal number of generators we may further assume that $E_i = k_{x_i}(z_i)$ for some elements z_i in E_i. Let

$$f_i(z) = z_i^d + a_{i1} z^{d_i - 1} + \ldots + a_{id_i}$$

be the minimal polinomial of z_i over k_{x_i} and

$$g_i = f_i^{d_1 \ldots d_n / d_i}.$$

Then g_i are of the same degree $d = d_1 \ldots d_n$. Let b_{ij}, $i = 1, \ldots, n$, $j = 1, \ldots, d$ be the coefficients of g_i. We will find then elements B_j in $\mathcal{O}(S)$ such that the reduction of B_j in the point x_i equals b_{ij}. We set $R = \mathcal{O}(S)[z]/(Z^d + B_1 z^{d-1} + \ldots + B_d)$. It is a finite algebra over $\mathcal{O}(S)$ and the morphism $Spec(R) \to S$ satisfies the required conditions.

Lemma 4.2.17 *Let $p : X \to S$ be a scheme of finite type over a Noetherian scheme S. Assume further that p is surjective. Then there exists a finite surjective morphism $S' \to S$ such that for any field k and any k-point $s' : Spec(k) \to S'$ there is a lifting of s' to a k-point of $X' = X \times_S S'$.*

Proof: Let Z be a closed subscheme of S. We say that our lemma holds outside Z if there is a finite surjective morphism $S_Z \to S$ such that for any field k and any k-point s' of $S_Z \times_S (S - Z)$ there is a lifting of s' to $S_Z \times_S X$.

Our lemma obviously holds for $Z = X$. By Noetherian induction it is sufficient to show that if our lemma holds for Z then it holds for a closed subset $Z' \subset Z$ such that $Z' \neq Z$.

Let $\eta_1, \ldots, \eta_k$ be all generic points of Z. Since the morphism p is of finite type there are finite extensions $E_1, \ldots, E_k$ of the fields $k_{\eta_1}, \ldots, k_{\eta_k}$ such that the points $Spec(E_i) \to S$ of S admit liftings to X. By Lemma 4.2.16 there is a finite surjective morphism $f : S' \to S$ such that for any field k and any k-point s' of S' over one of the points η_i there is a lifting of s' to $X \times_S S'$. Since p is finite type this condition also holds for any point s' of S' which belongs to $f^{-1}(U)$ for a dense open subset U of Z. Let $Z' = Z - U$. Setting $S_{Z'} = S_Z \times_S S'$ we conclude that our lemma holds for Z'.

Let us now prove Proposition 4.2.15. We will only consider the case of $z(X/S, r)$. Note first that by Proposition 4.2.2 both $z(X/S, r)_{qfh}$ and $z(X/S, r)_h$ are subpresheaves in $Cycl(X/S, r)_{\mathbf{Q}}$. We have only to show that if $\mathcal{Z}$ is an element of $Cycl(X/S, r) \otimes \mathbf{Q}$ and there is an h-covering $f : S' \to S$ such that $cycl(f)(\mathcal{Z})$ belongs to $z(X/S, r)(S')$ then there is a qfh-covering with the same property. It follows from 3.3.9 and Lemma 4.2.17.

Corollary 4.2.18 *Let $X \to S$ be a scheme of finite type over a Noetherian scheme S. Then for any $r \geq 0$ the qfh-sheaves $Cycl_{equi}(X/S, r)_{\mathbf{Q}}$,*

$PropCycl_{equi}(X/S, r)_{\mathbf{Q}}$ are isomorphic on the category of excellent Noetherian schemes over S to the qfh-sheaves of abelian groups associated in the obvious way with the qfh-sheaves of abelian monoids $Cycl^{eff}(X/S, r)_{\mathbf{Q}}$, $PropCycl^{eff}(X/S, r)_{\mathbf{Q}}$ and the same holds for the corresponding p-divisible sheaves if S is a scheme of exponential characteristic p.

Proof: Note first that one has by 3.1.7:

$$Cycl^{eff}(X/S, r)_{\mathbf{Q}} \subset Cycl_{equi}(X/S, r)_{\mathbf{Q}}$$

$$PropCycl^{eff}(X/S, r)_{\mathbf{Q}} \subset PropCycl_{equi}(X/S, r)_{\mathbf{Q}}$$

As in the proof of 4.2.8 it is sufficient to show that for any element $\mathcal{Z} = \sum n_i z_i$ in $Cycl_{equi}(X/S, r)_{\mathbf{Q}}$ there is a qfh-covering $f : S' \to S$ such that $cycl(f)(\mathcal{Z})$ is a linear combination of elements of $Cycl^{eff}(X/S, r)_{\mathbf{Q}}(S')$. Since S is an excellent scheme its normalization is a qfh-covering and our statement follows from Corollary 3.4.3.

Remark: Corollary 4.2.18 remains true without the excellency assumption.

4.3. Fundamental exact sequences for Chow sheaves

Theorem 4.3.1 Let $p : X \to S$ be a scheme of finite type over a Noetherian scheme S, $i : Z \to X$ be a closed subscheme of X and $j : U \to X$ be the complement to Z in X. Then for any $r \geq 0$ the following sequence of presheaves is left exact and it is also right exact as a sequence of cdh-sheaves:

$$0 \to z(Z/S, r) \xrightarrow{i_*} z(X/S, r) \xrightarrow{j^*} z(U/S, r) \to 0.$$

Proof: This sequence is obviously left exact as a sequence of presheaves. It is sufficient to show that the last arrow is a surjection in the cdh-topology. Let $\mathcal{Z} = \sum n_i z_i$ be an element of $z(U/S, r)$. By Lemma 4.1.4 it is sufficient to show that there is an abstract blow-up (2.2.4) of the form $f : S' \to S$ such that $cycl(f)(\mathcal{Z})$ belongs to the image of $z(X/S, r)(S')$ in $z(U/S, r)(S')$. Let Z_i be the closures of the points z_i in X wich we consider as closed integral subschemes. We may assume that S is reduced. Then by Theorem 2.2.2 there is a blow-up $f : S' \to S$ such that the proper transforms $\tilde{Z}_i$ of Z_i are flat over S'. We set $\mathcal{Z}' = \sum n_i cycl_{X \times_S S'}(\tilde{Z}_i)$. Then by Corollary 3.3.11 one has $\mathcal{Z}' \in z(X/S, r)(S')$ and its restriction to $U \times_S S'$ obviously equals $cycl(f)(\mathcal{Z})$.

Corollary 4.3.2 *Let $p : X \to S$ be a scheme of finite type over a Noetherian scheme S and $X = X_1 \cup X_2$ be an open covering of X. Denote the inclusions $X_i \subset X, X_1 \cap X_2 \subset X_i$ by f_i and g_i respectively. Then for any $r \geq 0$ the following sequence of presheaves is left exact and it is also right exact as a sequence of cdh-sheaves:*

$$0 \to z(X/S, r) \overset{(f_1)^* + (f_2)^*}{\to} z(X_1/S, r) \oplus z(X_2/S, r) \overset{(g_1)^* - (g_2)^*}{\to}$$

$$\to z((X_1 \cap X_2)/S, r) \to 0.$$

Remark: Note that the sequence of Theorem 4.3.1 is in fact already exact in the topology where coverings are proper cdh-coverings.

Proposition 4.3.3 *Let S be a Noetherian scheme, $p : X \to S$ be a scheme of finite type over S, $i : Z \to X$ be a closed subscheme of X and $f : X' \to X$ be a proper morphism such that the morphism $f^{-1}(X - Z) \to X - Z$ is an isomorphism.*

Consider the pull-back square

$$
\begin{array}{ccc}
f^{-1}(Z) & \overset{i'}{\longrightarrow} & X' \\
{\scriptstyle f_Z}\downarrow & & \downarrow{\scriptstyle f} \\
Z & \overset{i}{\dashrightarrow} & X
\end{array}
$$

Denote by $F(-, r)$ one of the cdh-sheaves $z(-, r)$, $c(-, r)$. Then the following sequence of presheaves is left exact and it is also right exact as a sequence of cdh-sheaves

$$0 \to F(f^{-1}(Z)/S, r) \overset{i'_* \oplus (f_Z)_*}{\to} F(X'/S, r) \oplus F(Z/S, r) \overset{f_* \oplus (-i_*)}{\to}$$

$$\to F(X/S, r) \to 0.$$

Proof: It is clearly sufficient to consider the case of the sheaves $z(-, -)$.

Lemma 4.3.4 *In the notations of Proposition 4.3.3 the following sequence of abelian groups is exact:*

$$0 \to Cycl_{equi}(f^{-1}(Z)/S, r) \overset{i'_* \oplus (f_Z)_*}{\to} Cycl_{equi}(X'/S, r) \oplus Cycl_{equi}(Z/S, r)$$

$$\overset{f_* \oplus (-i_*)}{\to} Cycl_{equi}(X/S, r).$$

Proof: The first arrow is a monomorphism since $i' : f^{-1}(Z) \to X'$ is a monomorphism.

Let us show that the sequence is exact in the middle term. Let $\mathcal{Y} = \sum n_i y_i$ be an element of $Cycl_{equi}(Z/S, r)$ and $\mathcal{W} = \sum m_j w_j$ be an element of $Cycl_{equi}(X'/S, r)$ such that $f_*(\mathcal{W}) = i_*(\mathcal{Y})$. The cycle $\mathcal{W}$ can be represented (uniquely) as a sum $\mathcal{W} = \mathcal{W}_0 + \mathcal{W}_1$ such that $supp(\mathcal{W}_0) \subset f^{-1}(Z)$ and $supp(\mathcal{W}_1) \cap f^{-1}(Z) = \emptyset$. By our condition on $\mathcal{W}$ we have

$$f_*(\mathcal{W}_0) = i_*(\mathcal{Y})$$

$$f_*(\mathcal{W}_1) = 0$$

and since f is an isomorphism outside $f^{-1}(Z)$ we conclude that $\mathcal{W}_1 = 0$ and hence $\mathcal{W} \oplus \mathcal{Y}$ belongs to the image of the homomorphism $i'_* \oplus (f_Z)_*$.

One can easily see that Lemma 4.3.4 implies that the sequence of abelian groups

$$0 \to z_{equi}(f^{-1}(Z)/S, r) \overset{i'_* \oplus (f_Z)_*}{\to} z_{equi}(X'/S, r) \oplus z_{equi}(Z/S, r) \overset{f_* \oplus (-i_*)}{\to}$$

$$\to z_{equi}(X/S, r).$$

is also exact. Hence, our sequence of sheaves is left exact and it is sufficient to show that the homomorphism

$$z_{equi}(X'/S, r) \oplus z_{equi}(Z/S, r) \overset{f_* \oplus (-i_*)}{\to} z_{equi}(X/S, r)$$

is surjective as a homomorphism of cdh-sheaves. By Lemma 4.1.4 we may assume that S is integral and it is sufficient to show that for any element $\mathcal{W}$ of the group $z_{equi}(X/S, r)$ there is a blow-up $g : S' \to S$ such that $cycl(g)(\mathcal{W})$ belongs to the image of $z_{equi}(X' \times_S S'/S', r) \oplus z_{equi}(Z \times_S S'/S', r)$. By Lemma 4.2.1 we may assume that $\mathcal{W} = cycl_X(W)$ where W is an integral closed subscheme of X which is equidimensional of relative dimension r over S. If $W \subset Z$ our statement is obvious. Let w be the generic point of W and W' be the closure of $f^{-1}(w)$ in X'. By Theorem 2.2.2 there is a blow-up $g : S' \to S$ such that the proper transform $\tilde{W}'$ of W' is flat over S'. Denote by f' the morphism $f \times_S S' : X' \times_S S' \to X \times_S S'$. Since g is an isomorphism in generic points and f is an isomorphism ouside Z we clearly have $f'_*(cycl(\tilde{W}')) = cycl(g)(\mathcal{W})$. To finish the proof it is sufficient to note that $\tilde{W}'$ belongs to $z(X' \times_S S'/S', r)$ by Corollary 3.3.11.

Corollary 4.3.5 *Let S be a Noetherian scheme, $p : X \to S$ be a scheme of finite type over S and $X = Z_1 \cup Z_2$ be a covering of X by two closed subschemes. Denote the inclusions $Z_1 \cap Z_2 \subset Z_i$, $Z_i \subset X$ by f_i and g_i respectively and let $F(-, r)$ be one of the cdh-sheaves $z(-, r)$, $c(-, r)$.*

Then the following sequence of cdh-sheaves is exact

$$0 \to F(Z_1 \cap Z_2, r) \xrightarrow{(f_1)_* + (f_2)_*} F(Z_1, r) \oplus F(Z_2, r)) \xrightarrow{(g_1)_* - (g_2)_*} F(X, r) \to 0.$$

Proof: It is sufficient to apply Proposition 4.3.3 in the case $X' = Z_1 \coprod Z_2$, $Z = Z_1 \cap Z_2$.

Proposition 4.3.6 *In the notations of Proposition 4.3.3 assume in addition that f is a finite morphism and S is a geometrically unibranch scheme. Then the following sequence of abelian groups is exact:*

$$0 \to Cycl_{equi}(f^{-1}(Z)/S, r) \xrightarrow{i'_* \oplus (f_Z)_*} Cycl_{equi}(X'/S, r) \oplus Cycl_{equi}(Z/S, r)$$

$$\xrightarrow{f_* \oplus (-i_*)} Cycl_{equi}(X/S, r) \to 0.$$

The same statement holds for the groups $PropCycl_{equi}(-, -)$.

Proof: It is clearly sufficient to consider the case of $Cycl_{equi}(-, -)$. By Lemma 4.3.4 our sequence is left exact.

Let $\mathcal{W}$ be an element of $Cycl_{equi}(X/S, r)$. By Corollary 3.4.3 we may assume that $\mathcal{W} = cycl_X(W)$ where W is an integral closed subscheme of X which is equidimensional of relative dimension r over S. If $W \subset Z$ then $\mathcal{W}$ belongs to the image of the homomorphism i_*. Otherwise let w be the generic point of W and let W' be the closure of $w' = f^{-1}(w)$ in X'. Since w belongs to $X - Z$ we have $f_*(cycl_{X'}(W')) = \mathcal{W}$. To finish the proof of our proposition it is sufficient to show that $cycl_{X'}(W') \in Cycl_{equi}(X'/S, r)$. It follows from the fact that f is finite and Theorem 3.4.2.

Remark: Note that in Proposition 4.3.6 one can not replace in general the groups $Cycl_{equi}(-, -)$ by the groups $z_{equi}(-, -)$. An example of the situation when the corresponding sequence is not right exact for the groups $z_{equi}(-, -)$ can be easily deduced from example 3.5.10(2).

Proposition 4.3.7 *Let S be a Noetherian scheme, $p : X \to S$ be a scheme of finite type over S and $X = U_1 \cup U_2$ be an open covering of X. Denote the inclusions $U_1 \cap U_2 \subset U_i$, $U_i \subset X$ by f_i and g_i respectively.*
Then the sequence of presheaves

$$0 \to c_{equi}(U_1 \cap U_2/S, 0) \xrightarrow{(f_1)_* + (f_2)_*} c_{equi}(U_1/S, 0) \oplus c_{equi}(U_2/S, 0) \xrightarrow{(g_1)_* - (g_2)_*}$$

$$\to c_{equi}(X/S, 0) \to 0.$$

is exact in the Nisnevich topology.

Proof: Note that our sequence is left exact as a sequence of presheaves by obvious reason. To prove that the last arrow is a surjection in the Nisnevich topology it is sufficient to show that the map

$$c_{equi}(U_1/S, 0) \oplus c_{equi}(U_2/S, 0) \overset{(g_1)_* - (g_2)_*}{\to} c_{equi}(X/S, 0)$$

is surjective if S is a local henselian scheme (see [12]). It follows trivially from the fact that for any element $\mathcal{Z}$ of $c_{equi}(X/S, 0)$ its support is finite over S and the existence of decomposition of schemes finite over local henselian schemes into a disjoint union of local schemes (see [11, I.4.2.9(c)]).

Corollary 4.3.8 *In the notations of Proposition 4.3.7 the sequence of cdh-sheaves*

$$0 \to c(U_1 \cap U_2, 0) \overset{(f_1)_* + (f_2)_*}{\to} c(U_1, 0) \oplus c(U_2, 0) \overset{(g_1)_* - (g_2)_*}{\to} c(X, 0) \to 0.$$

is exact.

Proof: It follows immediately from Proposition 4.3.7, Theorem 4.2.9 and exactness of the functor of associated sheaf.

Remark: Note that the exact sequence of Proposition 4.3.7 is quite different from the exact sequence of Corollary 4.3.2. In particular while the sequence for finite cycles requires only Nisnevich coverings to be exact the sequence for general cycles requires only abstract blow-ups, i.e. proper cdh-coverings to be exact.

Proposition 4.3.9 *Let $p : X \to S$ be a scheme of finite type over a Noetherian scheme S, $Z \to X$ be a closed subscheme of X and $f : X' \to X$ be an etale morphism such that the morphism $f^{-1}(Z) \to Z$ is an isomorphism. Let further U be the complement to Z in X and U' be the complement to $f^{-1}(Z)$ in X'. Then the canonical morphism of quotient sheaves in the Nisnevich topology*

$$c_{equi}(X'/S, 0)/c_{equi}(U'/S, 0) \to c_{equi}(X/S, 0)/c_{equi}(U/S, 0)$$

is an isomorphism.

Proof: It is sufficient to show that for a local henselian scheme S the morphism of abelian groups

$$c_{equi}(X'/S, 0)/c_{equi}(U'/S, 0) \overset{f_*}{\to} c_{equi}(X/S, 0)/c_{equi}(U/S, 0).$$

is an isomorphism. The following lemma is an easy corollary of the standard properties of henselian schemes (see [11]).

Lemma 4.3.10 *Let $q : X \to S$ be an etale morphism such that the scheme S is henselian and let x be a closed point of X over the closed point s of S such that the morphism $Spec(k_x) \to Spec(k_s)$ is an isomorphism. Then q is an isomorphism in a neighborhood of x.*

(**Injectivity.**) Let $\mathcal{W}'$ be an element of $c_{equi}(X'/S, 0)$ and assume that $f_*(\mathcal{W}')$ belongs to $c_{equi}(U/S, 0)$. Since S is henselian the support $W' = supp(\mathcal{W}')$ of $\mathcal{W}'$ is a disjoint union of local henselian schemes and we may assume that the closed points of W belong to $f^{-1}(Z)$ and $f_*(\mathcal{W}) = 0$.

Since f is an isomorphism on $f^{-1}(Z)$ we may further assume that W' is local. Then by Lemma 4.3.10 the morphism $supp(\mathcal{W}') \to supp(f_*(\mathcal{W}'))$ is an isomorphism and hence $\mathcal{W}' = 0$.

(**Surjectivity.**) Let $\mathcal{W}$ be an element of $c_{equi}(X/S, 0)$. As above we may assume that $W = supp(\mathcal{W})$ is local and its closed point belongs to Z. Let W' be the local scheme of the closed point of $f^{-1}(W)$ over the closed point of W. Then by Lemma 4.3.10 the morphism $W' \to W$ is an isomorphism. Denote by $\mathcal{W}'$ the cycle on W' (and hence on X') which corresponds to $\mathcal{W}$. We obviously have $f_*(\mathcal{W}') = \mathcal{W}$ and $\mathcal{W}' \in c_{equi}(X'/S, 0)$.

Corollary 4.3.11 *In notations of Proposition 4.3.9 there is a canonical isomorphism of quotient sheaves in the cdh-topology of the form:*

$$c(X'/S, 0)/c(U'/S, 0) \to c(X/S, 0)/c(U/S, 0).$$

Proof: It follows from Proposition 4.3.9 and Theorem 4.2.9.

4.4. Representability of Chow sheaves

In this section we consider representability of Chow sheaves of effective proper cycles on quasi-projective schemes over a Noetherian scheme S. Let us begin with the following definition.

Definition 4.4.1 *Let S be a Noetherian scheme and F be a presheaf of sets on the category of Noetherian schemes over S. We say that F is h-representable by a scheme X over S if there is an isomorphism $F_h \to L_h(X)$ of the h-sheaf associated with F with the h-sheaf associated with the presheaf represented by X.*

Note that the scheme X which h-represents a presheaf F is not uniquely defined up to isomorphism since the h-topology is not subcanonical. Nevertheless as was shown in [18] X is well defined up to a universal homeomorphism. In particular if for some X which h-represents F all generic points of

X are of characteristic zero then there exists a unique semi-normal scheme which h-represents F.

Let S be a Noetherian scheme and F be a presheaf on the category of Noetherian schemes over S. A k-point of F is a pair of the form $(Spec(k) \to S, \phi)$ where k is a field and $\phi \in F(Spec(k)/S)$ is a section of F over $Spec(k)$. We say that a k-point ϕ of F is equivalent to a k'-point ϕ' of F if there is a field k'' a morphism $Spec(k'') \to S$ and morphisms $u : Spec(k'') \to Spec(k)$, $u' : Spec(k'') \to Spec(k')$ over S such that the sections $u^*(\phi)$ and $(u')^*(\phi')$ of F over $Spec(k'')$ coincide. A point of F is by definition an equivalence class of k-points of F. Denote the set of points of F by $Top(F)$. One can easily see that for any morphism of presheaves $f : F \to G$ one has a map of sets $Top(f) : Top(F) \to Top(G)$ which is a monomorphism (resp. an epimorphism) if f is a monomorphism (resp. an epimorphism). In particular for any subpresheaf F_0 in F we get a subset $Top(F_0)$ in $Top(F)$.

The following lemma is trivial.

Lemma 4.4.2 *Let t be a Grothendieck topology on the category of Noetherian schemes over S such that for any algebraically closed field k any morphism $Spec(k) \to S$ and any t-covering $p : U \to Spec(k)$ there exists a section $s : Spec(k) \to U$ of p. For a presheaf F on the category of Noetherian schemes over S denote by F_t the associated t-sheaf. Then the map $Top(F) \to Top(F_t)$ is a bijection.*

Note that all topologies we use in this paper satisfty the condition of Lemma 4.4.2.

For any presheaf of sets F on the category of Noetherian schemes over S, any Noetherian scheme S' over S and any section ϕ of F over S' denote by $Top(\phi)$ the obvious map of sets $S' \to Top(F)$.

Let now $A \subset Top(F)$ be a subset in $Top(F)$. Denote by F_A the subpresheaf in F such that for any Noetherian scheme S' over S the subset $F_A(S') \subset F(S')$ consists of sections ϕ such that $Top(\phi)(S') \subset A$. If a subpresheaf F_0 in F is of the form F_A for a subset A in $Top(F)$ we say that F_0 is defined by a pointwise condition. One can easily see that F_0 is defined by a pointwise condition if and only if for any Noetherian scheme S' over S and a section ϕ of F over S' such that $Top(\phi)(S') \subset Top(F_0)$ one has $\phi \in F_0(S')$.

Let A be a subset in $Top(F)$. We say that A is open (resp. closed, constructible) if for any Noetherian scheme S' over S and any section ϕ of T on S' the set $Top(\phi)^{-1}(A)$ is open (resp. closed, constructible) in S'. We will say further that a subpresheaf F_0 of F is open (resp. closed, constructible) if F_0 is of the form F_A for an open (resp. closed, constructible) subset in $Top(F)$.

One can easily see that open subsets of $Top(F)$ form a topology on this set and that a subset is closed if and only if its complement is open. Note also that for any morphism of presheaves $f : F \to G$ the corresponding map $Top(f) : Top(F) \to Top(G)$ is a continuous map with respect to this topology. The following lemma is straightforward.

Lemma 4.4.3 *Let $X \to S$ be a Noetherian scheme over a Noetherian scheme S. Denote by $L(X/S)$ the presheaf of sets represented by X on the category of Noetherian schemes over S.*

1. *The map $Top(\phi) : X \to Top(L(X/S))$ defined by the tautological section of $L(X/S)$ over X is a homeomorphism of the corresponding topological spaces.*

2. *Let t be a topology on the category of Noetherian schemes over S satisfying the condition of Lemma 4.4.2 and such that for any Noetherian scheme S' over S the morphism $S'_{red} \to S'$ is a t-covering. Then for any open (resp. closed) subset $A \subset Top(L_t(X/S))$ the subpresheaf $L_t(X/S)_A$ is t-representable by the corresponding open (resp. closed) subscheme in X.*

Note that the h-topologies (h-,cdh-,qfh-) satisfy the conditions of Lemma 4.4.3(2).

We say that a presheaf F on the category of Noetherian schemes over a Noetherian scheme S is topologically separated if for any Noetherian scheme T over S and any dominant morphism $T' \to T$ the map $F(T) \to F(T')$ is injective. Note that all the Chow presheaves considered in this paper are topologically separated.

Lemma 4.4.4 *Let S be a Noetherian scheme, F be a topologically separated presheaf on the category of Noetherian schemes over S, $X \to S$ be a scheme of finite type over S and $f : z(X/S, r) \to F$ be a monomorphism of presheaves. Consider a closed subset A in $Top(z(X/S, r))$. Then the subsheaf $f(z(X/S, r)_A)_h$ of F_h is a closed subpresheaf of F if and only if $Top(f)(A)$ is a constructible subset in $Top(F) = Top(F_h)$.*

Proof: We obviously have $Top(f)(A) = Top(f(z(X/S, r)_A)) \subset Top(F)$ which proves the "only if" part.

Assume that $Top(f)(A)$ is a constructible subset in $Top(F)$. Let us show first that $f(z(X/S, r)_A)_h$ is a subpresheaf in F_h given by a pointwise condition. Let T be a Noetherian scheme over S and $\phi \in F_h(T)$ be a section of F_h such that for any geometrical point $x : Spec(k) \to T$ of T we have $x^*(\phi) \in z(X/S, r)_A$. We have to show that $\phi \in f(z(X/S, r)_A)_h(T)$. Our problem is h-local with respect to T. Replacing T by the union of its irreducible components we may assume that T is an integral scheme. Let η be the generic point of T, $\bar{k}_\eta$ be an algebraic closure of the function field of T and $\bar{\eta} : Spec(\bar{k}_\eta) \to T$ be the corresponding geometrical point of T. Then $\bar{\eta}^*(\phi)$ corresponds to an element $\mathcal{Z}$ in $z(X \times_T Spec(\bar{k}_\eta)/Spec(\bar{k}_\eta), r)$. Since X is of finite type over S and A is a closed subset of $Top(Z(X/S, r))$ there is a quasi-finite dominant morphism $p : U \to T$ and an elememt $\mathcal{Z}_U$ in $z(X/S, r)_A(U)$ such that the restriction of $\mathcal{Z}_U$ to $Spec(\bar{k}_\eta)$ equals $\mathcal{Z}$. Theorem 2.2.2 implies easily now that there is an h-covering $\bar{p} : \bar{U} \to T$ which is finite over the generic point of T and a section $\mathcal{Z}_{\bar{U}}$ of $z(X/S, r)_A$ over $\bar{U}$ such that its restriction to $Spec(\bar{k}_\eta)$ equals $\mathcal{Z}$. Since F is topologically separated we conclude that $\bar{p}^*(\phi) = \mathcal{Z}_{\bar{U}}$ and hence $f(z(X/S, A))_h$ is indeed defined by a pointwise condition. To prove that $Top(f(z(X/S, A))_h)$ is a closed subset in $Top(F) = Top(F_h)$ is trivial.

Proposition 4.4.5 *Let $X \to S$ be a scheme of finite type over a Noetherian scheme S and X_0 be a closed subscheme in X. Then the sheaf $z(X_0/S, r)$ is a closed subpresheaf in the sheaf $z(X/S, r)$.*

Proof: Note first that the sheaf $z(X_0/S, r)$ is a subpresheaf in $z(X/S, r)$ given by a pointwise condition. Moreover a section of $z(X/S, r)$ belongs to $z(X_0/S, r)$ if and only if it belongs to $z(X_0/S, r)$ over the generic points of S. It implies easily that the only thing we have to show is that for any cycle $\mathcal{Z}$ in $z(X/S, r)$ which does not belong to $z(X_0/S, r)$ in a generic point η of S there is a neighborhood U of η in S such that $\mathcal{Z}_s$ does not belong to $z((X_0)_s/Spec(k_s), r)$ for any point s in U. It is obvious.

Proposition 4.4.6 *Let $X \to S$ be a scheme of finite type over a Noetherian scheme S and X_0 be an open subscheme in X. Then the sheaf $c^{eff}(X_0/S, r)$ is an open subpresheaf in the sheaf $c^{eff}(X/S, r)$.*

Proof: Let us show first that the subpresheaf $c^{eff}(X_0/S, r)$ is given by a pointwise condition. Let $\mathcal{Z}$ be an element of $c^{eff}(X/S, r)$. Assume that for any geometrical point $x : Spec(k) \to S$ of S we have $cycl(x)(\mathcal{Z}) \in c^{eff}(X_0 \times_S Spec(k)/Spec(k), r)$. We have to show that $\mathcal{Z} \in c^{eff}(X_0/S, r)$. Since $c^{eff}(X_0/S, r)$ is a cdh-subsheaf in $c^{eff}(X/S, r)$ it is sufficient to show that there is a blow-up $f : S' \to S$ such that $cycl(f)(\mathcal{Z}) \in c^{eff}(X_0 \times_S S'/S', r)$. We may assume therefore that $\mathcal{Z} = \sum n_i cycl(Z_i)$ where Z_i are irreducible closed subschemes of X which are flat over S. It is sufficient to show that our condition on $\mathcal{Z}$ implies that $supp(\mathcal{Z}) \subset X_0$. It follows immediately from the fact that $\mathcal{Z}$ is effective since $supp(cycl(x)(cycl(Z_i))) = Z_i \times_S Spec(k)$ for any geometrical point $Spec(k) \to S$ of S.

Let now $\mathcal{Z}$ be an arbitrary element of $c^{eff}(X/S, r)$. We have to show that the set of points s of S such that $\mathcal{Z}_{\bar{k}_s} \in c^{eff}(X_0 \times_S Spec(\bar{k}_s)/Spec(\bar{k}_s), r)$ is open in S. Again we may replace S by its blow-up and assume that $\mathcal{Z} = \sum n_i cycl(Z_i)$ where Z_i are irreducible closed subschemes of X which are flat over S. Then clearly our subset is the intersection of subsets U_i where $s \in U_i$ if and only if $Z_i \times_S Spec(k_s) \subset X_0 \times_S Spec(k_s)$. Then $U_i = S - p(Z_i \cap (X - X_0))$ and since $p : Z_i \to S$ is proper we conclude that U_i are open subsets of S.

Example 4.4.7 The analog of Proposition 4.4.6 for the sheaves $c(X/S, r)$ is false. Let us consider the scheme $X = \mathbf{P}_k^2 \times \mathbf{A}^1 \to \mathbf{A}^1$ over the affine line. Let further Z_1 (resp. Z_2) be the closed subscheme given by the equation $x_0 + x_1 + tx_2 = 0$ (resp. $x_0 + x_1 - tx_2 = 0$) where x_i are the coordinates on $\mathbf{P}^2$ and t is the coordinate on $\mathbf{A}^1$. Let $\mathcal{Z} = Z_1 - Z_2$ and $X_0 = \mathbf{A}^2 \times \mathbf{A}^1 \subset \mathbf{P}_k^2 \times \mathbf{A}^1$. Then the set of points $t \in \mathbf{A}^1$ where $\mathcal{Z}_t \in (X_0)_t$ consists of one point $t = 0$.

Let S be a Noetherian scheme. For any multi-index $I = (i_1, \dots, i_k)$ denote by $\mathbf{P}_S^I$ the product of projective spaces $\mathbf{P}_S^{i_1} \times_S \dots \times_S \mathbf{P}^{i_k}$ over S. Let $\mathcal{Z}$ be an element of $Cycl^{eff}(\mathbf{P}_S^I/S, r)$. For any point s of S denote by $deg_s(\mathcal{Z})$ the (multi-)degree of the cycle $\mathcal{Z}_s$ on $\mathbf{P}_{k_s}^I$ (note that apriory $deg_s(\mathcal{Z})$ is sequence of rational numbers).

Proposition 4.4.8 *Let S be a Noetherian scheme and $\mathcal{Z}$ be an element of $Cycl^{eff}(\mathbf{P}_S^I/S, r)$. Then the function $s \to deg_s(\mathcal{Z})$ is locally constant on S.*

Proof: It is sufficient to show that if η is a generic point of S and s is a point in the closure of η then $deg_\eta(\mathcal{Z}) = deg_s(\mathcal{Z})$. Since for any cycle $\mathcal{W}$ on $\mathbf{P}_k^I$ and any field extension k'/k we have $deg(\mathcal{Z}) = deg(\mathcal{Z} \otimes_k k')$ it is sufficient to show that for some field extensions L, E of k_η and k_s respectively the cycles $\mathcal{Z}_{Spec(L)}$ and $\mathcal{Z}_{Spec(E)}$ have the same degree. Let (x_0, x_1, R) be a fat point on S such that the image of x_1 is $\{\eta, s\}$. Replacing S by $Spec(R)$ we may assume that S is the spectrum of a discrete valuation ring. In this case $\mathcal{Z} = \sum n_i cycl(Z_i)$ where Z_i are closed subschemes of $\mathbf{P}_S^I$ which are flat and equidimensional over S and our statement follows from the invariance of (multi-)degree in flat families.

Corollary 4.4.9 *Let S be a connected Noetherian scheme. Then for any cycle $\mathcal{Z}$ in $Cycl(\mathbf{P}_S^I/S, r)$ and any point s of S the degree $deg_s(\mathcal{Z})$ is a sequence of integers which does not depend on s.*

For a Noetherian scheme S, a multi-index $I = (i_1, \dots, i_k)$ and a sequence of nonnegative integers $D = (d_1, \dots, d_n)$ denote by $z_D^{eff}(\mathbf{P}_S^I/S, r)$ the subset in $z(\mathbf{P}_S^I/S, r)$ which consists of cycles $\mathcal{Z}$ such that for any point s of S one has $deg_s(\mathcal{Z}) = D$. One can easily see that $z_d^{eff}(\mathbf{P}_S^I/S, r)$ is in fact a cdh-subsheaf in $z^{eff}(\mathbf{P}_S^I/S, r)$. Proposition 4.4.8 implies further that for a connected Noetherian scheme one has

$$z^{eff}(\mathbf{P}_S^I/S, r) = \bigcup_D z_D^{eff}(\mathbf{P}_S^I/S, r).$$

The proof of the following lemma is standard.

Lemma 4.4.10 *Let S be a Noetherian scheme. Then for any multi-index $I = (i_1, \dots, i_k)$ and any sequence of nonnegative integers $D = (d_1, \dots, d_k)$ the sheaf $z_D^{eff}(\mathbf{P}_S^I/S, (\sum i_j) - 1)$ is h-representable by the projective space P_S^N for some $N = N(I, D)$.*

Denote by G the product $(\mathbf{P}^n)^* \times_{Spec(\mathbf{Z})} \cdots_{Spec(\mathbf{Z})} (\mathbf{P}^n)^*$ of $r + 1$-copies of the projective space dual to the standard projective space (i.e., $(\mathbf{P}^n)^*$ is the scheme which parameterizes hyperplanes in $\mathbf{P}^n$). Let further $L \in \mathbf{P}^n \times_{Spec(\mathbf{Z})} G$ be the closed subscheme of points $(x, H_1, \dots, H_{r+1})$ such that $x \in H_i$ for all $i = 1, \dots, r + 1$. It is smooth over $\mathbf{P}^n$ and fibers of the projection $f : L \to \mathbf{P}^n$ are isomorphic to $(\mathbf{P}^{n-1})^{r+1}$.

Denote further by $Div_d(G)$ the projective space which parametrizes cycles of codimension 1 and degree $(d, \dots, d)$ on $((\mathbf{P}^n)^*)^{r+1}$ (see Lemma 4.4.10). Let $Div_d^{irr}(G)$ be the open subspace in $Div_d(G)$ which parametrizes irreducible divisors on G and let $\Gamma \in G \times Div_d^{irr}(G)$ be the support of the corresponding relative cycle on $G \times Div_d^{irr}(G)$ over $Div_d^{irr}(G)$. Set

$$U = L \times Div_d^{irr}(G) - (L \times Div_d^{irr}(G) \cap \mathbf{P}^n \times \Gamma)$$

Let finally Φ be the subset $(pr : L \times Div_d^{irr}(G) \to \mathbf{P}^n \times Div_d^{irr}(G))(U)$. Since L is smooth (and hence universally open) over $\mathbf{P}^n$ this subset is open in $\mathbf{P}^n \times Div_d^{irr}(G)$. We define $C_{r,d}^{irr}$ to be the subset in $Div_d^{irr}(G)$ which consists of points s such that $dim((\mathbf{P}^n \times Div_d^{irr}(G) - U) \times_{Div_d^{irr}(G)} Spec(k_s)) \geq r$. Since the projection $\mathbf{P}^n \times Div_d^{irr}(G) \to Div_d^{irr}(G)$ is proper Chevalley theorem (2.1.1) implies that $C_{r,d}^{irr}$ is a closed subset in $Div_d^{irr}(G)$.

Theorem 4.4.11 *For any Noetherian scheme S and any $r, d \geq 0$ the sheaf of sets $z_d^{eff}(\mathbf{P}_S^n/S, r)$ is h-representable by a projective scheme $C_{r,d}$ over S.*

Proof: Note that it is obviously sufficient to prove our theorem for $S = Spec(\mathbf{Z})$. We will use the notations which we introduced in the construction of $C_{r,d}^{irr}$ above. Let $\mathcal{Z}$ be an element of $z^{eff}(\mathbf{P}_S^n/S, r)$. Let

$$f : L \to \mathbf{P}^n$$

$$p : L \to G$$

be the obvious morphisms. Since f is smooth and p is proper we get a homomorphism of presheaves

$$Chow = p_* f^* : z^{eff}(\mathbf{P}^n/Spec(\mathbf{Z}), r) \to z^{eff}(G/Spec(\mathbf{Z}), (n-1)(r+1) + r)$$

The following lemma is straighforward.

Lemma 4.4.12 1. *The homomorphism Chow is a monomorphism.*

2. *The homomorphism Chow takes the subsheaf $z_d^{eff}(\mathbf{P}_S^n/S, r)$ to the subsheaf $z_D^{eff}(((\mathbf{P}_S^n)^*)^{r+1}/S, (n-1)(r+1) + r)$ where $D = (d, \ldots, d)$.*

In view of Lemma 4.4.12 and Lemma 4.4.10 the homomorphism $Chow$ gives us an embedding of the sheaf $z_d^{eff}(\mathbf{P}_S^n/S, r)$ to the cdh-sheaf representable by the projective space $Div_d(G)$. Since $z_d^{eff}(\mathbf{P}^n/Spec(\mathbf{Z}), r)$ is clearly a closed subpresheaf in the sheaf $z(\mathbf{P}^n/Spec(\mathbf{Z}), r)$ it is sufficient by 4.4.3 and 4.4.4 to show that the subset $C_{r,d} = Im(Top(Chow))$ in $Div^d(G)$ is constructible.

Denote by F_d the subpresheaf in $z_d^{eff}(\mathbf{P}^n/Spec(\mathbf{Z}), r)$ such that for a Noetherian scheme S the subset $F_d(S)$ in $z_d^{eff}(\mathbf{P}_S^n/S, r)$ consists of cycles $\mathcal{Z}$ such that for any algebraically closed field k and a k-point $x : Spec(k) \to S$ the cycle $cycl(x)(\mathcal{Z})$ on $\mathbf{P}_k^n$ is of the form $cycl(Z)$ for a closed integral subscheme Z in $\mathbf{P}_k^n$ of dimension r and degree d. Let further

$$\mathbf{F}_d = \coprod_{k=1}^{d} \coprod_{d_1 + \ldots + d_k = d} \prod_{i=1,\ldots,k} F_{d_i}.$$

We have the following diagram of morphisms of presheaves

$$\begin{array}{ccc} \mathbf{F}_d & \overset{Chow}{\to} & \coprod_{k=1}^{d} \coprod_{d_1 + \ldots + d_k = d} \prod_{i=1,\ldots,k} L(Div_{d_i}^{irr}(G)) \\ \downarrow & & \downarrow \\ z_d^{eff}(\mathbf{P}^n/Spec(\mathbf{Z}), r) & \to & L(Div_d(G)) \end{array}$$

The first vertical arrow is a surjection on the corresponding topological spaces and the second vertical arrow being induced by a morphism of schemes of finite type takes constructible sets to constructible sets. It is sufficient therefore to show that the image of $Top(Chow^{irr})$ where $Chow^{irr}$ is the morphism

$$F_d \to L(Div_d^{irr}(G))$$

is constructible. Let us show that in fact $Im(Top(Chow^{irr})) = C_{r,d}^{irr}$.

(" $\mathbf{Im(Top(Chow^{irr}))} \subset \mathbf{C_{r,d}^{irr}}$") Let k be an algebraically closed field and $\mathcal{Z} \in F_d(\mathbf{P}_k^n)$. We have to show that the point x on $Div_d^{irr}(G)_k$ which corresponds to $Chow^{irr}(\mathcal{Z})$ belongs to $C_{r,d}^{irr}$. It follows immediately from our definition of $C_{r,d}^{irr}$ since in this case the fiber of the projection $\mathbf{P}^n \times Div_d^{irr}(G) - U \to Div_d^{irr}(G)$ over x contains the support of $\mathcal{Z}$.

(" $\mathbf{C_{r,d}^{irr}} \subset \mathbf{Im(Top(Chow^{irr}))}$ ") Let k be an algebraically closed field and $x : Spec(k) \to Div_d^{irr}(G)$ be a point of $Div_d^{irr}(G)$ which belongs to $C_{r,d}^{irr}$, i.e. such that $dim((\mathbf{P}^n \times Div_d^{irr}(G) - U) \times_{Div_d^{irr}(G)} Spec(k)) \geq r$. Let D_x be the divisor on G_k which corresponds to x. One can verify easily that fibers of $\mathbf{P}^n \times Div_d(G) - U$ over $Div_d(G)$ are of dimension $\leq r$. Denote by Z_i the irreducible components of the fiber of $\mathbf{P}^n \times Div_d(G) - U$ over x which have dimension r. Then $p(f^{-1}(Z_i))$ is an irreducible divisor in G which is obviously contained in $supp(D)$. Since $p(f^{-1}(Z_i)) \neq p(f^{-1}(Z_j))$ for $i \neq j$ (Lemma 4.4.12(1)) and $supp(D)$ is irreducible we conclude that there is only one component Z of dimension r and $p(f^{-1}(Z)) = supp(D)$. It implies easily that $D = Chow^{irr}(cycl(Z))$. Theorem is proven.

Let S be a Noetherian scheme and $i : X \to \mathbf{P}_S^n$ be a projective scheme over S. Denote by $z_d^{eff}((X, i)/S, r)$ the subpresheaf in the Chow presheaf $z^{eff}(X/S, r)$ such that for a Noetherian scheme T over S the subset $z_d^{eff}((X, i)/S, r)(T)$ in $z^{eff}(X/S, r)(T)$ consists of relative cycles of degree d with respect to the embedding $i \times_S Id_T$. If $U \subset X$ is an open subset of X we denote by $c_d^{eff}((U, i)/S, r)$ the preimage of $z_d^{eff}((X, i)/S, r)$ with respect to the obvious morphism $c^{eff}(U/S, r) \to z^{eff}(X/S, r)$.

Corollary 4.4.13 *For any Noetherian scheme S and a projective scheme $i : X \to \mathbf{P}_S^n$ over S the presheaf $z_d^{eff}((X, i)/S, r)$ is h-representable by a projective scheme $C_{r,d}(X, i)$ over S and*

$$z^{eff}(X/S, r) = \coprod_{d \geq 0} z_d^{eff}((X, i)/S, r).$$

If U is an open subscheme in X then $c_d^{eff}((U, i)/S, r)$ is representable by an open subscheme $C_{r,d}(U, i)$ in $C_{r,d}(X, i)$ and

$$c^{eff}(U/S, r) = \coprod_{d \geq 0} c_d^{eff}((U, i)/S, r).$$

Proof: It follows immediately from Theorem 4.4.11, Propositions 4.4.5, 4.4.6 and Lemma 4.4.3.

Let $X \to S$ be a scheme of finite type over a Noetherian scheme S. A (closed) equivalence relation on X is a closed subscheme $R \subset X \times_S X$ such that for any Noetherian scheme T over S the subset $Hom_S(T, R)$ of the set $Hom_S(T, X \times_S X) = Hom_S(T, X) \times Hom_S(T, X)$ is an equivalence relation on $Hom_S(T, X)$. An equivalence relation R is called a proper equivalence relation if the projections $R \to X$ are proper morphisms.

Let R be an equivalence relation on X and Y be a Noetherian scheme over S. Let $\Gamma \to Y \times_S X$ be a closed subset of $Y \times_S X$. Denote by $p_Y : \Gamma \to Y$, $p_X : \Gamma \to X$ the obvious morphisms. We call Γ a graph-like closed subset (with respect to R) if the following conditions hold:

1. The morphism $p_Y : \Gamma \to Y$ is a universal topological epimorphism (i.e. an h-covering).

2. For any algebraically closed field $\bar{k}$ and a $\bar{k}$-valued point $\bar{y} : Spec(\bar{k}) \to Y$ all the elements of the subset $p_X(p_Y^{-1}(\bar{y}))$ in $X(\bar{k})$ are equivalent with respect to $R(\bar{k})$.

Any such Γ defines for any algebraically closed field $\bar{k}$ a map of sets $f_\Gamma : Y(\bar{k}) \to X(\bar{k})/R(\bar{k})$. We say that two graph-like subsets Γ_1, Γ_2 are equivalent if for any algebraically closed field $\bar{k}$ the corresponding maps f_{Γ_1} and f_{Γ_2} coincide. A continuous algebraic map from Y to X/R (over S) is an equivalence class of graph-like closed subschemes of $Y \times_S X$ with respect to this equivalence relation[3]. We denote the set of continuous algebraic maps from Y to X/R over S by $Hom_S^{a.c.}(Y, X/R)$.

Let now R be a proper equivalence relation. Denote the projections $R \to X$ by pr_1 and pr_2 respectively. For any closed subset Γ in $Y \times_S X$ which is of finite type over Y consider the subset $\Gamma_R \subset Y \times_S X$ of the form $(Id_Y \times_S pr_2)(Id_Y \times_S pr_1)^{-1}(\Gamma)$. Since R is proper Γ_R is a closed subset. One can verify easily that if Γ is a graph-like subset with respect to R then Γ_R is graph-like and Γ is equivalent to Γ_R. Moreover two graph-like closed subsets Γ_1 and Γ_2 are equivalent if and only if $(\Gamma_1)_R = (\Gamma_2)_R$.

Lemma 4.4.14 *Let $X \to S$ be a scheme of finite type over a Noetherian scheme S and $R \subset X \times_S X$ be a proper equivalence relation on X. Let $L_h(X)$ be the h-sheaf represented by X on the category of Noetherian schemes over S and let $L_h(X)/R$ be the quotient sheaf (in the h-topology)*

---

[3]Our definition of continuous algebraic maps is a natural generalization of the definition given in [3].

of $L_h(X)$ with respect to the equivalence relation defined by R. Then for any Noetherian scheme Y over S there is a canonical bijection:

$$Hom_S^{a.c.}(Y, X/R) = (L_h(X)/R)(Y).$$

Proof: Clearly $Hom_S^{a.c.}(Y, X/R)$ is a presheaf on the category of Noetherian schemes over S with respect to Y. Let us show that it is in fact an h-sheaf. Consider an h-covering $p : U \to Y$ of Y. Note first that if Γ_1, Γ_2 are two graph-like closed subsets in $Y \times_S X$ such that $(p \times_S Id_X)^{-1}(\Gamma_1)$ is equivalent to $(p \times_S Id_X)^{-1}(\Gamma_2)$ then Γ_1 is equivalent to Γ_2.

Let now $\Gamma \subset U \times_S X$ be a graph-like closed subset such that $pr_1^{-1}(\Gamma)$ is equivalent to $pr_2^{-1}(\Gamma)$ where $pr_i : U \times_Y U \times_S X \to U \times_S X$ are the projections. We may assume that $\Gamma = \Gamma_R$. Then the same obviously holds for $pr_i^{-1}(\Gamma)$ and we conclude that $pr_1^{-1}(\Gamma) = pr_2^{-1}(\Gamma)$. Since p is a universal topological epimorphism it implies trivially that $\Gamma = (p \times_S Id_X)^{-1}(\Gamma_0)$ for a closed subset Γ_0 in $Y \times_S X$. Since Γ_0 is obviously graph-like with respect to R it proves that $Hom_S^{a.c.}(-, X/R)$ is indeed an h-sheaf.

Let us construct a morphism of sheaves

$$\phi : Hom_S^{a.c.}(-, X/R) \to L_h(X)/R.$$

Let Γ be a graph-like closed subset in $Y \times_S X$. We have a morphism $\Gamma \to X$. Since $\Gamma \to Y$ is an h-covering our definition of a graph-like subset implies trivially that the corresponding section of $L_h(X)/R$ on Γ can be descended to a section of $L_h(X)/R$ on Y which does not depend on the choice of Γ in its equivalence class.

The proof of the fact that ϕ is an isomorphism is trivial.

Let $X \to S$ be a quasi-projective scheme over a Noetherian scheme S and $i : \bar{X} \to \mathbf{P}_S^n$ be a projective scheme over S such that there is an open embedding $X \subset \bar{X}$ over S. Consider the scheme $C_{r,d}(\bar{X}, i)$ which h-represents the presheaf $z_d^{eff}(\bar{X}/S, r)$ by Corollary 4.4.13. Set

$$C_{r,\leq d}(\bar{X}, i) = \coprod_{i \leq d} C_{r,i}(\bar{X}, r).$$

Consider the canonical section $\mathcal{Z}$ of $z(\bar{X}/S, r)_h$ over $C_{r,\leq d}(\bar{X}, i)$ and let $\tilde{\mathcal{Z}}$ be the section $(pr_1^*(\mathcal{Z}) - pr_2^*(\mathcal{Z})) - (pr_3^*(\mathcal{Z}) - pr_4^*(\mathcal{Z}))$ of $z(\bar{X}/S, r)_h$ over the product $(C_{r,\leq d}(\bar{X}, i))_S^4$. Denote by R_d^X the closed subset of points of $(C_{r,\leq d}(\bar{X}, i))_S^4$ where $\tilde{\mathcal{Z}}$ belongs to $z((\bar{X} - X)/S, r) \subset z(\bar{X}/S, r)$ (see 4.4.5). One can easily see that R_d^X is an equivalence relation on $(C_{r,\leq d}(\bar{X}, i))_S^2$ and since $C_{r,\leq d}(\bar{X}, i)$ is proper over S it is a proper equivalence relation. Note further that the obvious embeddings $C_{r,\leq d}(\bar{X}, i) \to C_{r,\leq(d+1)}(\bar{X}, i)$ take R_d^X to R_{d+1}^X and hence for any Noetherian scheme T over S there is a family of maps

$$Hom_S^{a.c.}(T, (C_{r,\leq d}(\bar{X}, i))_S^2 / R_d^X) \to Hom_S^{a.c.}(T, (C_{r,\leq(d+1)}(\bar{X}, i))_S^2 / R_{d+1}^X).$$

Proposition 4.4.15 *For any Noetherian scheme T over S there is a canonical bijection*

$$colim_d Hom_S^{a.c.}(T, (C_{r,\leq d}(\bar{X}, i) \times_S C_{r,\leq d}(\bar{X}, i)) / R_d^X) = z(X/S, r)_h(T).$$

Proof: It follows immediately from Lemma 4.4.14 and our definition of the equivalence relations R_d^X.

References

1. M. Artin. *Grothendieck topologies*. Harvard Univ., Cambridge, 1962.
2. Eric M. Friedlander. Some computations of algebraic cycle homology. *K-theory*, 8:271–285, 1994.
3. Eric M. Friedlander and O. Gabber. Cycle spaces and intersection theory. *Topological methods in modern mathematics*, pages 325–370, 1993.
4. W. Fulton. *Intersection Theory*. Springer-Verlag, Berlin, 1984.
5. A. Grothendieck. *Revetements etale et groupe fondamental(SGA 1)*. Lecture Notes in Math. 224. Springer, Heidelberg, 1971.
6. A. Grothendieck and J. Dieudonne. *Etude Globale Elementaire de Quelques Classes de Morphismes (EGA 2)*. Publ. Math. IHES,8, 1961.
7. A. Grothendieck and J. Dieudonne. *Etude Cohomologique des Faisceaux Coherents (EGA 3)*. Publ. Math. IHES,11,17, 1961,1963.
8. A. Grothendieck and J. Dieudonne. *Etude Locale des Schemas et des Morphismes de Schemas (EGA 4)*. Publ. Math. IHES,20,24,28,32, 1964-67.
9. A. Grothendiek. Les schemas de Picard: theoremes d'existence. In *Seminaire Bourbaki*, n.232 1961/62.
10. H. Matsumura. *Commutative ring theory*. Cambridge University Press, 1986.
11. J.S. Milne. *Etale Cohomology*. Princeton Univ. Press, Princeton, NJ, 1980.
12. Y. Nisnevich. The completely decomposed topology on schemes and associated descent spectral sequences in algebraic K-theory. In *Algebraic K-theory: connections with geometry and topology*, pages 241–342. Kluwer Acad. Publ., Dordrecht, 1989.
13. M. Raynaud and L. Gruson. Criteres de platitude et de projectivite. *Inv. Math.*, 13:1–89, 1971.
14. P. Samuel. *Methodes d'algebre abstraite en geometrie algebrique*. Ergebnisse der Mathematik, N.F.4. Springer-Verlag, Berlin, 1955.

15. J.-P. Serre. Algebre locale. Multiplicities. *Lecture Notes in Math.*, 11, 1965.
16. Andrei Suslin and Vladimir Voevodsky. Singular homology of abstract algebraic varieties. *Invent. Math.*, 123(1):61–94, 1996.
17. V. Voevodsky. Cohomological theory of presheaves with transfers. *This volume.*
18. V. Voevodsky. Homology of schemes. *Selecta Mathematica, New Series,* 2(1):111–153, 1996.

3

Cohomological Theory of Presheaves with Transfers

Vladimir Voevodsky

Contents

1	**Introduction**	**88**
2	**Divisors with compact support on relative curves**	**89**
3	**Pretheories**	**94**
	3.1 Categories of pretheories	94
	3.2 Homotopy invariant pretheories	96
	3.3 Elementary properties of transfer maps	100
	3.4 Examples of pretheories	104
4	**Zariski sheaves associated with pretheories**	**105**
	4.1 Technical lemmas	105
	4.2 Pretheories in a neighborhood of a smooth subscheme	111
	4.3 Pretheories on curves over a field	114
	4.4 Pretheories on semi-local schemes	115
	4.5 Zariski cohomology with coefficients in pretheories	119
	4.6 The Gersten resolution	124
5	**Sheaves in Nisnevich and etale topologies associated with pretheories**	**125**
	5.1 Pretheoretical structures on Nisnevich cohomology	125
	5.2 Cohomology of affine space and the comparison theorem	127
	5.3 Applications to Suslin homology	129
	5.4 Cohomology of blowups	131
	5.5 Sheaves in the etale topology associated with pretheories	135

1. Introduction

Let k be a field and Sm/k be the category of smooth schemes over k. In this paper we study contravariant functors from the category Sm/k to additive categories equipped with *transfer maps*. More precisely we consider contravariant functors $F : (Sm/k)^{op} \to A$ together with a family of morphisms $\phi_{X/S}(Z) : F(X) \to F(S)$ given for any smooth curve $X \to S$ over a smooth scheme S over k and a relative divisor Z on X over S which is finite over S. If these maps satisfy some natural properties (see definition 3.1) such a collection of data is called a *pretheory* over k. Some examples of pretheories are given in §3.4.

A pretheory F is called homotopy invariant if for any smooth scheme X one has $F(X \times \mathbf{A}^1) = F(X)$. For any pretheory F with values in an abelian category we define a family of homotopy invariant pretheories $\underline{h}_i(F)$. Applying this construction to particular pretheories F one can get algebraic cycle homology and bivariant morphic homology of Friedlander and Gabber (see [1]) or algebraic singular homology and algebraic Lawson homology defined in [7].

In the first section we prove some elementary facts about divisors with proper support on relative curves (see also [7]). In Section 3 we define pretheories and prove some of their basic properties.

In the third section we study local behavior of homotopy invariant pretheories. In most cases we assume that our pretheory takes values in the category of abelian groups and study the sheaf F_{Zar} in the Zariski topology on Sm/k associated with the presheaf F. The main theorem (Theorem 4.27) of this section states that if F is a homotopy invariant pretheory over a perfect field k with values in the category of abelian groups then the functors $U \to H^i_{Zar}(U, F_{Zar})$ have canonical structures of homotopy invariant pretheories.

As an application we show in §4.6 that for any such pretheory F the sheaf F_{Zar} has the Gersten resolution.

In the next section we consider the sheaves in the Nisnevich topology (see [5]) associated with pretheories with values in the category of abelian groups. It turns out that the Nisnevich topology has several serious advantages over the Zariski topology in the context of pretheories. In particular for any reasonable (see Definition 5.1), not necessarily homotopy invariant, pretheory F with values in the category of abelian groups the functors $U \to H^i_{Nis}(U, F_{Nis})$ have canonical structures of pretheories (Theorem 5.3). On the other hand, using the results of Section 4 we show that for any homotopy invariant pretheory F over a perfect field with values in the category of abelian groups one has canonical isomorphisms

$$H^i_{Nis}(-, F_{Nis}) \cong H^i_{Zar}(-, F_{Zar})$$

for all $i \geq 0$ (Theorem 5.7). It allows us to reformulate all the results of the previous section for the Nisnevich topology. In the last paragraph we apply our technique to prove the existence of a long exact sequence of Mayer-Vietoris type for algebraic singular homology (Theorem 5.17). Though apparently we are only interested in open coverings the use of the Nisnevich topology in the proof is crucial.

In paragraph 5.4 we consider the behavior of homotopy invariant pretheories with respect to blow-ups. Proposition 5.21 has important applications to algebraic cycle cohomology (see [2]).

In the last paragraph we consider the etale topology. The first result is the rigidity Theorem 5.25 (which is just a version of the rigidity theorem proven in [7]). It says that the sheaf in the etale topology on Sm/k associated with a homotopy invariant pretheory with values in the category of torsion abelian groups of torsion prime to characteristic of the field k is locally constant. For rational coefficients we get in the case of a perfect field k isomorphisms (Theorem 5.28):

$$H^i_{et}(-, F_{et}) \otimes \mathbf{Q} \cong H^i_{Zar}(-, F_{Zar}) \otimes \mathbf{Q}.$$

Most results of this paper are formulated for smooth schemes over a perfect field. This restriction on the base scheme is important. In fact almost all the results of this paper are wrong even for smooth schemes over the spectrum of a discrete valuation ring of characteristic zero.

The idea to consider presheaves with transfer (instead of qfh-sheaves) which allows us to obtain the "correct" answers with integral coefficients is due completely to my conversations with A.Suslin. I am also very grateful to D. Kazhdan for his patient interest in my work and to Yu. Tschinkel who told me about the existence of the Nisnevich topology.

2. Divisors with compact support on relative curves

Let S be a smooth scheme. A curve $p : X \to S$ over S is an equi-dimensional morphism of finite type of relative dimension one. The morphism $\emptyset \to S$ will also be considered as a curve over S.

For any curve $X \to S$ over a smooth scheme S denote by $c_{equi}(X/S, 0)$ the free abelian group generated by closed integral subschemes of X which are finite over S and surjective over an irreducible component of S.

Let Z be any closed subscheme of X which is finite over S. Consider the cycle $Cycl_X(Z)$ of X in Z. It follows immediately from our definition that it belongs to $c_{equi}(X/S, 0)$ if and only if any irreducible component of Z dominates an irreducible component of S. In particular $Cycl_X(Z)$ belongs to $c_{equi}(X/S, 0)$ for any closed subscheme Z in X which is finite and flat over S.

As was shown in [8] for any morphism of smooth schemes $f : S' \to S$ one can define a homomorphism

$$cycl(f) : c_{equi}(X/S, 0) \to c_{equi}(X \times_S S'/S', 0)$$

such that the following conditions hold:

1. For a composable pair of morphisms $S'' \xrightarrow{g} S' \xrightarrow{f} S$ one has

$$cycl(f \circ g) = cycl(g) \circ cycl(f)$$

2. For any dominant morphism $f : S' \to S$ of smooth schemes and an integral closed subscheme Z in $c_{equi}(X/S, 0)$ one has

$$cycl(f)(Z) = Cycl_{X \times_S S'}(Z \times_S S')$$

3. For any morphism of smooth schemes $f : S' \to S$ and a closed subscheme Z in X which is finite and flat over S one has:

$$cycl(f)(Cycl_X(Z)) = Cycl_{X \times_S S'}(Z \times_S S').$$

It can be shown easily (using Platification Theorem, see [6] or [8, Th. 2.2.2]) that homomorphisms $cycl(f)$ are uniquely determined by the conditions (1)-(3) given above.

Let $g : X_1 \to X_2$ be a morphism of curves over a smooth scheme S and Z be an integral closed subscheme in X_1 which belongs to $c_{equi}(X_1/S, 0)$. Since it is proper over S its image $g(Z)$ in X_2 is a closed subscheme which belongs to $c_{equi}(X_2/S, 0)$. The morphism $Z \to g(Z)$ is a finite surjective morphism of integral schemes. Denote its degree (i.e. the degree of the corresponding function fields extension) by $n(Z, g)$.

We define the direct image homomorphism

$$g_* : c_{equi}(X_1/S, 0) \to c_{equi}(X_2/S, 0)$$

setting

$$g_*(Z) = n(Z, g)g(Z).$$

The following proposition is a particular case of [8, Prop. 3.6.2].

Proposition 2.1 *Let $X_1 \to S$, $X_2 \to S$ be a pair of curves over a smooth scheme S, $g : X_1 \to X_2$ be a morphism over S and $f : S' \to S$ be a morphism of smooth schemes. Then the following diagram commutes:*

$$
\begin{array}{ccc}
c_{equi}(X_1/S,0) & \xrightarrow{cycl(f)} & c_{equi}(X_1 \times_S S'/S',0) \\
g_* \downarrow & & \downarrow g_* \\
c_{equi}(X_2/S,0) & \xrightarrow{cycl(f)} & c_{equi}(X_2 \times_S S'/S',0).
\end{array}
$$

The following technique provides us with a method of construction of elements in the groups $c_{equi}(X/S,0)$ which will be used extensively in this paper.

Let D be a Cartier divisor on X. Denote by $\mathcal{L}_D$ the corresponding line bundle on X. There is a canonical trivialization $t_D : \mathcal{O}_U \to (\mathcal{L}_D)_{|U}$ of $\mathcal{L}_D$ on $U = X - D$. If D is effective this trivialization extends to a canonical section $s_D : \mathcal{O}_X \to \mathcal{L}_D$ of $\mathcal{L}_D$ on X. Suppose that X is normal and connected. Then for any line bundle $\mathcal{L}$ on X, open subset U of X and a trivialization t of $\mathcal{L}$ on U there exists a unique Cartier divisor $D(\mathcal{L},U,t)$ and an isomorphism $\mathcal{L} \cong \mathcal{L}_{D(\mathcal{L},U,t)}$ such that $t = t_{\mathcal{L}_D}$ (see [4, 21.1.4]). Note that $D(\mathcal{L},U,t)$ considered as a closed subset of X lies in $X - U$.

Let $\bar{p} : \bar{X} \to S$ be a proper normal curve over a smooth scheme S and $X \subset \bar{X}$ be an open subscheme in $\bar{X}$ which is quasi-affine and smooth over S. Denote by X_∞ the reduced scheme $\bar{X} - X$.

Proposition 2.2 *Let $\mathcal{L}$ be a line bundle on $\bar{X}$, U be an open neighborhood of X_∞ in $\bar{X}$ and $t : \mathcal{O}_U \to \mathcal{L}_{|U}$ be a trivialization of $\mathcal{L}$ on U. Consider the Cartier divisor $D(\mathcal{L},U,t)$ defined by the triple $(\mathcal{L},U,t)$. Then $Cycl_X(D(\mathcal{L},U,t))$ belongs to $c_{equi}(X/S,0)$.*

Proof: Let

$$
Cycl_X(D(\mathcal{L},U,t)) = \sum n_i Z_i
$$

where Z_i are integral closed subschemes of X. Since $\bar{p} : \bar{X} \to S$ is proper and X is quasi-affine over S the open subset U is dense in all fibers of $\bar{p}$ and $\bar{X} - U$ is finite over S. Since $D(\mathcal{L},U,t)$ as a closed subset in $\bar{X}$ belongs to $\bar{X} - U$ the closed subschemes Z_i are finite over S. On the other hand each of Z_i is of pure codimension one in X and being finite over S it must dominate an irreducible component of S.

Lemma 2.3 *Under the assumption of Proposition 2.2 consider a morphism of smooth schemes $f : S' \to S$. Then for any triple $(\mathcal{L},U,t)$ as above we have:*

$$
cycl(f)(Cycl_X(D(\mathcal{L},U,t))) = Cycl_{X \times_S S'}(D(f_X^*(\mathcal{L}), f_X^{-1}(U), f_X^*(t)))
$$

where f_X is the projection $X \times_S S' \to X$.

Proof: It follows from the fact that $D(\mathcal{L}, U, t)$ is flat over S by the third property of base change homomorphisms.

Let $\mathcal{Z}_0, \mathcal{Z}_1$ be two elements of $c_{equi}(X/S, 0)$. We say that they are equivalent if there is an element $\mathcal{Z}$ in $S \times \mathbf{A}^1$ such that

$$cycl(i_0)(\mathcal{Z}) = \mathcal{Z}_0$$

$$cycl(i_1)(\mathcal{Z}) = \mathcal{Z}_1$$

where $i_0, i_1 : S \to S \times \mathbf{A}^1$ are the closed embeddings which correspond to the points 0 and 1 of $\mathbf{A}^1$ respectively. We denote the group of equivalence classes of elements of $c_{equi}(X/S, 0)$ with respect to this equivalence realtion by $h_0(X/S)$. It follows immediately from the definition that homomorphisms $cycl(f)$ give us homomorphisms $h_0(X/S) \to h_0(X \times_S S'/S')$ and by Proposition 2.1 the groups $h_0(X/S)$ are covariantly functorial with respect to morphisms $X_1 \to X_2$ of curves over S.

Definition 2.4 *Let $p : X \to S$ be a curve over a smooth scheme S. A good compactification of X is a pair $(\bar{p} : \bar{X} \to S, j : X \to \bar{X})$ where $\bar{X}$ is a proper curve over S and j is an open embedding over S such that the closed subset $X_\infty = \bar{X} - X$ in $\bar{X}$ has an open neighborhood which is affine over S.*

Note that if $(\bar{p} : \bar{X} \to S, j : X \to \bar{X})$ is a good compactification of a curve $p : X \to S$ and $f : S' \to S$ is a morphism of smooth schemes then $(\bar{p}' : \bar{X} \times_S S' \to S', j \times_S Id_{S'} : X \times_S S' \to \bar{X} \times_S S')$ is a good compactification of the curve $p' : X \times_S S' \to S'$.

The following lemma shows that if a normal curve over a smooth scheme has a good compactification then we may always choose it in such a way that the corresponding proper curve is also normal.

Lemma 2.5 *Let $p : X \to S$ be a normal curve over a smooth scheme S and $(\bar{p} : \bar{X} \to S, j : X \to \bar{X})$ be a good compactification of X. Let further $\bar{X}'$ be the normalization of $\bar{X}$. Then the open embedding $j : X \to \bar{X}$ defines an open embedding $j' : X \to \bar{X}'$ and the pair $(\bar{p}' : \bar{X}' \to S, j' : X \to \bar{X}')$ is a good compactification of X.*

Proof: Since the morphism $\bar{X}' \to \bar{X}$ is finite it is also affine which implies that the preimage of the affine open neighborhood of X_∞ in $\bar{X}'$ is an affine open neighborhood of $X'_\infty = \bar{X}' - X$ in $\bar{X}'$.

Proposition 2.6 *Let* $p : X \to S$ *be a smooth quasi-affine curve over a smooth scheme* S *and* $(\bar{p} : \bar{X} \to S, j : X \to \bar{X})$ *be a good compactification of* X. *Let* X_∞ *be the reduced scheme* $\bar{X} - X$.

Let further $\mathcal{L}$ *be a line bundle on* $\bar{X}$ *and* $s : \mathcal{O}_{X_\infty} \to \mathcal{L}_{|X_\infty}$ *be a trivialization of* $\mathcal{L}$ *over* X_∞. *Then the following statements hold.*

1. *For any two extensions* $\tilde{s}_1, \tilde{s}_2$ *of the trivialization* s *to an open neighborhood* U *of* X_∞ *the cycles* $Cycl_X(D(\mathcal{L}, U, \tilde{s}_1))$, $Cycl_X(D(\mathcal{L}, U, \tilde{s}_2))$ *give the same element in* $h_0(X/S)$.

2. *If* S *is affine there exists an open neighborhood* U *of* X_∞ *in* $\bar{X}$ *and an extension* $\tilde{s} : \mathcal{O}_U \to \mathcal{L}_{|U}$ *of the trivialization* s *to a trivialization of* $\mathcal{L}$ *on* U.

Proof: The second statement follows trivially from the fact that X_∞ has an affine open neighborhood in $\bar{X}$.

To prove the first one consider the curve

$$\bar{X} \times \mathbf{A}^1 \to S \times \mathbf{A}^1$$

over $S \times \mathbf{A}^1$ and let $\mathcal{L}'$ be the pull-back of $\mathcal{L}$ with respect to the projection $\bar{X} \times \mathbf{A}^1 \to \bar{X}$. The trivializations $\tilde{s}_1, \tilde{s}_2$ define trivializations $\tilde{s}_1', \tilde{s}_2'$ of $\mathcal{L}'$ on $U \times \mathbf{A}^1$ which coincide on $X_\infty \times \mathbf{A}^1$. Consider the section $t\tilde{s}_1' + (1 - t)\tilde{s}_2'$ of $\mathcal{L}'$ on $U \times \mathbf{A}^1$ where t is the regular function given by the projection $U \times \mathbf{A}^1 \to \mathbf{A}^1$. Its restriction to $X_\infty \times \mathbf{A}^1$ is equal to the trivialization induced by s and hence it definies a trivialization h of $\mathcal{L}'$ on an open neighborhood V of $X_\infty \times \mathbf{A}^1$ in $\bar{X} \times \mathbf{A}^1$. The divisor $D(\mathcal{L}', V, h)$ associated with h gives us by Proposition 2.2 an element $\mathcal{Z}$ in $c_{equi}(X \times \mathbf{A}^1/S \times \mathbf{A}^1, 0)$. Consider the closed embeddings $i_0, i_1 : S \to S \times \mathbf{A}^1$ which correspond to the points 0 and 1 on $\mathbf{A}^1$ respectively. Lemma 2.3 implies that

$$cycl(i_0)(\mathcal{Z}) = Cycl_X(D(\mathcal{L}, U, \tilde{s}_2))$$

and

$$cycl(i_1)(\mathcal{Z}) = Cycl_X(D(\mathcal{L}, U, \tilde{s}_1)).$$

Let $p : X \to S$ be a smooth quasi-affine curve over an affine smooth scheme S and $(\bar{p} : \bar{X} \to S, j : X \to \bar{X})$ be a good compactification of X. Denote by $Pic(\bar{X}, X_\infty)$ the set of isomorphism classes of pairs of the form $(\mathcal{L}, s)$ where $\mathcal{L}$ is a line bundle on $\bar{X}$ and s is a trivialization of $\mathcal{L}$ on X_∞. Tensor multiplication of line bundles and their sections gives a structure of abelian group on $Pic(\bar{X}, X_\infty)$. Proposition 2.6 implies that there is a well defined homomorphism

$$Pic(\bar{X}, X_\infty) \to h_0(X/S).$$

Lemma 2.3 implies immediately that this homomorphism commutes in the obvious sense with base change homomorphisms $cycl(-)$.

Note that if we consider pairs of the form $(\mathcal{O}_{\bar{X}}, s)$ we get a homomorphism

$$\mathcal{O}^*(X_\infty) \to h_0(X/S).$$

Remark 2.7 It can be shown (see [7]) that for a smooth quasi-affine curve $X \to S$ over a smooth affine scheme S and a good compactification of X the homomorphism $Pic(\bar{X}, X_\infty) \to h_0(X/S)$ is in fact an isomorphism. We will not use this fact in the present paper.

3. Pretheories

3.1. Categories of pretheories

Let k be a field. We denote by Sm/k the category of smooth schemes of finite type over k.

Definition 3.1 *A pretheory (F, ϕ, A) over a field k is the following collection of data.*

1. *An additive category A.*

2. *A contravariant functor $F : Sm/k \to A$.*

3. *For any object S of Sm/k and any smooth curve $p : X \to S$ over S a homomorphism of abelian groups*

$$\phi_{X/S} : c_{equi}(X/S, 0) \to Hom(F(X), F(S)).$$

These data should satisfy the following conditions.

1. *For any object S of Sm/k, any smooth curve $p : X \to S$ over S and any S-point $i : S \to X$ of X one has $\phi_{X/S}(i(S)) = F(i)$.*

2. *Let $f : S_1 \to S_2$ be a morphism of smooth schemes of finite type over k and $p : X_2 \to S_2$ be a smooth curve over S_2. Consider the Cartesian square*

$$\begin{array}{ccc} X_1 & \xrightarrow{g} & X_1 \\ \downarrow & & \downarrow \\ S_1 & \xrightarrow{f} & S_2. \end{array}$$

Then for any $\mathcal{Z}$ in $c_{equi}(X_2/S_2, 0)$ one has

$$F(f) \circ \phi_{X_2/S_2}(\mathcal{Z}) = \phi_{X_1/S_1}(f^*(\mathcal{Z})) \circ F(g).$$

3. *For any pair X, Y of objects of Sm/k the canonical morphism*

$$F(X \coprod Y) \to F(X) \oplus F(Y)$$

is an isomorphism.

Definition 3.2 *Let (F_1, ϕ_1, A), (F_2, ϕ_2, A) be a pair of pretheories with values in an additive category A. A morphism of pretheories*

$$T : (F_1, \phi_1, A) \to (F_2, \phi_2, A)$$

is a morphism $T : F_1 \to F_2$ of presheaves on the category Sm/k such that for any smooth scheme S over k, any smooth curve $p : X \to S$ over S and any element $\mathcal{Z} \in c_{equi}(X/S, 0)$ the following diagram commutes:

$$
\begin{array}{ccc}
F_1(X) & \stackrel{\phi_1(\mathcal{Z})}{\longrightarrow} & F_1(S) \\
T_X \downarrow & & \downarrow T_S \\
F_2(X) & \stackrel{\phi_2(\mathcal{Z})}{\longrightarrow} & F_2(S).
\end{array}
$$

Denote by $Preth(k, A)$ the category of pretheories over k with values in an additive category A. It is obviously an additive category. Moreover if A is an abelian category one has the following obvious proposition.

Proposition 3.3 *Let k be a field and A be an abelian category. Then the category $Preth(k, A)$ is abelian and the forgetful functor*

$$Preth(k, A) \to Preshv(Sm/k, A)$$

to the category of A-valued presheaves on Sm/k is exact.

Let (F, ϕ, A) be a pretheory over a field k and U be an object of Sm/k. We define a pretheory (F_U, ϕ_U, A) as follows:

1. The presheaf F_U takes a smooth scheme U of finite type over k to the object $F(U \times_k X)$.

2. Let $p : X \to S$ be a smooth curve over a smooth scheme S of finite type over k and $\mathcal{Z}$ be an element of $c_{equi}(X/S, 0)$. Then we define $(\phi_U)_{X/S}(\mathcal{Z})$ to be the homomorphism $\phi(p^*(\mathcal{Z}))$ where p is the projection $S \times U \to S$.

One can easily see that (F_U, ϕ_U, A) is indeed a pretheory and this construction gives us a functor

$$(Sm/k)^{op} \times Preth(k, A) \to Preth(k, A).$$

For any pretheory (F, ϕ, A) with values in an abelian category A we denote by (F_{-1}, ϕ_{-1}, A) the pretheory

$$coker((F_{\mathbf{A}^1}, \phi_{\mathbf{A}^1}, A) \to (F_{\mathbf{A}^1 - \{0\}}, \phi_{\mathbf{A}^1 - \{0\}}, A)).$$

Consider a separable field extension $k \subset L$ and let $f : Spec(L) \to Spec(k)$ be the corresponding morphism of schemes. Then for any pretheory (F, ϕ, Ab) over k where Ab is the category of abelian groups the inverse image $p^*(F)$ of the presheaf F has a canonical structure of a pretheory over L. It gives us an exact functor $p^* : Preth(k, Ab) \to Preth(L, Ab)$. If U is a smooth scheme over k then one has a canonical isomorphism:

$$p^*((F_U, \phi_U, Ab)) \cong p^*((F, \phi, Ab))_{U_L}.$$

In particular there is a canonical isomorphism

$$p^*((F, \phi, Ab)_{-1}) = p^*((F, \phi, Ab))_{-1}.$$

3.2. Homotopy invariant pretheories

Definition 3.4 *Let k be a field and (F, ϕ, A) be a pretheory over k. It is called homotopy invariant if for any object X of Sm/k the projection $X \times \mathbf{A}^1 \to X$ induces an isomorphism $F(X) \to F(X \times \mathbf{A}^1)$.*

For any field k and an additive category A denote by $HIPreth(k, A)$ the full subcategory in the category $Preth(k, A)$ which consists of homotopy invariant pretheories. In the following proposition we summarize basic properties of homotopy invariant pretheories which follow directly from our definitions.

Proposition 3.5 *Let k be a field and A be an abelian category. Then the category $HIPreth(k, A)$ is abelian and the inclusion functor $HIPreth(k, A) \to Preth(k, A)$ is exact. Moreover any subpretheory of a homotopy invariant pretheory is a homotopy invariant pretheory and any extension of homotopy invariant pretheories is a homotopy invariant pretheory.*

Let F be a presheaf on the category Sm/k with values in an abelian category. Let $\Delta^\bullet = (\Delta^n, \partial_i^n, s_i^n)$ be the standard cosimplicial object in Sm/k (see [2, §4]). Denote by $\underline{C}_*(F)$ the complex of presheaves such that $\underline{C}_n(F)(U) = F(U \times \Delta^n)$ and the differencial is given by the alternated sums of morphisms induced by ∂_i^n. Denote by $\underline{h}_i(F)$ the cohomology presheaves $H^{-i}(\underline{C}_*(F))$ of this complex.

If (F, ϕ, A) is a pretheory with values in an abelian category A then $\underline{C}_*(F)$ is a complex of pretheories (or equivalently a pretheory with values in the category of complexes over A) and in particular the presheaves $\underline{h}_i(F)$ have canonical structures of pretheories.

Proposition 3.6 *For any pretheory* (F, ϕ, A) *over* k *where* A *is an abelian category the pretheories* $\underline{h}_i(F, \phi, A)$ *are homotopy invariant.*

Proof: Since a pretheory is homotopy invariant if and only if the underlying presheaf is it is sufficient to show that for any presheaf of abelian groups F on Sm/k the presheaves $\underline{h}_i(F)$ are homotopy invariant. Let U be a smooth scheme over k. We have to show that the morphisms $\underline{h}_i(F)(U) \to \underline{h}_i(F)(U \times \mathbf{A}^1)$ induced by the projection $U \times \mathbf{A}^1 \to U$ are isomorphisms.

Denote by $i_0, i_1 : U \to U \times \mathbf{A}^1$ the closed embeddings $Id_U \times \{0\}$ and $Id_U \times \{1\}$ respectively. Let us show first that the morphisms

$$i_0^* : \underline{h}_i(F)(U \times \mathbf{A}^1) \to \underline{h}_i(F)(U)$$

$$i_1^* : \underline{h}_i(F)(U \times \mathbf{A}^1) \to \underline{h}_i(F)(U)$$

coincide. It is sufficient to prove that the corresponding morphisms of complexes of abelian groups $\underline{C}_*(F)(U \times \mathbf{A}^1) \to \underline{C}_*(F)(U)$ are homotopic.

Define homomorphisms

$$s_n : F(U \times \mathbf{A}^1 \times \Delta^n) \to F(U \times \Delta^{n+1})$$

by the formula

$$s_n = \sum_{i=0}^{n} (-1)^i (Id_U \times \psi_i)^*$$

where $\psi_i : \Delta^{n+1} \to \Delta^n \times \mathbf{A}^1$ is the linear isomorphism taking v_j to $v_j \times 0$ if $j \leq i$ or to $v_{j-1} \times 1$ if $j > i$ (here $v_j = (0, \ldots, 1, \ldots, 0)$ is the j-th vertex of Δ^{n+1} (resp. Δ^n)). A straightforward computation shows that $sd + ds = i_1^* - i_0^*$.

Consider now the morphism

$$Id_U \times \mu : U \times \mathbf{A}^1 \times \mathbf{A}^1 \to U \times \mathbf{A}^1$$

where $\mu : \mathbf{A}^1 \times \mathbf{A}^1 \to \mathbf{A}^1$ is given by multiplication of functions. Applying the previous result to the embeddings

$$i_0, i_1 : U \times \mathbf{A}^1 \to (U \times \mathbf{A}^1) \times \mathbf{A}^1$$

we conclude that the homomorphism

$$\underline{h}_i(F)(U \times \mathbf{A}^1) \to \underline{h}_i(F)(U \times \mathbf{A}^1)$$

induced by the composition

$$U \times \mathbf{A}^1 \overset{pr_1}{\to} U \overset{i_0}{\to} U \times \mathbf{A}^1$$

is the identity homomorphism which implies immediately the assertion of the proposition.

Note that we have a canonical morphism:

$$(F, \phi, A) \to \underline{h}_0(F, \phi, A).$$

The following proposition follows easily from the definition of $\underline{h}_0(-)$.

Proposition 3.7 *For any pretheory (F, ϕ, A) and any homotopy invariant pretheory (G, ψ, A) the homomorphism*

$$Hom(\underline{h}_0(F, \phi, A), (G, \psi, A)) \to Hom((F, \phi, A), (G, \psi, A))$$

induced by the projection

$$(F, \phi, A) \to \underline{h}_0(F, \phi, A).$$

is an isomorphism.

Thus $\underline{h}_0(-)$ is a functor $Preth(k, A) \to HIPreth(k, A)$ which is left adjoint to the inclusion functor. In particular it is right exact. The functors $\underline{h}_i(-)$ are in some sense the left derived functors of $\underline{h}_0(-)$. More precisely one has:

Proposition 3.8 *Let*

$$0 \to (F_1, \phi_1, A) \to (F_2, \phi_2, A) \to (F_3, \phi_3, A) \to 0$$

be a short exact sequence of pretheories with values in an abelian category A. Then there is a canonical long exact sequence of homotopy invariant pretheories of the form:

$$\ldots \to \underline{h}_i(F_1, \phi_1, A) \to \underline{h}_i(F_2, \phi_2, A) \to \underline{h}_i(F_3, \phi_3, A) \to \underline{h}_{i-1}(F_1, \phi_1, A) \to \ldots$$

Proof: It follows immediately from the obvious fact that $F \to \underline{C}_*(F)$ is an exact functor from the category of presheaves to the category of complexes of presheaves.

Unfortunately one can not apply usual techniques from homological algebra to define $\underline{h}_i(-)$ as the left derived functors of $\underline{h}_0$ because even if A is the category of abelian groups there is not enough pretheories (F, ϕ, A) such that $\underline{h}_i(F, \phi, A) = 0$ for all $i > 0$.

We will often use below the following simple results.

Proposition 3.9 *Let (F, ϕ, A) be a pretheory such that A is an abelian category and U be a smooth scheme of finite type over k. Then there are canonical isomorphisms of pretheories*

$$(\underline{h}_i(F, \phi, A))_U \cong \underline{h}_i((F, \phi, A)_U)$$

in particular if (F, ϕ, A) is a homotopy invariant pretheory then $(F, \phi, A)_U$ is also homotopy invariant.

Proposition 3.10 *Let $k \subset L$ be a separable extension of fields and $f : Spec(L) \to Spec(k)$ be the corresponding morphism. Then the inverse image functor $f^* : Preth(k, Ab) \to Preth(L, Ab)$ takes homotopy invariant pretheories to homotopy invariant pretheories and commutes with the corresponding functors $\underline{h}_i$.*

Proposition 3.11 *Let (F, ϕ, A) be a pretheory over k. It is homotopy invariant if and only if for any smooth scheme S of finite type over k and any smooth curve $X \to S$ over S the morphism $\phi_{X/S}$ can be factored through the canonical epimorphism $c_{equi}(X/S, 0) \to h_0(X/S)$ (see the definition of $h_0(-)$ after Lemma 2.3).*

Proof: The "only if" part is obvious. Suppose that (F, ϕ, A) is a pretheory such that the morphisms $\phi_{X/S}$ can be factored through $h_0(X/S)$. Let us show that it is homotopy invariant.

Let $i : X \to X \times \mathbf{A}^1$ be the embedding which corresponds to the point $\{0\}$ of $\mathbf{A}^1$. It is obviously sufficient to show that the following composition is equal to the identity morphism:

$$F(X \times \mathbf{A}^1) \overset{F(i)}{\to} F(X) \overset{F(pr_1)}{\to} F(X \times \mathbf{A}^1).$$

Consider the curve $(X \times \mathbf{A}^1) \times \mathbf{A}^1$ over $X \times \mathbf{A}^1$ and let

$$\mathcal{Z} \in h_0((X \times \mathbf{A}^1) \times \mathbf{A}^1 / X \times \mathbf{A}^1)$$

be the element which corresponds to the divisor $(X \times \mathbf{A}^1) \times \{0\}$. Then by the properties (1) and (2) of pretheories our composition is equal to the composition:

$$F(X \times \mathbf{A}^1) \xrightarrow{F(pr_1 \times Id_{\mathbf{A}^1})} F((X \times \mathbf{A}^1) \times \mathbf{A}^1) \xrightarrow{\phi(\mathcal{Z})} F(X \times \mathbf{A}^1).$$

The element $\mathcal{Z}$ coincides in $h_0((X \times \mathbf{A}^1) \times \mathbf{A}^1 / X \times \mathbf{A}^1)$ with the element represented by the graph of the morphism

$$Id_X \times \Delta : X \times \mathbf{A}^1 \to X \times \mathbf{A}^1 \times \mathbf{A}^1$$

and hence by the property (1) of pretheories we have $\phi(\mathcal{Z}) = F(Id_X \times \Delta)$. Therefore our composition is equal to

$$F((pr_1 \times Id_{\mathbf{A}^1}) \circ (Id_X \times \Delta)) = F(Id_{X \times \mathbf{A}^1}) = Id_{F(X \times \mathbf{A}^1)}.$$

Proposition is proven.

3.3. Elementary properties of transfer maps

Proposition 3.12 *Let k be a field and (F, ϕ, A) be a homotopy invariant pretheory over k. Let S be an object of Sm/k, $j : U \to X$ be an open embedding of smooth curves over S and a be an element of $c_{equi}(U/S, 0)$. Then one has*

$$\phi_{X/S}(j_*(a)) = \phi_{U/S}(a) \circ F(j).$$

Proof: Let Z be the closed subset $X - U$ of X. Consider the smooth curve $W = X \times \mathbf{A}^1 - Z \times \{0\}$ over $S \times \mathbf{A}^1$. Let $p : S \times \mathbf{A}^1 \to S$ be the canonical projection. Then there exists a unique element $\tilde{a}$ in $c_{equi}(W/S \times \mathbf{A}^1, 0)$ such that its image in $c_{equi}(X \times \mathbf{A}^1 / S \times /\mathbf{A}^1, 0)$ equals $cycl(p)(j_*(a))$. Let $i_0, i_1 : S \to S \times \mathbf{A}^1$ be the morphisms which correspond to the points $\{0\}$ and $\{1\}$ of $\mathbf{A}^1$ respectively. Then the fiber of W over i_0 is canonically isomorphic to U and the fiber of W over i_1 is canonically isomorphic to X and with respect to these isomorphisms we have

$$cycl(i_0)(\tilde{a}) = a$$

$$cycl(i_1)(\tilde{a}) = j_*(a).$$

Let $q : W \to X$ be the obvious projection. Consider the following diagram:

$$
\begin{array}{ccccc}
& & F(X) & \xrightarrow{\phi_{X/S}(j_*(a))} & F(S) \\
& Id \nearrow & \uparrow & & \uparrow F(i_1) \\
F(X) & \xrightarrow{F(q)} & F(W) & \xrightarrow{\phi_{W/S \times \mathbf{A}^1}(\tilde{a})} & F(S \times \mathbf{A}^1) \\
& F(j) \searrow & \downarrow & & \downarrow F(i_0) \\
& & F(U) & \xrightarrow{\phi_{U/S}(a)} & F(S)
\end{array}
$$

It is commutative by the property (2) of pretheories. Since F is homotopy invariant $F(i_0), F(i_1)$ are isomorphisms and we have

$$\phi_{U/S}(a) \circ F(j) = \phi_{X/S}(j_*(a)).$$

The proposition is proven.

Let $k \subset E_1 \subset E_2$ be finite separable extensions of k and $f : Spec(E_2) \rightarrow Spec(E_1)$ be the corresponding morphism of smooth schemes over k. For any pretheory (F, ϕ, A) over k we define a morphism

$$F_*(f) : F(Spec(E_2)) \rightarrow F(Spec(E_1))$$

as follows. Consider the smooth curve $\mathbf{A}^1_{E_2}$ over $Spec(E_1)$. The point $\{0\}$ of $\mathbf{A}^1$ gives as an element in $c_{equi}(\mathbf{A}^1_{E_2}/Spec(E_1), 0)$ and we set $F_*(f)$ to be the composition

$$F(Spec(E_2)) \overset{F(p)}{\rightarrow} F(\mathbf{A}^1_{E_2}) \overset{\phi_{\mathbf{A}^1_{E_2}/Spec(E_1)}(\{0\})}{\rightarrow} F(Spec(E_1))$$

where $p : \mathbf{A}^1_{E_2} \rightarrow Spec(E_2)$ is the canonical projection.

Proposition 3.13 *Let C be a smooth curve over k and x be a closed point of C. Suppose that the residue field k_x of x is separable over k and consider the following diagram of smooth schemes over k*

$$Spec(k_x) \overset{i}{\rightarrow} C$$
$$p_0 \searrow \quad \downarrow p$$
$$Spec(k)$$

Then for any homotopy invariant pretheory (F, ϕ, A) over k the following diagram commutes

$$F(C) \overset{F(i)}{\rightarrow} F(Spec(k_x))$$
$$\phi_{C/Spec(k)}(x) \downarrow \quad \swarrow F_*(p_0)$$
$$F(Spec(k))$$

Proof: Since k_x is separable over k the embedding $Spec(k_x) \rightarrow C$ is a smooth pair over k and hence there exists an open neighborhood U of x in C and an etale morphism $f : U \rightarrow \mathbf{A}^1_k$ such that $f^{-1}(\{0\}) = x$. By Proposition 3.12 we may suppose that $U = C$.

Consider a Cartesian square of the form

$$X$$
$$\tilde{f} \swarrow \quad \searrow pr$$
$$\mathbf{A}^2_k \qquad C$$
$$g \searrow \quad \swarrow f$$
$$\mathbf{A}^1_k$$

where g is the morphism given by multiplication of functions (i.e $g(u,v) = uv$).

The composition

$$X \xrightarrow{\tilde{f}} \mathbf{A}_k^2 \xrightarrow{pr_1} \mathbf{A}_k^1$$

defines a smooth curve over $\mathbf{A}_k^1$ (here pr_1 is the projection $(u,v) \mapsto u$). Consider the reduced closed subscheme $Z = \tilde{f}^{-1}(\{v=0\})$ in X. One can easily see that it is integral and belongs to $c_{equi}(X/\mathbf{A}_k^1, 0)$. The fibers of $X/\mathbf{A}_k^1$ over the points $\{0\}$ and $\{1\}$ of $\mathbf{A}_k^1$ are isomorphic to $\mathbf{A}_{k_x}^1$ and C respectively and the intersections of these fibers with Z are the points $\{0\}$ (on $\mathbf{A}_{k_x}^1$) and x (on C).

Let a be an element of $F(C)$. Consider the element $\phi_{X/\mathbf{A}_k^1}(Z)(F(pr)(a))$. By the property (2) of pretheories its restriction to the fiber over $\{0\}$ is equal to $F_*(p_0)(F(i)(a))$ and its restriction to the fiber over $\{1\}$ is equal to $\phi_{S/Spec(k)}(x)(a)$. Therefore these two elements coincide since F is homotopy invariant.

Proposition 3.14 *Let $k \subset E_1 \subset E_2$ be finite separable extensions of a field k and $d = [E_2 : E_1]$. Consider the diagram*

$$Spec(E_2) \xrightarrow{f} Spec(E_1)$$
$$p_2 \searrow \quad \downarrow p_1$$
$$Spec(k)$$

Then for any homotopy invariant pretheory (F, ϕ, A) over k one has

$$F_*(p_2) \circ F(f) = d\, F_*(p_1).$$

Proof: Consider the smooth curve $\mathbf{A}_{E_1}^1$ over $Spec(k)$. Since E_2 is separable over k it has a primitive element and hence there exists a closed point x on $\mathbf{A}_{E_1}^1$ such that the field of functions of x is isomorphic to E_2. Let us identify $Spec(E_1)$ with the point $\{0\}$ of $\mathbf{A}_{E_1}^1$. Then we have two morphisms $f_1, f_0 : Spec(E_2) \to \mathbf{A}_{E_1}^1$ where f_0 corresponds to the morphism f with respect to our identification and f_1 corresponds to the point x. Since the compositions of f_1 and f_0 with the canonical projection $\mathbf{A}_{E_1}^1 \to Spec(E_1)$ On the other hand as an element in $h_0(\mathbf{A}_{E_1}^1/Spec(k))$ the point x is equal to $d\{0\}$. Since F is homotopy invariant we conclude by Proposition 3.13 that

$$dF_*(p_1) = d\phi_{\mathbf{A}_{E_1}^1/Spec(k)}(\{0\}) = \phi_{\mathbf{A}_{E_1}^1/Spec(k)}(x) =$$
$$= F_*(p_2) \circ F(f_1) = F_*(p_2) \circ F(f).$$

The proposition is proven.

Corollary 3.15 *Let $k \subset E_1 \subset E_2$ be finite separable extensions of a field k, $d = [E_2 : E_1]$ and $f : Spec(E_2) \to Spec(E_1)$ be the corresponding morphism. Then for any homotopy invariant pretheory (F, ϕ, A) over k one has*

$$F_*(f)F(f) = d\,Id_{F(Spec(E_1))}.$$

Lemma 3.16 *Let k be a field and C be a smooth curve over k. Then the group $h_0(C/Spec(k))$ is generated by closed points x of C whose residue fields k_x are separable over k.*

Proof: Let z be a closed point of C such that the field k_z is not separable over k. Let further $\bar{C}$ be a smooth compactification of C. Denote by $x_1, \ldots, x_k$ the closed points $\bar{C} - C$. To show that the element of $h_0(C/Spec(k))$ which corresponds to z is a sum of elements which correspond to closed points of C with separable residue fields it is sufficient to find a rational function f on $\bar{C}$ such that the following conditions hold:

1. $f = 1$ at the points $x_1, \ldots, x_k$.

2. The divisor $D(f)$ of f is of the form $\sum q_i - z - \sum p_j$ where q_i, p_j are closed points of C such that the fields k_{q_i}, k_{p_j} are separable extensions of k.

Note first that since C is smooth over k there is infinitely many closed points on C with separable residue fields. Therefore there exists a finite set $p_1, \ldots, p_n$ of such points and a meromorphic function g on $\bar{C}$ such that $g(x_i) = 0$ for all $i = 1, \ldots, k$ and g has poles in the points $z, p_1, \ldots, p_n$ with multiplicities $[k_z : k], [k_{p_1}, k], \ldots, [k_{p_k} : k]$ respectively.

Consider the function g as a morphism $\bar{C} \to \mathbf{P}_k^1$. It is smooth in the points $p_1, \ldots, p_k$ and hence it is smooth everywhere but in a finite set of closed points $s_1, \ldots, s_t$ (note that since the extension k_z/k is not separable and $g(z) = \infty$ is a point with separable residue field on $\mathbf{P}^1$ we have $z \in \{s_1, \ldots, s_t\}$). We may assume that k is infinite (since finite fields are perfect). Then there is a k-point y on $\mathbf{P}_k^1$ such that $y \neq 0, \infty$ and g is smooth over y. We set $f = (y - g)/y$. One can easily see that this function satisfies the conditions (1)-(2) from above.

Proposition 3.17 *Let $f : X_1 \to X_2$ be a morphism of smooth curves over a field k and (F, ϕ, A) be a homotopy invariant pretheory over k. Then the* *following diagram commutes*

$$
\begin{array}{ccc}
c_{equi}(X_1/Spec(k),0) & \overset{\phi_{X_1/Spec(k)}}{\to} & Hom(F(X_1),F(Spec(k))) \\
f_* \downarrow & & \downarrow \\
c_{equi}(X_2/Spec(k),0) & \overset{\phi_{X_2/Spec(k)}}{\to} & Hom(F(X_2),F(Spec(k))).
\end{array}
$$

Proof: It follows immediately from Propositions 3.13, 3.14 and Lemma 3.16.

Definition 3.18 *Let k be a field and (F,ϕ,A) be a pretheory over k. It is said to be of homological type if for any smooth scheme S over k, any morphism $f : X_1 \to X_2$ of smooth curves over S and any element a in $c_{equi}(X_1/S,0)$ one has*

$$
\phi_{X_2/S}(f_*(a)) = \phi_{X_1/S}(a) \circ F(f).
$$

3.4. Examples of pretheories

Etale cohomology. For any field k consider the functor $\mathcal{H}_{et}(-,F)$ from the category Sm/k to the derived category of $Gal(\bar{k}/k)$-modules which takes a smooth scheme X to the object $\mathbf{R}p_*(p^*(F))$ where F is a sheaf on the small etale site $Spec(k)_{et}$ and $p : X_{et} \to Spec(k)_{et}$ is the morphism of sites defined by the structural morphism of X. This functor has a canonical structure of a pretheory. If F is a sheaf of $\mathbf{Z}[1/char(k)]$-modules this pretheory is homotopy invariant. A fortiori, for any $i \geq 0$ the functor from Sm/k to the category of abelian groups which takes X to the group $H^i_{et}(X,F)$ has a canonical structure of pretheory which is homotopy invariant if F is a sheaf of $\mathbf{Z}[1/char(k)]$-modules.

If S is the shift functor on the derived category of $Gal(\bar{k}/k)$-modules then for any locally constant n-torsion sheaf F such that n is prime to characteristic of k we have (by Kunnet formula for the etale cohomology) a canonical isomorphism

$$
(\mathcal{H}_{et}(-,F))_{-1} \cong S^{-1} \circ \mathcal{H}_{et}(-,F \otimes \mu_n^{-1}).
$$

In particular

$$
(H^i_{et}(-,F))_{-1} \cong H^{i-1}_{et}(-,F \otimes \mu_n^{-1})
$$

where μ_n is the sheaf of n-th roots of unit.

Algebraic K-theory. For any field k consider the functor K_i from Sm/k to the catgeory of abelian groups which takes a smooth scheme X to the Quillen's K-group $K_i(X)$. Since for any smooth scheme S over k and any smooth curve $p : X \to S$ over S an element of $c_{equi}(X/S, 0)$ is a formal linear combination of closed subschemes in X which are flat and finite over S one can define transfer homomorphisms

$$\phi_{X/S} : c_{equi}(X/S, 0) \to Hom(K_i(X), K_i(S)).$$

Unfortunately, as was shown by Mark Walker these homomorphisms do not satisfy in general the second axiom of the definition of pretheory. It can be proven though that if we consider the separated in the Zariski topology presheaves $\tilde{K}_i(-)$ which correspond to $K_i(-)$ we get a pretheory which is clearly homotopy invariant.

Note also that we have canonical isomorphisms $(\tilde{K}_i)_{-1} \cong \tilde{K}_{i-1}$.

Algebraic De Rham cohomology. Let k be a field of characteristic zero. One can define a structure of pretheory on the functor which takes a smooth scheme $p : X \to Spec(k)$ over k to the object $\mathbf{R}p_*(\Omega^\bullet)$ of the derived category of k-vector spaces. In particular we get structures of pretheories on the functors $X \mapsto H^k_{DR}(X)$.

qfh-sheaves. Let F be a qfh-sheaf of abelian groups on the category Sch/k of schemes of finite type over a field k (see [10], [7]). Then one can easily see that the restriction of F to the category Sm/k has a canonical structure of a pretheory and the same holds for all functors of the form $X \mapsto \mathbf{H}^i_{qfh}(X, K)$ where K is a complex of sheaves in the qfh-topology.

4. Zariski sheaves associated with pretheories

4.1. Technical lemmas

Definition 4.1 *Let S be a smooth scheme. A standard triple*

$$(\bar{p} : \bar{X} \to S, X_\infty, Z)$$

over S is the following collection of data:

1. *a proper normal curve $\bar{p} : \bar{X} \to S$ over S,*

2. *a pair of reduced closed subschemes Z, X_∞ in X.*

These data should satisfy the following conditions:

1. *the closed subset $Z \cup X_\infty$ has an open neighborhood in $\bar{X}$ which is affine over S,*

2. $Z \cap X_\infty = \emptyset$,

3. *the scheme $X = \bar{X} - X_\infty$ is quasi-affine and smooth over S.*

One can easily see that if $(\bar{p} : \bar{X} \to S, X_\infty, Z)$ is a standard triple then for any closed reduced subscheme Z' of Z the triple $(\bar{p} : \bar{X} \to S, X_\infty, Z')$ is also a standard triple.

Lemma 4.2 *Let S be a smooth scheme and $(\bar{p} : \bar{X} \to S, X_\infty, Z)$ be a standard triple over S. Denote the scheme $\bar{X} - X_\infty$ by X and let $p : X \to S$ be the corresponding morphism.*

Let $f : X' \to X$ be an etale morphism. Denote by W_f the closed subspace in X which consists of points x such that f is not finite over x and let Z_0 be a closed subspace in Z. Suppose that the following conditions hold:

1. $W_f \subset Z$

2. *the morphism $f^{-1}(Z_0) \to Z_0$ is finite.*

Denote by $\bar{f} : \bar{X}' \to \bar{X}$ the finite morphism with normal $\bar{X}'$ associated with f. Then the triple

$$(\bar{p} \circ \bar{f} : \bar{X}' \to S, \bar{X}' - X', f^{-1}(Z_0))$$

is a standard triple over S.

Proof: Note first that since the morphism $f^{-1}(Z_0) \to Z_0 \to S$ is finite the subspace $f^{-1}(Z_0)$ is closed in $\bar{X}'$. On the other hand $f^{-1}(Z_0)$ lies in X' and hence $f^{-1}(Z_0) \cap (\bar{X}' - X') = \emptyset$.

The scheme $X' \to X \to S$ is smooth and quasi-affine over S since etale morphisms are smooth and quasi-affine. It is sufficient to verify that $f^{-1}(Z_0) \cup (\bar{X}' - X')$ has an affine open neighborhood in $\bar{X}'$. Since $W_f \subset Z$ the image $\bar{f}(f^{-1}(Z_0) \cup (\bar{X}' - X'))$ of this subscheme lies in $X_\infty \cup Z$ and hence has an affine open neighborhood since $\bar{f}$ is affine morphism.

Lemma 4.3 *Let $S = Spec(\mathcal{O})$ be a smooth affine semi-local scheme over a field k, $\bar{p} : \bar{X} \to S$ be a projective curve over S and Z be a closed subscheme in $\bar{X}$. Then Z has an affine open neighborhood in $\bar{X}$ if and only if the fibers of Z over the closed points of S are finite.*

Proof: The "only if" part is obvious. Suppose that the fibers of Z over the closed points of S are finite and let $z_1, \ldots, z_n$ be the corresponding points of X. Since X is projective over S and S is affine there exists a quasi-projective scheme $\tilde{X}$ over k and an affine morphism $X \to \tilde{X}$. Our lemma follows now from the fact that any finite set of points in a quasi-projective scheme has an affine open neighborhood.

Let S be a smooth affine scheme and $T = (\bar{p} : \bar{X} \to S, X_\infty, Z)$ be a standard triple over S. Denote by $p : X \to S$ the scheme $\bar{X} - X_\infty$ over S. Consider the smooth curve $X \times_S X \to X$ over X. The diagonal $\Delta : X \to X \times_S X$ gives is a relative Cartier divisor on $X \times_S X$ over X. Denote by $\mathcal{L}_\Delta$ be the corresponding line bundle.

Definition 4.4 *A splitting of T over an open subset U of X is a trivialization of the restriction of $\mathcal{L}_\Delta$ to $U \times_S Z$.*

Lemma 4.5 *Let k be a field, S be a smooth scheme over k, $T = ((\bar{p} : \bar{X} \to S, X_\infty, Z)$ a standard triple over S and $j : U \to X$ be an affine open subset in X such that T splits over U. Then for any homotopy invariant pretheory (F, ϕ, A) over k there is a morphism $F(X - Z) \to F(U)$ such that the composition $F(X) \to F(X - Z) \to F(U)$ coincides with the morphism $F(j) : F(X) \to F(U)$.*

Proof: Consider the line bundle $\mathcal{M}$ on $U \times_S \bar{X}$ which corresponds to the obvious U-point of $U \times_S \bar{X}$ and let s be its section which defines our divisor. Its restriction to $U \times_S X_\infty$ gives a trivialization $\tilde{s}$ of $\mathcal{M}$ there. For a splitting t of T over U let $\tilde{t}$ be the trivialization of $\mathcal{M}$ on $U \times_S (X_\infty \coprod Z)$ equal to $\tilde{s}$ on $U \times_S X_\infty$ and to t on $U \times_S Z$.

Suppose now that U is affine. Then by Proposition 2.6 the pair $(\mathcal{M}, \tilde{t})$ defines an element A_t in $h_0(U \times_S (X - Z)/U)$ such that the direct image of A_t to $h_0(U \times_S X)$ coincides with the class of tautological U-point of $U \times_S X$. In particular if (F, ϕ, A) is a homotopy invariant pretheory over k then any splitting t of T over an affine open subset U gives us a morphism $F(X - Z) \to F(U)$ and the composition $F(X) \to F(X - Z) \to F(U)$ coincides by Proposition 3.12 with the morphism $F(j)$.

Lemma 4.6 *Let k be a field, S be a smooth scheme over k, $T = ((\bar{p} : \bar{X} \to S, X_\infty, Z)$ be a standard triple over S and U be an affine open subset in X such that T splits over U. Suppose further that $Z = Z_1 \coprod Z_2$ where Z_1, Z_2 are reduced closed subschemes of X such that $Z_1 \cap Z_2 = \emptyset$.*

Then for any homotopy invariant pretheory (F, ϕ, A) *over* k *the canonical morphism of complexes*

$$
\begin{array}{ccccccc}
0 \to F(X) \to & & F(X-Z_1) \oplus F(X-Z_2) & & F(X-Z) & \to 0 \\
\downarrow & & \downarrow & & \downarrow & \\
0 \to F(U) \to & F((X-Z_1) \cap U) \oplus F((X-Z_2) \cap U) & \to & F((X-Z) \cap U) & \to 0
\end{array}
$$

is homotopic to zero.

Proof: We have to define morphisms

$$
f_i : F(X-Z) \to F((X-Z_i) \cap U)
$$
$$
g_i : F(X-Z_i) \to F(U)
$$

which satisfy the usual axioms of a homotopy. As in the proof of Lemma 4.5 we will first construct some elements

$$
A_i \in h_0(((X-Z_i) \cap U) \times_S (X-Z)/((X-Z_i) \cap U))
$$
$$
B_i \in h_0(U \times_S (X-Z_i)/U)
$$

and then will use the transfer maps with respect to these elements to define the morphisms we need.

Consider the line bundle $\mathcal{M}$ on $U \times_S \bar{X}/U$ which corresponds to the obvious U-point of $U \times_S \bar{X}$ and let s be its section which defines our divisor. Its restriction to $U \times_S X_\infty$ gives a trivialization s_∞ of $\mathcal{M}$ there. By our assumption the restriction of $\mathcal{M}$ to the closed subscheme $U \times_S Z$ is trivial. Since U is affine there is an affine open neighborhood V of the subscheme $U \times_S (X_\infty \amalg Z)$ in $U \times_S \bar{X}$ and a trivialization t of $\mathcal{M}$ on V whose restriction to $U \times_S X_\infty$ is equal to s_∞.

Suppose for simplicity of the notations that $i = 1$. Consider the smooth curve $((X - Z_1) \cap U) \times_S (X - Z)$ over $(X - Z_1) \cap U$. Consider now the restriction of $\mathcal{M}$ to the subscheme $((X - Z_1) \cap U) \times_S (X_\infty \amalg Z)$. Since the corresponding divisor on $((X - Z_1) \cap U) \times_S \bar{X}$ does not intersect $((X-Z_1) \cap U) \times_S (X_\infty \amalg Z_1)$ we have the canonical trivialization s_1 of $\mathcal{M}$ there. Consider now the trivialization r of $\mathcal{M}$ on $((X-Z_1) \cap U) \times_S (X_\infty \amalg Z)$ which is equal to s_1 on $((X-Z_1) \cap U) \times_S (X_\infty \amalg Z_1)$ and to the restriction of t on $((X - Z_1) \cap U) \times_S Z_2$.

We claim that r can be extended to an open neighborhood V_0 of $((X - Z_1) \cap U) \times_S (X_\infty \amalg Z)$ in $((X - Z_1) \cap U) \times_S \bar{X}$ (of course the only problem is that if $codim(Z_1) \geq 1$ the scheme $(X - Z_1) \cap U$ is not affine). Since V is affine there exists a regular function h on V whose restriction to $U \times_S (X_\infty \amalg Z_2)$ is equal to one and whose restriction to $U \times_S Z_1$ is equal to s/t. Let Y be the divisor of h. Then the open subset $V_0 = ((X - Z_1) \cap U) \times_S \bar{X} \cap (V - Y)$ is an open neighborhood of

$((X - Z_1) \cap U) \times_S (X_\infty \coprod Z)$. Consider now the trivialization th of $\mathcal{M}$ on V_0. It is obviously an extension of the form we need.

We define now A_1 to be the element which corresponds to r by Proposition 2.6(a). The definition of A_2 is similar.

Let us now construct the elements B_i. Since U is affine it is sufficient to define a trivialization of $\mathcal{M}$ on $U \times_S (X_\infty \coprod Z_i)$. We set it to be equal to the restriction of t.

Denote by C the element in $h_0(U \times_S (X - Z)/U)$ which corresponds to the trivialization t of $\mathcal{M}$.

We got morphisms:

$$\tilde{C} : F(X - Z) \to F(U)$$

$$\tilde{A}_i : F(X - Z) \to F((X - Z_i) \cap U)$$

$$\tilde{B}_i : F(X - Z_i) \to F(U).$$

Let us define the morphism g_2 to be equal to the morphism $\tilde{B}_2$, the morphism g_1 to be equal to zero, the morphism f_1 to be equal to $\tilde{A}_1$ and the morphism f_2 to be equal to the difference $\delta - \tilde{A}_2$ where δ is the composition

$$F(X - Z) \xrightarrow{\tilde{C}} F(U) \to F((X - Z_2) \cap U).$$

One can verify easily now using Proposition 3.12 that we indeed constructed a homotopy of our morphism to the zero morphism.

Proposition 4.7 *Let S be a smooth scheme over a field k and $T_i = (\bar{p}_i : \bar{X}_i \to S, X_{\infty,i}, Z_i)$, $i = 1, 2$ be two standard triples over S. Denote the schemes $\bar{X}_i - X_{\infty,i}$ by X_i.*

Let $\bar{f} : \bar{X}_1 \to \bar{X}_2$ be a finite morphism such that the following conditions hold.

1. $\bar{f}^{-1}(X_{\infty,2}) \subset X_{\infty,1}$.

2. *Let $f : X_1 \to X_2$ be the morphism defined by $\bar{f}$ then $f^{-1}(Z_2) = Z_1$ and the morphism $f^{-1}(Z_2) \to Z_2$ is an isomorphism.*

Let (F, ϕ, A) be a homotopy invariant pretheory over k of homological type such that A is an abelian category and V be an open affine subset of X_2 such that T_2 splits over V. Then there is a morphism

$$\psi : F(X_1 - Z_1) \to F(V - V \cap Z_2)$$

satisfying the following conditions.

1. *The composition*

$$F(X_1) \to F(X_1 - Z_1) \to F(V - V \cap Z_2)$$

   is zero.

2. *The following diagram is commutative:*

$$
\begin{array}{ccc}
F(X_2 - Z_2) & & \\
\downarrow & \searrow & \\
F(X_1 - Z_1) & \to & F(V - V \cap Z_2)/F(V) \\
& \searrow & \downarrow \\
& & F(f^{-1}(V) - f^{-1}(V) \cap Z_1)/F(f^{-1}(V))
\end{array}
$$

Proof: Let $\mathcal{M}$ be the line bundle on $X_2 \times_S \bar{X}_1$ which is the preimage of the diagonal line bundle on $X_2 \times_S \bar{X}_2$ with respect to the morphism $Id \times_S \bar{f}$ and let s be a section of $\mathcal{M}$ which defines the corresponding divisor $\tilde{X}_2 \to X_2 \times_S \bar{X}_1$ where $\tilde{X}_2 = \bar{f}^{-1}(X_2)$. Since the restriction of $\bar{f}$ to Z_1 is an isomorphism a splitting of T_2 over V gives us a trivialization t of $\mathcal{M}$ on $V \times_S Z_1$.

Let U be an affine open neighborhood of $V \times_S (Z_1 \coprod X_{1,\infty})$ in $V \times_S \bar{X}$. There exists a regular function h on U which is equal to 1 on $V \times_S X_{1,\infty}$ and to the regular function s/t on $V \times_S Z_1$. Let Y be the divisor of h. Its intersection with $V \times_S X_{1,\infty}$ is empty and its intersection with $V \times_S Z_1$ is $(V \cap Z_2) \times_S Z_1$. Hence the open subset $V_0 = ((V - V \cap Z_2) \times_S \bar{X}_1) \cap (U - Y)$ is a neighborhood of $(V - V \cap Z_2) \times_S (Z_1 \coprod X_{1,\infty})$ in $(V - V \cap Z_2) \times_S \bar{X}_1$ and h gives an invertible function on U_0.

Let C be the corresponding element of $H_0((V - V \cap Z_2) \times_S (X_1 - Z_1)/(V - V \cap Z_2))$. We define our morphism ψ to be the composition

$$F(X_1 - Z_1) \overset{F(pr_2)}{\to} F((V - V \cap Z_2) \times_S (X_1 - Z_1)) \overset{\phi(C)}{\to} F(V - V \cap Z_2).$$

One can easily verify now that it satisfies the conditions of our proposition (we need the fact that F is of homological type to prove the commutativity of the upper triangle in our diagram).

Corollary 4.8 *In the notations of Proposition 4.7 suppose that A is the category of abelian groups. Then the following statments hold.*

1. *Let $a \in F(X_2 - Z_2)$ be an element such that its restriction to $F(X_1 - Z_1)$ can be extended to $F(X_1)$. Then for any closed point z of Z_2 there exists an open neighborhood V of z in X_2 such that the restriction of a to $V - V \cap Z_2$ can be extended to V.*

2. *Let a be an element it $F(X_1 - Z_1)$. Then for any closed point z of Z_2 there exists an open neighborhood V of z in X_2 and an element $a_0 \in F(V - V \cap Z_2)$ such that the images of a and a_0 in $F(f^{-1}(V) - f^{-1}(V) \cap Z_1)/F(f^{-1}(V))$ coincide.*

Proposition 4.9 *Let W be a smooth quasi-projective variety over a field k. Let $Y \subset W$ be a closed reduced subscheme in W such that $Y \neq W$ and $\{y_1, \dots, y_n\}$ be a finite set of closed points of Y. Then there exists an affine open neighborhood U of $\{y_1, \dots, y_n\}$ in W and a standard triple $(\bar{p} : \bar{X} \to S, X_\infty, Z)$ over a smooth affine variety S such that the pair $(U, U \cap Y)$ is isomorphic to the pair $(X = \bar{X} - X_\infty, Z)$.*

Proof: See [11, Remark 4.13].

4.2. Pretheories in a neighborhood of a smooth subscheme

Consider the inclusion of categories $i : Sm/k \to Sch/k$ where Sch/k is the category of Noetherian schemes over k. Then for any presheaf F on Sm/k with values in the category of abelian groups the restriction of the presheaf $i^*(F)$ on Sch/S to Sm/k is equal to F.

For any pretheory (F, ϕ, A) over k such that A is the category of abelian groups and any Noetherian scheme X over k we will use the notation $F(X)$ instead of $i^*(F)(X)$.

We denote by F_{Zar} the sheaf in the Zariski topology on Sch/k associated with the presheaf $i^*(F)$.

Let k be a field, $i : Z \to X$ be a smooth pair over k and F be a presheaf of abelian groups on Sm/k. Denote the open subscheme $X - Z$ by U and let $j : U \to X$ be the corresponding open embedding. Consider the exact sequence of presheaves on X_{Zar}:

$$F \to j_* j^*(F) \to coker \to 0.$$

Let $coker_{Zar}$ be the sheaf in the Zariski topology associated with the presheaf $coker$. One can easily see that the canonical morphism

$$coker_{Zar} \to i_* i^*(coker_{Zar})$$

is an isomorphism. We denote by $F_{(X,Z)}$ the sheaf $i^*(coker_{Zar})$ on Z_{Zar}.

Remark 4.10 Since the associated sheaf functor is exact we have an exact sequence of the form

$$F_{Zar} \to (j_* j^* F)_{Zar} \to (coker)_{Zar} \to 0$$

but in general the canonical morphism $(j_* j^* F)_{Zar} \to j_* j^* (F_{Zar})$ is not an isomorphism since the associated sheaf functor does not commute with the direct image functor. For example if F is a sheaf on X_{Zar} we have $(j_* j^* (H_{Zar}^n(-, F)))_{Zar} \cong R^n j_*(F)$ but $(H_{Zar}^n(-, F))_{Zar} = 0$.

Proposition 4.11 *Let Z be a smooth scheme over a field k and (F, ϕ, A) be a homotopy invariant pretheory over k such that A is the category of abelian groups. Then there is a canonical isomorphism of sheaves on Z_{Zar} of the form*

$$(F_{-1})_{Zar} \to F_{(Z \times \mathbf{A}^1, Z \times \{0\})}.$$

Proof: Let us first define a morphism

$$(F_{-1})_{Zar} \to F_{(Z \times \mathbf{A}^1, Z \times \{0\})}.$$

Let $i : Z \to Z \times \mathbf{A}^1$, $j : Z \times (\mathbf{A}^1 - \{0\}) \to Z \times \mathbf{A}^1$ be the obvious closed and open embeddings. Since the associated sheaf functor commutes with the inverse image functor it is sufficient to define a morphism of presheaves

$$F_{-1} \to i^* (coker(F \to j_* j^* (F)))$$

on Z_{Zar}. Almost by definition a section of the right hand side presheaf over an open subset U of Z is an equivalence class of pairs of the form (V, a) where V is an open subset in $Z \times \mathbf{A}^1$ such that $U \subset Z \cap V$ and a is an element of $F(U - U \cap (Z \times \{0\}))$. Two pairs (V_1, a_1), (V_2, a_2) correspond to the same section of our presheaf if there exists an open subset V_0 in $V_1 \cap V_2$ such that $U \subset V_0 \cap (Z \times \{0\})$ and the images of a_1 and a_2 in $F(V_0 - V_0 \cap (Z \times \{0\}))/F(V_0)$ coincide.

Since a section of F_{-1} over an open subset U of Z is by definition an element of $F(U \times (\mathbf{A}^1 - \{0\}))/F(U \times \mathbf{A}^1)$ the definition of our morphism is obvious.

To prove that the corresponding morphism of associated sheaves in the Zariski topology on Z is an isomorphism it is sufficient to show that if Z is a local smooth scheme over k and V is an open neighborhood of $Z \times \{0\}$ in $Z \times \mathbf{A}^1$ then the morphism

$$F(Z \times (\mathbf{A}^1 - \{0\}))/F(Z \times \mathbf{A}^1) \to F(U - U \cap (Z \times \{0\}))/F(U)$$

is an isomorphism.

Note that since U is a neighborhood of $Z \times \{0\}$ and we may assume Z to be local the triple $T_1 = (\mathbf{P}_Z^1 \to Z, \mathbf{P}_Z^1 - U, Z \times \{0\})$ is a standard triple over Z by Lemma 4.3 and we may apply Proposition 4.7 to the standard triples $T_1, T_2 = (\mathbf{P}_Z^1 \to Z, \mathbf{P}_Z^1 - \mathbf{A}_Z^1, Z \times \{0\})$ and the identity morphism

$Id_{\mathbf{P}^1_Z}$. Since the standard triple T_2 splits over $\mathbf{A}^1_Z$ it implies the result we need.

Up to the end of this section (F, ϕ, A) denote a homotopy invariant pretheory of homological type over a field k such that A is the category of abelian groups.

Proposition 4.12 *Let $f : X_1 \to X_2$ be an etale morphism of smooth schemes over k and Z be a closed subscheme in X_2 such that the morphism $f^{-1}(Z) \to Z$ is an isomorphism. Then for any (F, ϕ, A) as above the following conditions hold:*

1. *Let $a \in F(X_2 - Z)$ be an element such that the preimage of a to $X_1 - f^{-1}(Z)$ can be extended to X_1. Then for any closed point x of X_2 there exists an open neighborhood U in X_2 such that the restriction of a to $U - U \cap Z$ can be extended to U.*

2. *For any element $a \in F(X_1 - f^{-1}(Z))$ and any closed point x of X_2 there exists an open neighborhood U of x on X_2 and an element $a' \in F(U - U \cap Z)$ such that the images of a and a' in $F(f^{-1}(U - U \cap Z))/F(f^{-1}(U))$ coincide.*

Proof: Denote by $\bar{f} : X_1' \to X_2$ the finite morphism with normal X_1' associated with f and let Y be the closed subspace $\bar{f}(X_1' - X_1)$ in X_2. By Proposition 4.9 for any closed point x of X_2 there is an open neighborhood U and a standard triple $(\bar{p}_1 : \bar{W}_1 \to S, W_{1,\infty}, R_1)$ over a smooth scheme S such that the pair $(U, U \cap (Z \cup Y))$ is isomorphic to the pair $(W_1 = \bar{W}_1 - W_{1,\infty}, R_1)$.

Let g be the morphism $X_1 \times_{X_2} U \to X_2$. Lemma 4.2 implies that there is a standard triple $(\bar{p}_2 : \bar{W}_2 \to S, W_{2,\infty}, R_2)$ such that the pair $(W_2 = \bar{W}_2 - W_{2,\infty}, R_2)$ is isomorphic to the pair $(X_1 \times_{X_2} U, (X_1 \times_{X_2} U) \cap f^{-1}(Z))$. To finish the proof it is sufficient now to apply Corollary 4.8 to the standard triples $(\bar{p}_1 : \bar{W}_1 \to S, W_{1,\infty}, U \cap Z)$, $(\bar{p}_2 : \bar{W}_2 \to S, W_{2,\infty}, R_2)$.

Corollary 4.13 *Let $f : X_1 \to X_2$ be an etale morphism of smooth schemes over k and Z be a closed subscheme in X_2 such that the morphism $f^{-1}(Z) \to Z$ is an isomorphism. Then for any (F, ϕ, A) as above the canonical morphism of sheaves*

$$F_{(X_2, Z)} \to F_{(X_1, f^{-1}(Z))}$$

on $Z_{Zar} = f^{-1}(Z)_{Zar}$ is an isomorphism.

Theorem 4.14 *Let k and (F, ϕ, A) be as above, X be a smooth scheme over k and $Z \subset X$ be a smooth closed subscheme of X. Let further x be a closed point of X and d be the codimension of Z in x. Then there exists an open neighborhood U and a family of isomorphisms*

$$F_{(U \times Y, (U \cap Z) \times Y)} \cong F_{(\mathbf{A}^d \times (U \cap Z) \times Y, (U \cap Z) \times Y)}$$

of sheaves on $((U \cap Z) \times Y)_{Zar}$ given for all smooth schemes Y over k which are natural with respect to Y (here $(U \cap Z) \times Y$ is considered as a closed subscheme of $\mathbf{A}^d \times (U \cap Z) \times Y$ by means of the embedding which corresponds to the point $(0, \ldots, 0)$ of $\mathbf{A}^d$).

In particular for $d = 1$ we have isomorphisms

$$F_{(U \times Y, (U \cap Z) \times Y)} \cong (F_{-1})_{Zar}.$$

Proof: The second statement follows from the first one by Proposition 4.11.

Since Z is a smooth subscheme on a smooth variety X there exists an open neighborhood U of x and an etale morphism $p : U \to \mathbf{A}^n$ such that $Z \cap U = p^{-1}(\mathbf{A}^{n-d} \times \{0\})$. We may suppose that $U = X$. Let $p_0 : Z \to \mathbf{A}^{n-d}$ be the restriction of p to Z. Consider the Cartesian square

$$
\begin{array}{ccc}
W_0 & \longrightarrow & Z \times \mathbf{A}^d \\
q \downarrow & & \downarrow p_0 \times Id_{\mathbf{A}^d} \\
X & \xrightarrow{p} & \mathbf{A}^n = \mathbf{A}^{n-d} \times \mathbf{A}^d.
\end{array}
$$

Then $q^{-1}(Z) \cong Z \times_{\mathbf{A}^{n-d}} Z$ and since p_0 is etale the complement to the diagonal $q^{-1}(Z(-\Delta$ is a closed subset in $q^{-1}(Z)$ and, hence in W_0. Let W be the complement $W_0 - (q^{-1}(Z) - \Delta)$.

To finish the proof it is sufficient now to apply Corollary 4.13 to both arrows in the diagram

$$
\begin{array}{ccc}
W \times Y & \to & Z \times \mathbf{A}^d \times Y \\
\downarrow & & \\
U \times Y. & &
\end{array}
$$

4.3. Pretheories on curves over a field

Theorem 4.15 *Let k be a field and (F, ϕ, A) be a homotopy invariant pretheory over k such that A is the category of abelian groups. Then for any open subscheme U in $\mathbf{A}^1_k$ one has:*

$$H^i_{Zar}(U, F_{Zar}) = \begin{cases} F(U) & \text{for } i = 0 \\ 0 & \text{for } i \neq 0 \end{cases}$$

Proof: Consider an open covering of the form $U = U_1 \cup U_2$. Let Z_i be the reduced closed subscheme $U - U_i$. Then

$$T = (\mathbf{P}_k^1 \to Spec(k), \mathbf{P}_k^1 - U, Z_1 \coprod Z_2)$$

is a standard triple over $Spec(k)$ which obviously splits over U. Hence the sequence

$$0 \to F(U) \to F(U_1) \oplus F(U_2) \to F(U_1 \cap U_2) \to 0$$

is exact by Lemma 4.6. Therefore, F considered as a functor on the category of Zariski open subsets of U is a sheaf which implies the statement of the theorem for $i = 0$. To prove the statement for $i > 0$ note that using induction on the number of open subsets in a covering we can show that Cech cohomology of any open U' in U with coefficients in F are zero in positive dimensions. Then the standard spectral sequence which relates usual cohomology and Cech cohomology shows that $H_{Zar}^i(U, F) = 0$ for $i > 0$.

Corollary 4.10 *Let k be a field and (F, ϕ, A) be a homotopy invariant pretheory over k such that A is the category of abelian groups. Then the morphism $\mathbf{A}_k^1 \to Spec(k)$ induces an isomorphism*

$$F_{Zar}(Spec(k)) \to F_{Zar}(\mathbf{A}_k^1).$$

4.4. Pretheories on semi-local schemes

Let W be an object of Sch/k. We say that W is a smooth semi-local scheme over k if there exists a smooth affine scheme X of finite type over k and a finite set $\{x_1, \ldots, x_n\}$ of points of X such that W is the inverse limit of open neighborhoods of this set.

Proposition 4.17 *Let k be a field , (F, ϕ, A) be a homotopy invariant pretheory over k, W be a smooth quasi-projective variety over k and $\{x_1, \ldots, x_n\}$ be a finite set of points of W. Then for any nonempty open subset U of W there exists an open neighborhood V of $\{x_1, \ldots, x_n\}$ and a morphism $F(U) \to F(V)$ such that the following diagram commutes:*

$$\begin{array}{c} F(W) \to F(U) \\ \downarrow \quad \swarrow \\ F(V). \end{array}$$

Proof: Let N be the reduced closed subscheme $W - U$ of W. By Lemma 4.9 there exists an open neighborhood V_0 of the set $\{x_1, \ldots, x_n\}$ and a standard triple $T = (\bar{p} : \bar{X} \to S, X_\infty, Z)$ over a smooth affine variety S such that the pair $(V_0, V_0 \cap N)$ is isomorphic to the pair $(X = \bar{X} - X_\infty, Z)$. Since the Picard group of an affine semi-local scheme is zero there exists an open neighborhood V of the set $\{x_1, \ldots, x_n\}$ in X such that T splits over V. The statement of our proposition follows now from Lemma 4.5.

Corollary 4.18 *Let k be a field, (F, ϕ, A) be a homotopy invariant pretheory over k such that A is the category of abelian groups and W be a smooth semi-local scheme over k. Then for any nonempty open subscheme U of W the morphism $F(W) \to F(U)$ is a splitting monomorphism.*

Corollary 4.19 *Let k be a field, (F, ϕ, A) be a homotopy invariant pretheory over k such that A is the category of abelian groups and X be a smooth variety over k. Then for any nonempty open subscheme U of X the morphism $F_{Zar}(X) \to F_{Zar}(U)$ is a monomorphism.*

Proposition 4.20 *Let k be a field, (F_i, ϕ_i, A), $i = 1, 2$ be two homotopy invariant pretheories over k such that A is the category of abelian groups and $f : F_1 \to F_2$ be a morphism of pretheories such that for any field extension $k \subset E$ the morphism $f_E : F_1(Spec(E)) \to F_2(Spec(E))$ is an isomorphism. Then the morphism of associated sheaves $f_{Zar} : (F_1)_{Zar} \to (F_2)_{Zar}$ is an isomorphism.*

Proof: Consider the exact sequence of presheaves

$$0 \to ker(f) \to F_1 \to F_2 \to coker(f) \to 0.$$

By Proposition 3.5 the presheaves $ker(f)$ and $coker(f)$ have canonical structures of homotopy invariant pretheories. Since the associated sheaf functor is exact it is sufficient to show that $ker(f)_{Zar} = coker(f)_{Zar} = 0$. It follows from our assumption and Corollary 4.18.

Proposition 4.21 *Let k be a field and (F, ϕ, A) be a homotopy invariant pretheory over k such that A is the category of abelian groups. Then for any smooth scheme X over k one has:*

$$F_{Zar}(X \times \mathbf{A}^1) = F_{Zar}(X).$$

Proof: It follows from Corollary 4.16 and Corollary 4.19.

For a presheaf of abelian groups F' on Sch/k we denote by $s(F)$ the separated in the Zariski topology presheaf associated with F. By definition it means that for any noetherian scheme X over k we have

$$s(F)(X) = F(X)/\cup(ker : F(X) \to \oplus F(U_i))$$

where the union is taken over the partially ordered set of the Zariski open coverings $\{U_i\}$ of X.

Proposition 4.22 *Let k be a field and (F, ϕ, A) be a homotopy invariant pretheory over k such that A is the category of abelian groups. Then the presheaf $s(F)$ has a unique structure of a homotopy invariant pretheory such that the canonical morphism of presheaves $F \to s(F)$ is a morphism of pretheories.*

Proof: Note first that $s(F)$ is obviously homotopy invariant. Let S be a smooth scheme over k, $p : X \to S$ be a smooth curve over S and Z be a closed reduced subscheme of X which belongs to $c_{equi}(X/S, 0)$. To prove our proposition it is sufficient to show that if S is local and $a \in F(X)$ is an element such that the image of a in $s(F)(X)$ is equal to zero then $\phi_{X/S}(Z)(a) = 0$.

By Corollary 4.18 our condition implies that there is an open neighborhood U of Z in X such that the restriction of a to U is equal to zero. Thus $\psi_{X/S}(Z)(a) = 0$ by Proposition 3.12.

Proposition 4.23 *Let W be a smooth semi-local scheme over a field k and $W = U \cup V$ be an open covering of W. Then for any pretheory (F, ϕ, A) over k such that A is the category of abelian groups the following sequence is exact:*

$$0 \to F(W) \to F(U) \oplus F(V) \to F(U \cap V) \to 0.$$

Proof: It follows from Lemma 4.6 in exactly the same way as in the proof of Proposition 4.17.

Proposition 4.24 *Let k be a field , (F, ϕ, A) be a homotopy invariant pretheory over k such that A is the category of abelian groups and W be a smooth semi-local scheme over k. Then one has $F_{Zar}(W) = F(W)$.*

Proof: By Proposition 4.22 and Corollary 4.19 we may suppose that F is a separated presheaf. Let $W = \cup U_i$ be an open covering of W. It is sufficient to show that for any collection of elements $a_i \in F(U_i)$ such that the restrictions of a_i and a_j to $U_i \cap U_j$ coincide there is an element a in $F(W)$ such that its restrictions to U_i are equal to a_i.

We use induction on the number k of open subsets. For $k = 2$ our statement follows from Proposition 4.23.

For an arbitrary finite covering $W = \cup U_i$ let J be the subset in the set of closed points of W which consists of points which belong to $U_1 \cup U_2$. Proposition 4.23 implies that there is a closed subscheme Z in W such that Z does not contain points from J and an element $b \in F(U_1 \cup U_2 - (U_1 \cup U_2) \cap Z)$ such that the restriction of b to $U_l - U_l \cap Z_l$ coincides with the restrictions of a_l for $l = 1, 2$.

One can easily see that the collection of open subsets $((U_1 \cup U_2 - (U_1 \cup U_2) \cap Z), U_3, \dots, U_k)$ is a covering of W and we can use the inductive assumption to assign an element in $F(W)$ to the collection of sections $(b, a_3, \dots, a_k)$ on this open covering. Since F is separated Corollary 4.18 implies that this element satisfies the conditions we need. Proposition is proven.

Proposition 4.25 *Let k be a field and (F, ϕ, A) be a homotopy invariant pretheory over k such that A is the category of abelian groups. Then the restriction of the sheaf F_{Zar} to the category Sm/k has a unique structure of pretheory such that the morphism of presheaves $F \to F_{Zar}$ is a morphism of pretheories.*

Proof: By Proposition 4.22 we may assume that F is separated. Let S be a smooth scheme over k, $p : X \to S$ be a smooth curve over S and Z be a reduced irreducible subscheme of X which belongs to $c_{equi}(X/S, 0)$. Consider a section $a \in F_{Zar}(X)$. Since F is separated, Proposition 4.24 implies easily that there is an open covering $\{U_i \to S\}$ of S and open neighborhoods V_i of closed subschemes $Z \cap p^{-1}(U_i)$ in $p^{-1}(U_i)$ such that the restrictions a_i of a to V_i belong to the subgroups $F(V_i)$ of $F_{Zar}(V_i)$. Let b_i be the elements $\phi_{V_i/U_i}(Z \cap V_i)(a_i)$ in $F(U_i)$. Since F is separated Corollary 4.19 implies easily that the restrictions of b_i and b_j to $U_i \cap U_j$ coincide and hence there is a global element $b \in F(S)$. We set $\phi_{X/Z}(Z)(a) = b$. One can verify easily (using again the fact that F is separated and Corollary 4.19) that it gives a structure of pretheory on F_{Zar}.

Proposition 4.26 *Let k be a field and (F, ϕ, A) be a homotopy invariant pretheory over k such that A is the category of abelian groups. Then the pretheory F_{Zar} is a homotopy invariant pretheory of homological type.*

Proof: It follows easily from Proposition 3.17, Proposition 4.21 and Corollary 4.19.

4.5. Zariski cohomology with coefficients in pretheories

Theorem 4.27 *Let k be a perfect field and (F, ϕ, A) be a homotopy invariant pretheory over k such that A is the category of abelian groups. Then for all $i \geq 0$ the presheaves $H^i_{Zar}(-, F_{Zar})$ on Sm/k have canonical structures of homotopy invariant pretheories of homological type.*

Proof: We use induction on i. For $i = 0$ our statement is proven in Proposition 4.26. Suppose that the theorem is proven for all $i < n$.

Lemma 4.28 *Let k be a perfect field and (F, ϕ, A) be a homotopy invariant pretheory over k such that A is the category of abelian groups. Then under the inductive assumption for any smooth semi-local scheme W over k one has*

$$H^n_{Zar}(W, F_{Zar}) = 0.$$

Proof: We use induction on the number j of closed points of W. For $j = 1$ our statement is trivial. Suppose that W is a smooth semi-local scheme, $\{x_1, \ldots, x_j\}$ the set of closed points of W and a an element of $H^n_{Zar}(W, F_{Zar})$.

By the inductive assumption on j there is a closed subscheme Z_1 of W which does not contain x_1 and a closed subscheme Z_2 of W which does not contain any of the points $x_2, \ldots, x_j$ such that a becomes zero on $U_1 = W - Z_1$ and $U_2 = W - Z_2$. One can easily see that $Z_1 \cap Z_2 = \emptyset$ and hence we have a long exact sequence

$$\ldots \to H^{n-1}(U_1, F_{Zar}) \oplus H^{n-1}(U_2, F_{Zar}) \to H^{n-1}(U_1 \cap U_2, F_{Zar}) \to$$

$$\to H^n(W, F_{Zar}) \to H^n(U_1, F_{Zar}) \oplus H^n(U_2, F_{Zar}) \to \ldots.$$

But since $H^{n-1}(-, F_{Zar})$ has a structure of homotopy invariant pretheory the homomorphism

$$H^{n-1}(U_1, F_{Zar}) \oplus H^{n-1}(U_2, F_{Zar}) \to H^{n-1}(U_1 \cap U_2, F_{Zar})$$

is surjective by Proposition 4.23 which implies that $a = 0$.

Lemma 4.29 *Let k be a perfect field and (F, ϕ, A) be a homotopy invariant pretheory over k such that A is the category of abelian groups. Then under the inductive assumption the presheaf $U \to H^n_{Zar}(U, F_{Zar})$ on Sm/k has a canonical structure of pretheory.*

Proof: Let S be a smooth variety over k and $p : X \to S$ be a smooth curve over S. Consider an element A of $c_{equi}(X/S, 0)$ and let $Z = Supp(A)$ be the support of A which is a closed subset in X finite and surjective over S.

Let $\mathcal{C}_A$ be the full subcategory in the category X_{Zar} of open subsets of X which consists of open subsets U such, that

$$p^{-1}(p((X - U) \cap Z)) \cap Z = (X - U) \cap Z.$$

One can easily see that $\mathcal{C}_A$ is closed under finite intersections. Consider the Grothendieck topology on $\mathcal{C}_A$ such that coverings are families of the form $\{U_i \to U\}$ satisfying $U = \cup U_i$ and denote by $\alpha : X_{Zar} \to \mathcal{C}_A$ the obvious morphism of sites.

For any object U of $\mathcal{C}_A$ let $\beta^{-1}(U)$ be the open subset $S - p((X - U) \cap Z)$ of S. One can easily see that $\beta^{-1}(U \cap V) = \beta^{-1}(U) \cap \beta^{-1}(V)$ and that if $U = \cup U_i$ then $\beta^{-1}(U) = \cup \beta^{-1}(U_i)$. Therefore we have a morphism of sites $\beta : S_{Zar} \to \mathcal{C}_A$.

Lemma 4.28 implies easily that $R^i \alpha_*(F_{Zar}) = 0$ for $i \leq n$. Therefore by the Leray spectral sequence for the morphism α we have

$$H^n_{Zar}(X, F_{Zar}) = H^n(\mathcal{C}_A, \alpha_*(F_{Zar})).$$

On the other hand we have a canonical morphism

$$H^n(\mathcal{C}_A, \alpha_*(F_{Zar})) \to H^n_{Zar}(S, \beta^* \alpha_*(F_{Zar})).$$

To define a homomorphism

$$\phi^n_{X/S}(A) : H^n_{Zar}(X, F_{Zar}) \to H^n_{Zar}(S, F_{Zar})$$

it is sufficient now to define a homomorphism $\beta^* \alpha_* F_{Zar} \to F_{Zar}$ of sheaves on S_{Zar} or, equivalently, a homomorphism $\alpha_* F_{Zar} \to \beta_* F_{Zar}$ of sheaves on $\mathcal{C}_A$. It means that for any open subset U of X which belongs to $\mathcal{C}_A$ we have to define a homomorphism

$$F_{Zar}(U) \to F_{Zar}(\beta^{-1}(U)).$$

We obviously have $U \cap Z = p^{-1}(\beta^{-1}(U)) \cap Z$ and hence $U \cap Z \cap p^{-1}(\beta^{-1}(U))$ is finite over $\beta^{-1}(U)$. Therefore A gives us an element A_U in $c_{equi}(p^{-1}(\beta^{-1}(U)) \cap U/\beta^{-1}(U), 0)$ and we can define our homomorphism to be the composition

$$F_{Zar}(U) \to F_{Zar}(p^{-1}(\beta^{-1}(U)) \cap U) \to F_{Zar}(\beta^{-1}(U))$$

where the last arrow is the morphism which corresponds to A_U with respect to the canonical structure of pretheory on F_{Zar} defined in Proposition 4.25.

One can easily verify now that our construction indeed provides a structure of a pretheory on the presheaf $H^n_{Zar}(-, F_{Zar})$ on Sm/k.

Lemma 4.30 *Under the assumptions of Lemma 4.29 the pretheory* $H^n_{Zar}(-, F_{Zar})$ *is of homological type.*

Proof: Let S be a smooth scheme over k, $f : X_1 \to X_2$ be a morphism of smooth curves over S and A be an element of $c_{equi}(X_1/S, 0)$. Then f defines a morphism of sites $\psi : \mathcal{C}_A \to \mathcal{C}_{f_*(A)}$ such that the following diagram is commutative

$$(X_1)_{Zar} \xrightarrow{f} (X_2)_{Zar}$$

$$\alpha_1 \downarrow \qquad\qquad \downarrow \alpha_2$$

$$\mathcal{C}_A \xrightarrow{\psi} \mathcal{C}_{f_*(A)}$$

$$\beta_1 \nwarrow \qquad \uparrow \beta_2$$

$$S_{Zar}.$$

We only have to verify that the homomorphisms

$$\beta_1^* \psi^* F_{Zar} \twoheadrightarrow F_{Zar}$$

$$\beta_2^* F_{Zar} \twoheadrightarrow F_{Zar}$$

coincide with respect to the canonical isomorphism $\beta_1^* \psi^* \cong \beta_2^*$. One can easily see that it means exactly that F_{Zar} is a pretheory of homological type which was proven in Proposition 4.26.

We will need the following technical result.

Lemma 4.31 *Let X be a smooth scheme of finite type over a perfect field k, $Z \subset X$ be a closed subset of X such that $codim(Z) \geq 1$ and x be a point of X. Then there is an open neighborhood U of x in X and a sequence of reduced closed subschemes $\emptyset \subset Y_0 \subset Y_1 \subset \ldots \subset Y_k$ in U satisfying the following conditions:*

1. *The schemes $Y_i - Y_{i-1}$ are smooth divisors on $U - Y_{i-1}$.*

2. *$U \cap Z$ is a subset of Y_k.*

Proof: We may suppose that x is a closed point of X and that X is connected.

We proceed by induction on $dim(X)$. Since X is smooth over k there is an open neighborhood U of x and a smooth morphism $p : U \to V$ of relative dimension one such that V is a smooth scheme of dimension $dim(X) - 1$ over k. Consider the closure Z' of $p(Z \cap U)$ in V.

If $codim(Z') \geq 1$ then by the inductive assumption there is a sequence of closed subschemes in V of the form $\emptyset \subset Y_0' \subset \ldots \subset Y_n'$ such that the conditions of our lemma hold for this sequence with respect to Z'. One can easily see that these conditions also hold for the sequence $\emptyset \subset p^{-1}(Y_0') \subset \ldots \subset p^{-1}(Y_n')$ of closed subschemes of U with respect to Z.

Suppose now that $codim(Z') = 0$ i.e. $Z' = V$. Let Z_0 be the closed subset of $Z \cap U$ which is the union of irreducible components of Z of codimension > 1 and the closed subset of singular points of Z. Then the codimension of the closure of $p(Z_0)$ is greater than zero hence there is a sequence $\emptyset \subset Y_0 \subset \ldots \subset Y_n$ of closed subschemes of U satisfying the conditions of the lemma with respect to Z_0. One can easily see now that the sequence

$$\emptyset \subset Y_0 \subset \ldots \subset Y_n \subset Y_n \cup (Z \cap U)$$

satisfies the conditions of the lemma with respect to Z. Lemma is proven.

To finish the proof of the theorem it is sufficient to show now that under the inductive assumption for any smooth scheme X over k the canonical projection $X \times \mathbf{A}^1 \to X$ induces an isomorphism

$$H_{Zar}^n(X, F_{Zar}) \to H_{Zar}^n(X \times \mathbf{A}^1, F_{Zar}).$$

Since the projection $p : X \times \mathbf{A}^1 \to X$ splits our morphism is a monomorphism. The Leray spectral sequence for p together with the inductive assumption shows that it is sufficient to verify that any element $a \in H_{Zar}^n(X \times \mathbf{A}^1, F_{Zar})$ becomes zero on an open covering $X = \cup U_i$ of X. By Theorem 4.15 there is a nonempty open subscheme U of X such that a equals zero on $U \times \mathbf{A}^1$. Together witth Lemma 4.31 it implies that one has to show that for a local smooth X an a smooth divisor Z on X the morphism $H_{Zar}^n(X \times \mathbf{A}^1, F_{Zar}) \to H_{Zar}^n((X - Z) \times \mathbf{A}^1, F_{Zar})$ is a monomorphism.

Lemma 4.32 *Consider the open embedding $j : (X - Z) \times \mathbf{A}^1 \to X \times \mathbf{A}^1$ and suppose that X is a local scheme. Then under our assumptions we have*

$$R^m j_*(F_{Zar}) = 0$$

for $m < n$.

Proof: Let $i : Z \times \mathbf{A}^1 \to X \times \mathbf{A}^1$ be the complimentary closed embedding. By definition $R^m j_*(F_{Zar})$ is the sheaf associated with the presheaf $j_* j^*(H_{Zar}^m(-, F_{Zar}))$. Since the sheaf associated with the presheaf $H_{Zar}^m(-, F_{Zar})$ is zero we have $R^m j_*(F_{Zar}) = i_* i^*(R^m j_*(F_{Zar}))$ and

$$i^*(R^m j_*(F_{Zar})) = (H_{Zar}^m(-, F_{Zar}))_{(X \times \mathbf{A}^1, Z \times \mathbf{A}^1)}.$$

By our assumption and Lemma 4.30 the presheaf $H^m_{Zar}(-, F_{Zar})$ has a structure of a homotopy invariant pretheory of homological type. Therefore since X is local we have (by Theorem 4.14) an isomorphism

$$i^*(R^m j_*(F_{Zar})) \cong ((H^m_{Zar}(-, F_{Zar}))_{-1})_{Zar}.$$

Denote by $p_0 : Z \times (\mathbf{A}^1 - \{0\}) \to Z$ the canonical projection. The right hand side sheaf is isomorphic by definition to $R^m(p_0)_*(F_{Zar})$. Again by the inductive assumption and Corollary 4.19 this sheaf on Z_{Zar} is separated. On the other hand it is zero in the generic point by Theorem 4.15. Lemma is proven.

Consider the exact sequence of sheaves on $X \times \mathbf{A}^1$

$$0 \to F_{Zar} \to j_* j^*(F_{Zar}) \to coker \to 0$$

(the first morphism is a monomorphism by Corollary 4.19). Since the restriction of a to $(X - Z) \times \mathbf{A}^1$ is equal to zero Lemma 4.32 and the Leray spectral sequence for the open embedding j imply that the image of a in $H^n_{Zar}(X \times \mathbf{A}^1, j_* j^*(F_{Zar}))$ is equal to zero. Therefore a belongs to the image of the homomorphism

$$H^{n-1}_{Zar}(X \times \mathbf{A}^1, coker) \to H^n_{Zar}(X \times \mathbf{A}^1, F_{Zar}).$$

Since $coker \simeq i_* i^* coker$ and the higher direct images for closed embeddings are zero the left hand side group is isomorphic to $H^{n-1}_{Zar}(Z \times \mathbf{A}^1, i^* coker)$. We may suppose that X is local. Then by Proposition 4.16 and Theorem 4.14 the sheaf $i^*(coker)$ is isomorphic to the sheaf $(F_{-1})_{Zar}$.

If $n > 1$ it implies immediately that $a = 0$ by the induction. For $n = 1$ we have a homomorphism

$$H^0(X \times \mathbf{A}^1, j_* j^*(F_{Zar})) \to H^0(X \times \mathbf{A}^1, coker)$$

The left hand side group is equal to $F_{Zar}(X - Z)$ since F_{Zar} is homotopy invariant and the right hand side group is isomorphic to $F_{Zar}(Z \times (\mathbf{A}^1 - \{0\}))/F_{Zar}(F)$. Therefore our homomorphism is surjective by Theorem 4.14 since X and Z are local. Theorem is proven.

Corollary 4.33 *Let k be a perfect field and (F, ϕ, A) be a homotopy invariant pretheory over k such that A is the category of abelian groups. Let further X be a smooth scheme over k and $j : U \to X$ be an open subscheme of X such that $X - U$ is a smooth divisor on X. Then $R^n j_*(F_{Zar}) = 0$ for $n > 0$.*

Proof: See Lemma 4.32.

4.6. The Gersten resolution

Proposition 4.34 *Let k be a perfect field and (F, ϕ, A) be a homotopy invariant pretheory over k such that A is the category of abelian groups. Then for any $i \geq 0$ one has a canonical isomorphism of pretheories*

$$(H^i_{Zar}(-, F_{Zar}))_{-1} \cong H^i_{Zar}(-, (F_{-1})_{Zar}).$$

Proof: Let us first consider the case $i = 0$. We have to show that the canonical morphism of presheaves $(F_{-1})_{Zar} \to (F_{Zar})_{-1}$ is an isomorphism. By Proposition 4.26 both presheaves have canonical structures of homotopy invariant pretheories and obviously this morphism is a morphism of pretheories. Our statement follows now from Theorem 4.15 and Proposition 4.20.

Suppose now that $i > 0$. One can easily see that it is sufficient to show that for any smooth scheme X over k one has $R^i p_*(F_{Zar}) = 0$ for $i > 0$ where p is the projection $X \times (\mathbf{A}^1 - \{0\}) \to X$. It follows from Theorem 4.27, Proposition 4.18 and Theorem 4.15.

Lemma 4.35 *Let k, (F, ϕ, A) be as in Proposition 4.34 and S be a smooth local scheme over k then one has:*

$$H^i_{Zar}(S \times (\mathbf{A}^n - \{0\}), F_{Zar})/H^i_{Zar}(S, F_{Zar}) = \begin{cases} F_{-n}(S) & for \quad i = n - 1 \\ 0 & for \quad i \neq n - 1 \end{cases}$$

Proof: It follows easily by induction from Proposition 4.34, Theorem 4.27 and the Mayer-Vietoris long exact sequence for the Zariski cohomology.

For any point x of a smooth scheme X over k we denote by $H^i_x(X, F)$ the group $\lim_{\to} H^i_{\{\bar{x}\} \cap U}(U, F)$ where limit is taken over the partially ordered set of open neighborhoods of x in X.

Lemma 4.36 *Let k be a perfect field, X be a smooth scheme over k and x be a point of codimension p on X. Then for any F as above one has:*

$$H^i_x(X, F) = \begin{cases} F_{-p}(Spec(k_x)) & for \quad i = p \\ 0 & for \quad i \neq p \end{cases}$$

Proof: Let us recall that for any open embedding $j : U \to X$ and any presheaf F on X_{Zar} we defined in §4.2 a sheaf $F_{(X, X-U)}$ on $(X - U)_{Zar}$ as $i^*(coker : F \to j_* j^*(F))_{Zar}$ where $i : X - U \to X$ is the corresponding closed embedding.

Let X_x be the local scheme of x in X. For $p = 0$ or $i = 0$ our statement follows from Corollary 4.17. Assume that $i, p > 0$. Consider the sheaf

$H^{i-1}(-, F)_{(X_x, x)}$ on $Spec(k_x)_{Zar}$. Clearly it is the constant sheaf which corresponds to the abelian group $H_x^i(X, F)$. Applying Theorem 4.14 to $H^{i-1}(-, F)$ we conclude that

$$H_x^i(X, F) = (H^{i-1}(-, F))_{(\mathbf{A}_{k_x}^p, (0, ..., 0))}(Spec(k_x)) = H_{(0, ..., 0)}^i(\mathbf{A}_{k_x}^p, F).$$

We have the long exact sequence for the cohomology with supports

$$.. \to H_{(0)}^i(\mathbf{A}_{k_x}^p, F) \to H^i(\mathbf{A}_{k_x}^p, F) \to H^i(\mathbf{A}_{k_x}^p - \{0\}, F) \to H_{(0)}^{i+1}(\mathbf{A}_{k_x}^p, F) \to ..$$

By Theorem 4.27 we have isomorphisms $H^i(\mathbf{A}_{k_x}^p, F) \to H^i(Spec(k_x), F)$ and therefore homomorphisms $H^i(\mathbf{A}_{k_x}^p, F) \to H^i(\mathbf{A}_{k_x}^p - \{0\}, F)$ are injective. Our statement follows now from Lemma 4.35.

Theorem 4.37 *Let k be a perfect field, (F, ϕ, A) be a homotopy invariant pretheory over k such that A is the category of abelian groups and F is a sheaf in the Zariski topology. Let further X be a smooth scheme of dimension n over k. Denote by $X^{(k)}$ the set of points of X of codimension k. Then there is a canonical exact sequence of sheaves on X_{Zar} of the form:*

$$0 \to F \to \oplus_{x \in X^{(0)}}(i_x)_*(F) \to \ldots \to \oplus_{x \in X^{(n)}}(i_x)_*(F_{-n}) \to 0.$$

Proof: For any sheaf F we have a spectral sequence

$$(E_1^{pq} = \oplus_{x \in X^{(p)}} H_x^{p+q}(X, F)) \Rightarrow H^{p+q}(X, F).$$

Consider the corresponding spectral sequence of sheaves on X_{Zar}. It converges to the sheaves in the Zariski topology associated with the presheaves $H^{p+q}(-, F)$ which equal zero for $p + q \neq 0$ and F for $p + q = 0$. On the other hand Lemma 4.36 implies that it degenerates to a complex of the form

$$\oplus_{x \in X^{(0)}}(i_x)_*(F) \to \ldots \to \oplus_{x \in X^{(n)}}(i_x)_*(F_{-n}) \to 0$$

which immediately implies the statement of our theorem.

5. Sheaves in Nisnevich and etale topologies associated with pretheories

5.1. Pretheoretical structures on Nisnevich cohomology

Definition 5.1 *Let k be a field and (F, ϕ, A) be a pretheory over k such that A is the category of abelian groups. It is called a reasonable pretheory if the following condition holds. Let S be a smooth local scheme over k,*

$f : X_1 \to X_2$ be an etale morphism of smooth curves over S and Z be a reduced irreducible subscheme of X_1 which belongs to $c_{equi}(X/S, 0)$ such that $Z \to f(Z)$ is an isomorphism. Then one has

$$\phi_{X_2/S}(f(Z)) = \phi_{X_1/S}(Z) \circ F(f).$$

Proposition 5.2 Let k be a field and (F, ϕ, A) be a homotopy invariant pretheory over k such that A is the category of abelian groups. Then F is reasonable.

Proof: It follows from Corollary 4.19 and Proposition 3.17.

Theorem 5.3 Let k be a field and (F, ϕ, A) be a reasonable pretheory over k. Then the presheaves $U \to H_{Nis}^i(U, F_{Nis})$ have canonical structures of (reasonable) pretheories.

Proof: Let us consider first the case $i = 0$. Denote by $s_{Nis}(F)$ the separated in the Nisnevich topology presheaf associated with F. Let us show that it has a structure of a reasonable pretheory.

It is sufficient to show that for any smooth henselian local scheme S over k, smooth curve $p : X \to S$ over S and a reduced irreducible subscheme Z in $c_{equi}(X/S, 0)$ we have $\phi_{X/S}(Z)(a) = 0$ for any $a \in ker(F(X) \to s_{Nis}(F)(X))$. It follows immideately from the fact any scheme finite over S is a local henselian scheme and the fact that F is reasonable.

We may assume now that F is a separated presheaf in the Nisnevich topology. Let $p : X \to S$ be a smooth curve over a smooth scheme S over k, Z be a closed subscheme in X which belongs to $c_{equi}(X/S, 0)$ and a be an element in $F_{Nis}(X)$. We want to construct an element $b = \phi_{X/S}(Z)(a)$ in $F_{Nis}(S)$.

Let $\{U_i \to X\}$ be a finite Nisnevich covering of X such that the restrictions of a to U_i belong to $F(U_i)$. Since a scheme finite over a local henselian scheme is a local henselian scheme there is a Nisnevich covering $\{V_j \to S\}_{1 \leq j \leq m}$ of S such that for any $j = 1, \ldots, m$ the covering

$$\{U_i \times_S V_j \to X \times_S V_j\}$$

splits over $Z_j = Z \times_S V_j$.

Let $\tilde{Z}_j$ be the image of such a splitting, i.e. a closed subscheme in $\coprod_i U_i \times_S V_j$ which maps isomorphically on Z_j. Then $\coprod_i U_i \times_S V_j \to V_j$ is a smooth curve over V_j and $\tilde{Z}_j$ belongs to $c_{equi}((\coprod_i U_i \times_S V_j)/V_j, 0)$. Consider the elements $b_j = \phi(\tilde{Z}_j)(F(pr_1)(\coprod a_i))$.

One can easily see that b_j define a global element b in $F_{Nis}(S)$ and that this construction indeed provides us with a structure of a reasonable pretheory on F_{Nis}.

Consider now the case $i > 0$. Let $p : X \to S$ be a smooth curve over a smooth scheme S over k and Z be a reduced closed subscheme of X which belongs to $c_{equi}(X/S, 0)$. Consider the following diagram of morphisms of sites:

$$Z_{Nis} \xrightarrow{i} X_{Nis}$$
$$p_0 \searrow \quad \downarrow p$$
$$S_{Nis}$$

Since p_0 is finite the higher direct images in the Nisnevich topology with respect to p_0 are zero. It implies that one has only to construct a morphism of sheaves $(p_0)_* i^*(F_{Nis}) \to F_{Nis}$ on S_{Nis}. It can be easily done using our theorem for $i = 0$ and the fact that a scheme finite over a henselian local scheme is a disjoint union of henselian local schemes.

5.2. Cohomology of affine space and the comparison theorem

Proposition 5.4 *Let k be a field, (F, ϕ, A) be a homotopy invariant pretheory over k such that A is the category of abelian groups and U be an open subscheme in $\mathbf{A}^1_k$. Then one has:*

$$H^i_{Nis}(U, F_{Nis}) = \begin{cases} F(U) & \text{for } i = 0 \\ 0 & \text{for } i \neq 0 \end{cases}$$

Proof: Let us consider first the case $i = 0$. First of all by Theorem 4.15 we have $F \cong F_{Zar}$ on U_{Zar}. Hence the morphism $F(U) \to F_{Nis}(U)$ is a monomorphism by Corollary 4.18 since any Nisnevich covering contains at least one non-empty open subset. It implies that we may replace F by $s_{Nis}(F)$ and assume that F is a separated presheaf in the Nisnevich topology.

One can easily see now that it is sufficient to show that for any element a in $F_{Nis}(U)$ and any closed point x of U there is an open neighborhood V such that the restriction of a to V belongs to $F(V) \subset F_{Nis}(V)$. It follows immediately from Proposition 4.12.

To prove our proposition for $i > 0$ it is obviously sufficient to consider the case $i = 1$. The corresponding result again follows easily from Proposition 4.12.

Proposition 5.5 *Let k be a field and (F, ϕ, A) be a homotopy invariant pretheory over k such that A is the category of abelian groups. Then the canonical morphism of presheaves $F_{Zar} \to F_{Nis}$ is an isomorphism.*

Proof: Note first that Theorem 5.3 together with Proposition 5.4 imply that F_{Nis} is a homotopy invariant pretheory and our morphism is obviously a morphism of pretheories. To finish the proof it is sufficient to apply Proposition 4.20 to this morphism.

Theorem 5.6 *Let k be a perfect field, (F, ϕ, A) be a homotopy invariant pretheory over k such that A is the category of abelian groups and X be a smooth scheme over k. Then the canonical projection $X \times \mathbf{A}^1 \to X$ induces isomorphisms*

$$H^i_{Nis}(X, F_{Nis}) \to H^i_{Nis}(X \times \mathbf{A}^1, F_{Nis}).$$

Proof: The proof is strictly parallel to the proof of Theorem 4.27 with considerable simplification due to Theorem 5.3.

Theorem 5.7 *Let k be a perfect field, (F, ϕ, A) be a homotopy invariant pretheory over k such that A is the category of abelian groups and X be a smooth scheme over k. Then the canonical morphisms $H^i_{Zar}(X, F_{Zar}) \to H^i_{Nis}(X, F_{Nis})$ are isomorphisms.*

Proof: The case $i = 0$ is proven in Proposition 5.5. Suppose that $i > 0$. By Theorems 5.6 and 5.3 the presheaves $H^i_{Nis}(-, F_{Nis})$ are homotopy invariant pretheories. It is obviously sufficient to show that the sheaves in the Zariski topology associated with these presheaves are zero. It follows immediately from Corollary 4.18 since the Nisnevich and the Zariski cohomology coincide for fields.

Remark 5.8 Unfortunately the analogs of the previous results do not hold for the flat version of the Nisnevich topology. Consider the following example. Let F be the cokernel of the morphism of sheaves $\mathbf{G}_m \xrightarrow{z^2} \mathbf{G}_m$ in the Zariski topology. Our results imply that F has a canonical structure of a homotopy invariant pretheory of homological type.

Consider the Nis-flat covering of $\mathbf{A}^1$ by the open subset $U_0 = \mathbf{A}^1 - \{0\}$ and the flat morphism $U_1 = \mathbf{A}^1 \xrightarrow{z^2} \mathbf{A}^1$. Let a be the canonical section of F on U_0. Together with the zero section on U_1 they give us a nontrivial global section of F_{Nis-fl} on $\mathbf{A}^1$. It implies immediately that our theory does not work in this situation.

5.3. Applications to Suslin homology

We will show in this section that our results on cohomology of sheaves associated with pretheories imply some important vanishing properties for the sheaves $\underline{h}_i(-)_{Zar}$ (see the end of §3.2). As an application we will prove the Mayer-Vietoris theorem for the Suslin homology.

Theorem 5.9 *Let k be a perfect field and (F, ϕ, A) be a pretheory over k with values in the category of abelian groups. Then the following conditions are equivalent.*

1. *The sheaves $(\underline{h}_i(F))_{Zar}$ are zero for all $i \le n$.*

2. *For any homotopy invariant pretheory G which is a sheaf in the Nisnevich topology one has $Ext^i(F_{Nis}, G) = 0$ for all $i \le n$.*

Proof: Note that by Proposition 5.5 one has $(\underline{h}_i(F))_{Zar} = (\underline{h}_i(F))_{Nis}$ and that (by 4.26) these presheaves are homotopy invariant pretheories. $(1- > 2)$. Consider the reduced singular simplicial complex $\tilde{\underline{C}}_*(F)$ of F where

$$\tilde{\underline{C}}_0(F)(X) = F(X) = \underline{C}_0(F)(X)$$

$$\tilde{\underline{C}}_k(F)(X) = (coker : F(X) \to F(X \times \mathbf{A}^k)) = (\underline{C}_k(F)/F)(X) \text{ for } k > 0.$$

Clearly the obvious morphism $\underline{C}_*(F) \to \tilde{\underline{C}}_*(F)$ is a quasi-isomorphism. The reason why $\tilde{\underline{C}}_*$ is sometimes more convinient to work with is the following lemma.

Lemma 5.10 *Let G be a Nisnevich sheaf which is a homotopy invariant pretheory. Then for any $j > 0$ and any $i \ge 0$ one has*

$$Ext^i((\tilde{\underline{C}}_j(F))_{Nis}, G) = 0.$$

Proof: It follows easily from the fact that the presheaf $\tilde{\underline{C}}_j(F)$ is contractible (and hence so is the associated sheaf) and Theorem 5.6.

Lemma 5.10 implies that for any G as above and any i the obvious morphism

$$Hom_D((\underline{C}_*(F))_{Nis}, G[i]) = Hom_D((\tilde{\underline{C}}_*(F))_{Nis}, G[i]) \to Ext^i(F_{Nis}, G)$$

is an isomorphism (here Hom_D are morphisms in the derived category of Nisnevich sheaves on Sm/k). If $(\underline{h}_j(F))_{Nis} = 0$ for $j \le n$ then the left hand side groups are obviously zero for all $i \le n$.

($2- > 1$) Let $(\underline{h}_k(F))_{Nis}$ be the first nontrivial cohomology sheaf of $\underline{C}_*(F)$. Then there is a nonzero morphism in the derived category

$$(\underline{C}_*(F))_{Nis} \to (\underline{h}_k(F))_{Nis}[k].$$

Since $(\underline{h}_k(F))_{Nis}$ is a homotopy invariant pretheory it gives us as above a nontrivial element in $Ext^k(F_{Nis}, (\underline{h}_k(F))_{Nis})$ and hence by our condition $k > n$.

Corollary 5.11 *Let k be a perfect field and F be a pretheory over k with values in the category of abelian groups such that $F_{Nis} = 0$. Then for any $i \geq 0$ one has $\underline{h}_i(F)_{Zar} = 0$.*

Corollary 5.12 *Let k be a perfect field and*

$$0 \to F \to G \to H \to 0$$

be a sequence of reasonable pretheories such that the corresponding sequence of associated sheaves in the Nisnevich topology is exact. Then one has a canonical long exact sequence of sheaves of the form

$$\ldots \to (\underline{h}_i(F))_{Zar} \to (\underline{h}_i(G))_{Zar} \to (\underline{h}_i(H))_{Zar} \to (h_{i-1}(F))_{Zar} \to \ldots.$$

Proof: It follows immediately from Corollary 5.11 and Theorem 5.3.

Let k be a field, X be a scheme of finite type over k and U be a smooth scheme over k. Denote by $c_{equi}(X/Spec(k), 0)(U)$ the free abelian group generated by integral closed subschemes of $X \times U$ which are finite over U and surjective over an irreducible component of U. As was shown in [8] for any morphism of smooth schemes $f : U' \to U$ one can define a homomorphism

$$cycl(f) : c_{equi}(X/Spec(k), 0)(U) \to c_{equi}(X/Spec(k), 0)(U')$$

such that $c_{equi}(X/Spec(k), 0)(-)$ becomes a presheaf on the category Sm/k and the obvious analogs of properties (1)-(3) given at the beginning of §2 hold.

Proposition 5.13 *Let k be a field and X be a scheme of finite type over k. Then the presheaf $c_{equi}(X/Spec(k), 0)$ on Sm/k is a sheaf in the etale (and a fortiori in the Nisnevich) topology.*

Proof: It follows immediately from our definition of $c_{equi}(X/Spec(k), 0)$ and property (2) of the base change homomorphisms.

Proposition 5.14 *For any scheme of finite type X over a field k the presheaf of abelian groups $c_{equi}(X/Spec(k), 0)$ on Sm/k has a canonical structure of a pretheory of homological type.*

Proof: We have to define for any smooth curve C/S over a smooth scheme S over k a pairing of the form

$$c_{equi}(C/S, 0) \otimes c_{equi}(X/Spec(k), 0)(C) \to c_{equi}(X/Spec(k), 0)(S).$$

This is a special case of the correspondence pairings considered in [8, Section 3.7]. The fact that these pairing give us a structure of a pretheory follows from [8, Th. 3.7.3].

Proposition 5.15 *Let k be a field, X a separated scheme over k and $X = U \cup V$ an open covering of X. Then the sequence*

$$0 \to c_{equi}(U \cap V) \to c_{equi}(U) \oplus c_{equi}(V) \to c_{equi}(X) \to 0$$

is exact in the Nisnevich topology (we abbreviated $c_{equi}(-/Spec(k), 0)$ to $c_{equi}(-)$).

Proof: It is a special case of [8, Prop. 4.3.7].

Definition 5.16 *Let k be a field and X a scheme of finite type over k. The groups $\underline{h}_i(c_{equi}(X/Spec(k), 0))(Spec(k))$ are called the Suslin homology of X. We denote them by $H_i^S(X/k)$.*

Theorem 5.17 *Let k be a perfect field and X a scheme of finite type over k. Then for any open covering $X = U \cup V$ of X there is a canonical long exact sequence of the form*

$$\ldots \to H_i^S(U \cap V) \to H_i^S(U) \oplus H_i^S(V) \to H_i^S(X) \to H_{i-1}^S(U \cap V) \to \ldots.$$

Proof: It follows from Corollary 5.12 and Proposition 5.15.

5.4. Cohomology of blowups

For any smooth scheme X over a field k denote by $\mathbf{Z}_{Nis}(X)$ the sheaf of abelian groups on $(Sm/k)_{Nis}$ freely generated by the sheaf of sets representable by X. Let $Ab((Sm/k)_{Nis})$ be the category of sheaves of abelian groups on $(Sm/k)_{Nis}$. Then for any object F of this category we have canonical isomorphisms:

$$Ext^i_{Ab((Sm/k)_{Nis})}(\mathbf{Z}_{Nis}(X), F) \cong H^i_{Nis}(X, F).$$

Proposition 5.18 *Let k be a field, $f : X \to Y$ an etale morphism of smooth schemes over k and Z a closed subscheme in Y such that $f^{-1}(Z) \to Z$ is an isomorphism. Then the canonical morphism of sheaves*

$$\mathbf{Z}_{Nis}(X)/\mathbf{Z}_{Nis}(X - f^{-1}(Z)) \to \mathbf{Z}_{Nis}(Y)/\mathbf{Z}_{Nis}(Y - Z)$$

on $(Sm/k)_{Nis}$ is an isomorphism.

Proof: Note first that $(Y - Z) \coprod X \to Y$ is a Nisnevich covering of Y. Therefore our morphism is an epimorphism. To prove that it is a monomorphism it is sufficient to prove that it is a monomorphism on the corresponding freely generated presheaves (since the associated sheaf functor is exact). Consider the diagram:

$$
\begin{array}{ccccccccc}
0 \to & \mathbf{Z}(X - f^{-1}(Z)) & \xrightarrow{i} & \mathbf{Z}(X) & \to & \mathbf{Z}(X)/\mathbf{Z}(X - f^{-1}(Z)) & \to 0 \\
& \downarrow & & \downarrow & & \downarrow \\
0 \to & \mathbf{Z}(Y - Z) & \to & \mathbf{Z}(Y) & \to & \mathbf{Z}(Y)/\mathbf{Z}(Y - Z) & \to 0
\end{array}
$$

where $\mathbf{Z}(-)$ is the presheaf which sends a smooth scheme U to the direct sum of free abelian groups generated by sets of morphisms from the connected components of U to $-$. We have to show that $ker(\mathbf{Z}(f))$ lies in $Im(i)$. Let W be a connected scheme. Then $ker(\mathbf{Z}(f))$ is the group of formal sums of the form $\sum_{i \in I} n_i g_i$, where g_i are morphisms from W to X such that there exists a decomposition $I = \coprod I_k$ satisfying $f \circ g_i = f \circ g_j$ for $i, j \in I_k$ and $\sum_{i \in I_k} n_i = 0$ for any k. Therefore, we have to prove only that if $f \circ g = f \circ h$ for some $g, h : W \to X$ then either $g = h$ or g and h can be factored through $X - f^{-1}(Z)$. Let g, h be such morphisms. Then there exists a morphism $g \times h : W \to X \times_Y X$ whose compositions with the projections are the morphisms g and h respectively. To finish the proof it is sufficient to notice that since f is an etale morphism the diagonal in $X \times_Y X$ is a connected component i.e. there is a decomposition of the form $X \times_Y X = \Delta(X) \coprod X_0$. But since $f^{-1}(Z) \to Z$ is an isomorphism the projections $pr_1, pr_2 : X_0 \to X$ can be factored through $X - f^{-1}(Z)$. Proposition is proven.

Proposition 5.19 *Let X be a smooth scheme over a field k and $Z \to X$ be a smooth subscheme of X. Denote by $p_Z : X_Z \to X$ the blowup of X in Z and consider the following Cartesian square:*

$$
\begin{array}{ccc}
p_Z^{-1}(Z) & \xrightarrow{i} & X_Z \\
q \downarrow & & \downarrow p_Z \\
Z & \longrightarrow & X
\end{array}
$$

Then the canonical morphism of sheaves $ker(\mathbf{Z}_{Nis}(q)) \to ker(\mathbf{Z}_{Nis}(p_Z))$ on Sm/k is an isomorphism (and a fortiori the same holds for any stronger topology).

Proof: Obviously our morphism is a monomorphism. It is sufficient to show that it is an epimorphism. For any $X \in ob(Sm/k)$ let $L(X)$ be the sheaf of sets representable by X. Note that there is an obvious long exact sequence of the form

$$\ldots \to \mathbf{Z}_{Nis}(L(X_Z) \times_{L(X)} L(X_Z)) \to \mathbf{Z}_{Nis}(L(X_Z)) \to \mathbf{Z}_{Nis}(L(X))$$

It is sufficient to show that the morphism

$$L(i) \times L(i) \coprod L(\Delta) : L(p_Z^{-1}(Z) \times_Z p_Z^{-1}(Z)) \coprod L(X_Z) \to L(X_Z) \times_{L(X)} L(X_Z)$$

is an epimorphism. It is obvious since as a scheme $X_Z \times_X X_Z$ is a union of the diagonal and the image of $p_Z^{-1}(Z) \times_Z p_Z^{-1}(Z)$.

Remark 5.20 The statement of the proposition above is false for sheaves in the Nisnevich (or Zariski) topology on Sch/k which gives in particular an example of the fact that functor of the inverse image on the categories of sheaves associated with the canonical embedding of categories $Sm/k \to Sch/k$ is not left exact. Nevertheless the proposition is still valid for all schemes if we consider qfh- or cdh-topology (see [10], [8]).

Proposition 5.21 *In the notations of Proposition 5.19 suppose that k is a perfect field and let (F, ϕ, A) be a homotopy invariant pretheory over k such that A is the category of abelian groups. Then for any $i \geq 0$ one has:*

$$Ext^i(coker(\mathbf{Z}_{Nis}(p_Z)), F_{Nis}) = 0$$

Proof: One can easily see that our problem is local with respect to X. Denote by U the open subscheme $X - Z$ of X and consider the diagram of sheaves on $(Sm/k)_{Nis}$:

$$
\begin{array}{ccccccccc}
0 \to & \mathbf{Z}_{Nis}(U) & \to & \mathbf{Z}_{Nis}(X_Z) & \to & coker_1 & \to 0 \\
& \downarrow & & \downarrow & & \downarrow & \\
0 \to & \mathbf{Z}_{Nis}(U) & \to & \mathbf{Z}_{Nis}(X) & \to & coker_2 & \to 0
\end{array}
$$

One can easily see that $coker(\mathbf{Z}(p_Z)) = coker(coker_1 \to coker_2)$. Together with Proposition 5.18 and the fact that Z is smooth it implies that locally with respect to X we have:

$$coker(\mathbf{Z}(p_Z)) \cong coker(\mathbf{Z}(Z \times \mathbf{A}^d_{(Z \times \{0\})}) \to \mathbf{Z}(Z \times \mathbf{A}^d))$$

where d is the codimension of Z. We have a projection $Z \times \mathbf{A}^d_{(Z \times \{0\})} \to Z \times \mathbf{P}^{d-1}$ which is a locally trivial in the Zariski topology fibration with the fiber $\mathbf{A}^1$. By Theorem 5.6 it implies that we have:

$$Ext^i(\mathbf{Z}(Z \times \mathbf{A}^d_{(Z \times \{0\})}), F) = H^i(Z \times \mathbf{P}^{d-1}, F)$$
$$Ext^i(\mathbf{Z}(Z \times \mathbf{A}^d), F) = H^i(Z, F).$$

Our statement follows now trivially from Proposition 5.19.

Corollary 5.22 *Let k be a perfect field, X be a smooth scheme over k and Z be a smooth closed subscheme in X. Denote by $p : X_Z \to X$ the blow-up of Z in X. Then for any homotopy invariant pretheory F with values in the category of abelian groups there is a long exact sequence of the form:*

$$\ldots \to H^i(X, F_{Zar}) \to H^i(X_Z, F_{Zar}) \oplus H^i(Z, F_{Zar}) \to$$

$$\to H^i(p^{-1}(Z), F_{Zar}) \to H^{i+1}(X, F_{Zar}).$$

(Note that it is true for both the Zariski and the Nisnevich cohomology groups by Theorem 5.7)

Proof: Consider the following diagram of morphisms of sheaves in the Nisnevich topology:

$$
\begin{array}{ccc}
\mathbf{Z}_{Nis}(p_Z^{-1}(Z)) & \overset{i}{\to} & \mathbf{Z}_{Nis}(X_Z) \\
q\downarrow & & \downarrow \mathbf{Z}_{Nis}(p_Z) \\
\mathbf{Z}_{Nis}(Z) & \to & \mathbf{Z}_{Nis}(X)
\end{array}
$$

The left vertical arrow is an epimorphism since $p^{-1}(Z) \to Z$ is a locally trivial in Zarski topology bundle over Z. This fact and Proposition 5.19 imply immediately that we have an exact sequence of the form:

$$0 \to \mathbf{Z}_{Nis}(p_Z^{-1}(Z)) \to \mathbf{Z}_{Nis}(X_Z) \oplus \mathbf{Z}_{Nis}(Z) \to \mathbf{Z}_{Nis}(X)$$

$$\to coker(\mathbf{Z}_{Nis}(p_Z)) \to 0$$

and the corollary follows from Proposition 5.21.

Remark 5.23 It can be shown (see [9]) that the exact sequence from Corollary 5.22 splits canonically and one has:

$$H^i(X_Z, F_{Zar}) = H^i(X, F_{Zar}) \oplus \bigoplus_{j=1}^{c-1} H^{i-2j}(Z, (F_{-j})_{Zar}).$$

5.5. Sheaves in the etale topology associated with pretheories

Proposition 5.24 *Let k be a field and (F, ϕ, A) be a homotopy invariant pretheory over k such that A is the category of $\mathbf{Q}$-vector spaces. Then the canonical morphism of presheaves on Sm/k*

$$F_{Nis} \to F_{et}$$

is an isomorphism.

Proof: It follows trivially from Corollary 4.18 and the results of §3.3.

The corresponding result with intergral coefficients is wrong. Instead we have the following "rigidity" theorem.

Theorem 5.25 *Let k be a field of exponential characteristic p and (F, ϕ, A) be a homotopy invariant pretheory over k such that A is the category of torsion abelian groups of torsion prime to p. Then the sheaf F_{et} on Sm/k is locally constant.*

Proof: The proof is strictly parallel to the proof of the rigidity theorem in [7].

Corollary 5.26 *Let k be a field of exponential characteristic p and (F, ϕ, A) be a homotopy inavriant pretheory over k such that A is the category of $\mathbf{Z}[1/p]$-modules. Consider the exact sequence of sheaves*

$$0 \to (F_{tors})_{et} \to F_{et} \to F_{et} \otimes \bar{\mathbf{Q}} \to (F_{cotors})_{et} \to 0.$$

Then the sheaves $(F_{tors})_{et}, (F_{cotors})_{et}$ are locally constant.

Proof: It follows from Theorem 5.25 and Proposition 3.5.

Proposition 5.27 *Let X be a scheme and F be a sheaf of $\mathbf{Q}$-vector spaces on X_{et} then the canonical morphisms*

$$H^i_{Nis}(X, F) \to H^i_{et}(X, F)$$

are isomorphisms.

Proof: It follows immediately from the Leray spectral sequence and the fact that the etale cohomology of any henselian local scheme are torsion groups.

Proposition 5.28 *Let k be a perfect field and (F, ϕ, A) be a homotopy invariant pretheory over k such that A is the category of abelian groups. Then the canonical morphisms of pretheories*

$$H^i_{Zar}(-, F_{Zar} \otimes \mathbf{Q}) \to H^i_{et}(-, F_{et} \otimes \mathbf{Q})$$

are isomorphisms.

Proof: It follows from Theorem 5.7, Proposition 5.28 and Proposition 5.24.

Corollary 5.29 *Let k be a perfect field of exponential characteristic p, (F, ϕ, A) a homotopy invariant pretheory over k such that A is the category of $\mathbf{Z}[1/p]$-modules and X a smooth scheme over k. Then the canonical morphisms*

$$H^i_{et}(X, F_{et}) \to H^i_{et}(X \times \mathbf{A}^1, F_{et})$$

are isomorphisms.

Proof: It follows from our comparison theorem for the rational coefficients, Theorem 5.25 and homotopy invariance of the etale cohomology with locally constant torsion coefficients.

Proposition 5.30 *Let k be a perfect field and $F : (Sm/k)^{op} \to A$ a contravariant functor from the category Sm/k to a $\mathbf{Q}$-linear additive category A. Suppose that there are two pretheory structures ϕ^1, ϕ^2 of on F. Then ϕ^1 and ϕ^2 coincide semi-locally. In particular if F is a separated presheaf of $\mathbf{Q}$-vector spaces in the Zariski topology, then ϕ^1 and ϕ^2 coincide.*

Proof: Let S be a smooth affine variety over k and $p : X \to S$ a smooth curve over S. We have to show that for any finite set $\{s_1, \ldots, s_n\}$ of points of S and any element A in $c_{equi}(X/S, 0)$ there is an open neighborhood U of $\{s_1, \ldots, s_n\}$ such that the following two compositions coincide

$$F(X) \overset{\phi^1_{X/S}(A)}{\to} F(S) \to F(U)$$

$$F(X) \overset{\phi^2_{X/S}(A)}{\to} F(S) \to F(U).$$

By Proposition 4.17 it is sufficient to show that there exists a nonempty open subset U satisfying this property. It means that we may suppose that $S = Spec(K)$ where K is a field, X is a curve over K and A is a closed point of X. In this case our statement follows easily from the results of §3.3.

References

1. Eric M. Friedlander and O. Gabber. Cycle spaces and intersection theory. *Topological methods in modern mathematics*, pages 325–370, 1993.
2. Eric M. Friedlander and V. Voevodsky. Bivariant cycle cohomology. *This volume*.
3. A. Grothendieck, M. Artin, and J.-L. Verdier. *Theorie des topos et cohomologie etale des schemas (SGA 4)*. Lecture Notes in Math. 269, 270, 305. Springer, Heidelberg, 1972-73.
4. A. Grothendieck and J. Dieudonne. *Etude Locale des Schemas et des Morphismes de Schemas (EGA 4)*. Publ. Math. IHES, 20, 24, 28, 32, 1964-67.
5. Y. Nisnevich. The completely decomposed topology on schemes and associated descent spectral sequences in algebraic K-theory. In *Algebraic K-theory: connections with geometry and topology*, pages 241–342. Kluwer Acad. Publ., Dordrecht, 1989.
6. M. Raynaud and L. Gruson. Criteres de platitude et de projectivite. *Inv. Math.*, 13:1–89, 1971.
7. Andrei Suslin and Vladimir Voevodsky. Singular homology of abstract algebraic varieties. *Invent. Math.*, 123(1):61–94, 1996.
8. A. Suslin and V. Voevodsky. Relative cycles and Chow sheaves. *This volume*.
9. V. Voevodsky. Triangulated categories of motives over a field. *This volume*.
10. V. Voevodsky. Homology of schemes. *Selecta Mathematica, New Series*, 2(1):111–153, 1996.
11. Mark F. Walker. The primitive topology of a scheme. *J. Algebra*, 201(2):656–685, 1998.

4

Bivariant Cycle Cohomology

Eric M. Friedlander* and Vladimir Voevodsky

Contents

1 Introduction   138

2 Presheaves of relative cycles   141

3 The cdh-topology   145

4 Bivariant cycle cohomology   150

5 Pretheories   155

6 The moving Lemma   165

7 Duality   171

8 Properties and pairings   176

9 Motivic cohomology and homology   181

1. Introduction

The precursor of our bivariant cycle cohomology theory is the graded Chow group $A_*(X)$ of rational equivalence classes of algebraic cycles on a scheme X over a field k. In [1], S. Bloch introduced the higher Chow groups $CH^*(X, *)$ in order to extend to higher algebraic K-theory the relation established by A. Grothendieck between $K_0(X)$ and $A_*(X)$. More recently,

---

*Partially supported by the N.S.F. and NSA Grant #MDA904-

Lawson homology theory for complex algebraic varieties has been developed (cf. [13], [3]), in which the role of rational equivalence is replaced by algebraic equivalence, and a bivariant extension $L^*H^*(Y,X)$ [7] has been introduced. This more topological approach suggested a duality relating covariant and contravariant theories as established in [8].

Our bivariant cycle cohomology groups $A_{r,i}(Y,X)$, defined for schemes Y, X of finite type over a field k, satisfy $A_{r,i}(Spec(k),X)) = CH^{n-r}(X,i)$ whenever X is an affine scheme over k of pure dimension n. This bivariant theory is a somewhat more sophisticated version of a theory briefly introduced by the first author and O. Gabber in [6]. The added sophistication enables us to prove numerous good properties of this theory, most notably localization (Theorem 5.11) and duality (Theorem 8.2), which suggest that $A_{*,*}(Y,X)$ should be useful in various contexts. For example, we envision that the bivariance and duality of $A_{*,*}(Y,X)$ will be reflected in a close relationship to the algebraic K-theories of coherent sheaves and of locally free sheaves on schemes of finite type over a field k. Moreover, we establish pairings for our theory whose analogues in Lawson homology have proved useful in studying algebraic cycles (cf. [10],[5]).

The final section presents a formulation of motivic cohomology and homology in terms of $A_{*,*}(Y,X)$.

The development of our theory requires machinery developed recently by the second author, partially in joint work with A. Suslin. As we incorporate the functoriality of cycles [22], pretheories [23], sheaves for the cdh-topology [22], and homology vanishing theorems [23], we endeavor to prodvide an introduction to these ideas by summarizing relevant results from these papers and by providing complete proofs of modified results which we require.

A quick overview of this paper can be obtained from a glance at the table of contents. To conclude this introduction, we give a somewhat more detailed summary of the contents of the various sections of our paper. Section 2 recalls from [22] the functoriality of the presheaf $z_{equi}(X,r)$ which sends a smooth scheme U of finite type over k to the group of cycles on $X \times U$ equidimensional of relative dimension r over k. For X projective, this is closely related to the functor represented by the Chow monoid $C_r(X)$ of effective r-cycles on X. Despite the proliferation of notation, we find it very convenient to consider the related presheaf $z_{equi}(Y,X,r)$ defined by $z_{equi}(Y,X,r)(U) = z_{equi}(X,r)(Y \times U)$.

Following [21], we consider for a presheaf F the complex of presheaves $\underline{C}_*(F)$, sending a scheme U of finite type over k to $F(U \times \Delta^\bullet)$, and the associated homology presheaves $\underline{h}_i(F)$. In particular, the "naive" bivariant theory $\underline{h}_i(z_{equi}(X,r)(Y)$ (closely related to a construction of [6]) is modified to satisfy "cdh-descent", resulting in the definition (Definition 4.3)

$$A_{r,i}(Y,X) = \mathbf{H}_{cdh}^{-i}(Y,\underline{C}_*(z_{equi}(X,r)_{cdh}).$$

The role of the cdh-topology, more flexible than the Zariski topology, is to permit as coverings surjective maps which arise in resolving the singularities of Y.

As developed by the second author in [23], a pretheory is a presheaf F privided with well-behaved transfer maps. Such a structure has the following remarkable property (cf. Theorem 5.5): if the associated cdh-sheaf F_{cdh} vanishes, then $\underline{C}_*(F)_{Zar} = 0$. This result requires that "k admit resolution of singularities" (a precise definition of what it means is given in Definition 3.4). The importance of this property is seen by the ease with which the difficult property of localization for $A_{*,*}(Y,X)$ with respect to the second argument follows: one applies this vanishing result to the pretheory $coker(i^* : z_{equi}(X,r) \to z_{equi}(X - W,r))$ associated to a closed subscheme $W \subset X$ of X over such a field k.

In Section 6, we recast in functorial language the recent "moving lemma for families" established by the first author and H.B. Lawson in [9], leading to the duality theorems of Section 7. For projective, smooth schemes Y, X over an arbitrary field k, Theorem 7.1 provides isomorphisms of presheaves $\underline{h}_i(z_{equi}(Y,X,r)) \simeq \underline{h}_i(z_{equi}(X \times Y, r + n))$. Theorem 8.2 asserts that if k admits resolution of singularities, then there are canonical isomorphisms

$$A_{r,i}(Y \times U, X) \to A_{r+n,i}(X \times U))$$

for any smooth scheme U of pure dimension n over k. Other desired properties established in Section 8 include homotopy invariance, suspension and cosuspension isomorphisms, and a Gysin exact triangle. We also provide three pairings for our theory: the first is the direct analogue of the pairing that gives operations in Lawson homology [11], the second is a multiplicative pairing inspired by the multiplicative structure in morphic cohomology [7], and the final one is a composition product viewed in the context of Lawson homology as composition of correspondences.

Because of its good properties and its relationship to higher algebraic K-theory, one frequently views the higher Chow groups $CH^{n-r}(X,i)$ as at least one version of "motivic cohomology". Indeed, these higher Chow groups are a theory of "homology with locally compact supports." In the final section, we formulate four theories—cohomology/homology with/without compact supports—in terms of our bivariant cycle cohomology.

2. Presheaves of relative cycles

Let k be a field. For a scheme of finite type Y over k we denote by $Cycl(Y)$ the group of cycles on Y (i.e. the free abelian group generated by closed integral subschemes of Y). For a closed subscheme Z of Y we define a cycle $cycl_Y(Z)$ of Z in Y as a formal linear combination $\sum m_i Z_i$ where Z_i are (reduced) irreducible components of Z and m_i is the length of the local ring $\mathcal{O}_{Z,Z_i}$ (note that in our case this is just the dimension of the finite algebra $\mathcal{O}(Z) \otimes_{\mathcal{O}(Z_i)} F(Z_i)$ over the field $F(Z_i)$ of functions on Z_i).

For a smooth scheme U over k , a scheme of finite type X over k and an integer $r \geq 0$ we denote by $z_{equi}(X,r)(U)$ the subgroup in the group of cycles $Cycl(X \times U)$ generated by closed integral subschemes Z of $X \times U$ which are equidimensional of relative dimension r over U (note that this means in particular that Z dominates a connected component of U). We denote further by $z_{equi}^{eff}(X,r)(U)$ the submonoid of effective cycles in $z_{equi}(X,r)(U)$.

The groups $z_{equi}(X,r)(U)$, $z_{equi}^{eff}(X,r)(U)$ were considered in more general context of schemes of finite type over a Noetherian scheme in [22]. We will recall now some of the results obtained there.

For any morphism $f : U' \to U$ there is a homomorphism $cycl(f)$: $z_{equi}(X,r)(U) \to z_{equi}(X,r)(U')$ such that the following conditions hold:

1. For a composable pair of morphisms $U'' \xrightarrow{g} U' \xrightarrow{f} U$ one has $cycl(fg) = cycl(g)cycl(f)$.

2. For a dominant morphism $f : U' \to U$ one has
$$cycl(f)(\sum n_i Z_i) = \sum n_i cycl_{X \times U'}(Z_i \times_U U')$$
where $cycl_{X \times U'}(Z_i \times_U U')$ is the cycle associated with the closed subscheme $Z_i \times_U U'$ in $X \times U'$.

3. For any morphism $f : U' \to U$ and a closed susbscheme Z of $X \times U$ which is flat and equidimensional of relative dimension r over U one has
$$cycl(f)(cycl_{X \times U}(Z)) = cycl_{X \times U'}(Z \times_U U').$$

One can verify easily that homomorphisms $cycl(f)$ are uniquely determined by the conditions (1)-(3) above. Note that the condition (1) implies that we may consider $z_{equi}(X,r)$ as a presheaf of abelian groups on the category Sm/k of smooth schemes over k.

Since for an effective cycle $\mathcal{Z}$ the cycle $cycl(f)(\mathcal{Z})$ is effective, the submonoids $z_{equi}^{eff}(X,r)(U)$ of effective cycles in $z_{equi}(X,r)(U)$ form a subpresheaf $z_{equi}^{eff}(X,r)$ of $z_{equi}(X,r)(U)$.

Proposition 2.1 *For a cycle $\mathcal{Z}$ in $z_{equi}(X,r)(U)$, let $p_*(\mathcal{Z})$ denote the push-forward of $\mathcal{Z}$ to a cycle on $Y \times U$ for a proper morphism $p : X \to Y$; let $q^*(\mathcal{Z})$ denote the flat pull-back of $\mathcal{Z}$ to a cycle on $W \times U$ for a flat equidimensional morphism $q : W \to X$ of relative dimension n of schemes of finite type over k. Then*

$$p_*(\mathcal{Z}) \in z_{equi}(Y,r)(U) \quad , \quad q^*(\mathcal{Z}) \in z_{equi}(W, r+n)(U).$$

Moreover, for any morphism of smooth schemes $f : U' \to U$ one has

$$cycl(f)(p_*(\mathcal{Z})) = p_*(cycl(f)(\mathcal{Z})) \; , \; cycl(f)(q^*(\mathcal{Z})) = q^*(cycl(f)(\mathcal{Z})).$$

Proof: The assertions for proper push-forward are a special case of [22, Prop. 3.6.2(2)], whereas that for flat pull-back follow from [22, Lemma 3.6.4].

Proposition 2.1 implies that for any proper morphism $p : X \to Y$ of schemes of finite type over k there is a homomorphism of presheaves $p : z_{equi}(X,r) \to z_{equi}(Y,r)$. Clearly for a composable pair of morphisms $X \xrightarrow{p_1} Y \xrightarrow{p_2} Z$ one has $(p_2 p_1)_* = p_{2*} p_{1*}$. Similarly, for a flat equidimensional morphism $q : W \to X$, q^* is a morphism of presheaves $z_{equi}(Y,r) \to z_{equi}(X, r+n)$ and for a composable pair of morphisms $X \xrightarrow{q_1} Y \xrightarrow{q_2} Z$ we have $(q_2 q_1)^* = q_1^* q_2^*$.

Let X be a scheme of finite type over k, U be a smooth equidimensional scheme of finite type over k and V be a smooth scheme over k. Let $\mathcal{Z}$ be an element of $z_{equi}(X,r)(U \times V)$. Considered as a cycle on $X \times U \times V$ it clearly belongs to $z_{equi}(X \times U, r + dim(U))(V)$. By [22, Th. 3.7.3], we have for any morphism $f : V' \to V$ of smooth schemes over k the following equality of cycles on $X \times U \times V'$:

$$cycl(f)(\mathcal{Z}) = cycl(Id_U \times f)(\mathcal{Z}).$$

For any X, U, V with U, V smooth denote by $z_{equi}(U, X, r)(V)$ the group $z_{equi}(X,r)(U \times V)$. Then $z_{equi}(U, X, r)$ is a presheaf on Sm/k which is contravariantly functorial with respect to U. The above equality asserts that we have a canonical embedding of presheaves

$$\mathcal{D} : z_{equi}(U, X, r) \to z_{equi}(X \times U, r + dim(U)).$$

One can verify easily that homomorphisms $\mathcal{D}$ are consistent with covariant (resp. contravariant) functoriality of both presheaves with respect to proper (resp. flat equidimensional) morphisms $X \to X'$.

Consider now the case of a projective scheme $i_X : X \subset \mathbf{P}^m$ over k. For an effective cycle $\mathcal{Z} = \sum n_i Z_i \in z_{equi}^{eff}(X,r)(Spec(k))$ on X we define its degree $deg(\mathcal{Z}) = deg_{i_X}(\mathcal{Z})$ as the sum $\sum n_i deg_{i_X}(Z_i)$ where $deg_{i_X}(Z_i)$ is the degree of the reduced closed subscheme $i_X(Z_i)$ in $\mathbf{P}^m$. The lemma below follows easily from the invariance of degree in flat families.

Lemma 2.2 *Let U be a smooth connected scheme over k and $i_X : X \subset \mathbf{P}^m$ be a projective scheme over k. Let further $\mathcal{Z}$ be an element of $z_{equi}(X,r)(U)$ and*

$$u_1 : Spec(k_1) \to U$$

$$u_2 : Spec(k_2) \to U$$

be two points of U. Then one has

$$deg(cycl(u_1)(\mathcal{Z})) = deg(cycl(u_2)(\mathcal{Z})).$$

where the degree on the right hand side (resp. left hand side) is considered with respect to the closed embedding $i_X \times_{Spec(k)} Spec(k_2)$ (resp. $i_X \times_{Spec(k)} Spec(k_1)$).

In the situation of Lemma 2.2 we will use the notation $deg(\mathcal{Z})$ for the degree of the fibers of $\mathcal{Z}$. Denote by $z_{equi}^{eff}(X,r,d)$ (resp. $z_{equi}^{eff}(X,r,\leq d)(U)$) the subset in $z_{equi}^{eff}(X,r)(U)$ which consists of cycles of degree d (resp. $\leq d$). By Lemma 2.2 $z_{equi}^{eff}(X,r,\leq d)$ and $z_{equi}^{eff}(X,r,d)$ are subpresheaves in $z_{equi}^{eff}(X,d)$.

We may now relate our presheaves $z_{equi}^{eff}(X,r)$ to the presheaves represented by Chow varieties. Let us recall briefly the construction of the Chow variety of effective cycles of dimension r and degree d on a projective scheme $X \subset \mathbf{P}^m$ over a field k.

Let $(\mathbf{P}^m)^{r+1}$ be the be the product of projective spaces which we consider as the space which parametrizes families $(L_0, \ldots, L_r)$ of $r+1$ hyperplanes in $\mathbf{P}^m$. Consider the incidence correspondence

$$\Gamma \subset \mathbf{P}^m \times (\mathbf{P}^m)^{r+1}$$

which consists of pairs of the form $(x, (L_0, \ldots, L_r))$ such that $x \in L_0 \cap \ldots \cap L_r$. One can easily see that the projection $pr_1 : \Gamma \to \mathbf{P}^m$ is smooth. Therefore for any cycle $\mathcal{Z}$ of dimension r on $\mathbf{P}^m$ there is defined a cycle $Chow(\mathcal{Z}) = (pr_2)_*(pr_1)^*(\mathcal{Z})$ on $(\mathbf{P}^m)^{r+1}$. One can verify easily that $Chow(\mathcal{Z})$ is a cycle of codimension 1 and if $deg(\mathcal{Z}) = d$ then $Chow(\mathcal{Z})$ is a cycle of degree $(d, \ldots, d)$. Since effective cycles of codimension 1 and given degree on a product of projective spaces are parametrized by a projective space, this gives us a map

$$z_{equi}^{eff}(\mathbf{P}^m, r, d)(Spec(\bar{k})) \to \mathbf{P}^{N(m,r,d)}(\bar{k}).$$

It is well known in the classical theory of algebraic cycles (see [19] for a very classical approach or [22] for a very modern one) that this map is injective and its image coincides with the set of $\bar{k}$-points of a closed (reduced) subscheme $C_{r,d}(\mathbf{P}^m)$ in $\mathbf{P}^{N(m,r,d)}$. Moreover for any closed subscheme X in $\mathbf{P}^m$ the set of $\bar{k}$-points of $C_{r,d}(\mathbf{P}^m)$ which correspond to cycles with support on X coincides with the set of $\bar{k}$-points of a closed (reduced) subscheme $C_{r,d}(X)$ in $C_{r,d}(\mathbf{P}^m)$ which is called the Chow variety of cycles of degree d and dimension r on X.

Since the construction of the Chow map above only uses the operations of flat pull-back and proper push-forward on cycles, Proposition 2.1 implies that the obvious relative version of this map is well defined and gives a homomorphism of presheaves

$$z_{equi}^{eff}(\mathbf{P}^m, r, d) \to z_{equi}^{eff}((\mathbf{P}^m)^{r+1}, mr + m - 1, (d, \dots, d)).$$

The right hand side presheaf is representable on the category Sm/k by the projective space $\mathbf{P}^{N(m,r,d)}$ and we conclude that for any smooth scheme U over k there is a map

$$z_{equi}^{eff}(X, r, d)(U) \to Hom(U, C_{r,d}(X))$$

and that these maps give us a monomorphism from the presheaf $z_{equi}^{eff}(X, r, d)$ to the presheaf represented by $C_{r,d}(X)$ (this construction was explored in [3] and considered in the more general context of presheaves on the category of normal schemes in [7]).

We let $\mathcal{C}_r(X)$ denote the Chow monoid

$$\mathcal{C}_r(X) \equiv \coprod_{d \geq 0} C_{r,d}(X)$$

of effective r-cycles on X. Up to weak normalization, $\mathcal{C}_r(X)$ does not depend upon the projective embedding $X \subset \mathbf{P}^m$.

Proposition 2.3 *Let X be a projective scheme and U a smooth quasi-projective scheme over k. Then every $\mathcal{Z} \in z_{equi}(X, r)^{eff}(U)$ determines a morphism $f_{\mathcal{Z}} : U \to \mathcal{C}_r(X)$. If $f : U \to \mathcal{C}_r(X)$ is a morphism and if $u \in U$, then the effective cycle $\mathcal{Z}_{f(u)}$ with Chow point $f(u)$ is defined over a field $k(\mathcal{Z}_{f(u)})$ which is a finite radiciel extension of the residue field $k(f(u))$.*

Such a morphism $f : U \to \mathcal{C}_r(X)$ determines $\mathcal{Z}_f \in z_{equi}(X, r)^{eff}(U)$ if and only if $k(\mathcal{Z}_{f(\eta)}) = k(f(\eta))$ for every generic point $\eta \in U$. Finally, for any $\mathcal{Z} \in z_{equi}(X, r)^{eff}(U)$,

$$\mathcal{Z} = \mathcal{Z}_{f_\mathcal{Z}} \quad , \quad f = f_{\mathcal{Z}_f}.$$

Proof: This is proved in [3, 1.4]

In particular, if k is a field of characteristic zero, then Proposition 2.3 implies that effective cycles on $U \times X$ equidimensional over the smooth scheme U of relative dimension r correspond exactly to maps from U to $\mathcal{C}_r(X)$. The following example shows that in arbitrary charcteristic this is no longer true.

Example 2.4 *(cf. [19])* Let $k = F(t_1, t_2)$ where F is an algebraically closed field of characteristic $p > 0$. Consider the 0-cycle $\mathcal{Z}$ on $\mathbf{P}_F^2$ which corresponds to the closed reduced subscheme given by the equations

$$x^p = t_1 z^p$$

$$y^p = t_2 z^p$$

where x, y, z are homogeneous coordinates on $\mathbf{P}^2$. One can verify easily that $Chow(\mathcal{Z})$ is the cycle in $\mathbf{P}_F^2$ (with homogeneous coordinates a, b, c) of the form pD where D is the cycle of the closed subscheme given by the equation $t_1 a^p + t_2 b^p + c^p = 0$. It follows immediately from definition of Chow varieties that D is an F-rational point of $C_{0,p}(\mathbf{P}_F^2)$. On the other hand there is no cycle $\mathcal{W}$ on $\mathbf{P}_F^2$ such that $p\mathcal{W} = \mathcal{Z}$.

Remark 2.5 Another construction of a presheaf of r-cycles on X for a quasi-projective scheme X was given in [6]. We will discuss its relations to our definition at the end of the next section.

3. The cdh-topology

In this section we recall the definition of cdh-topology given in [22]. In order to work with possibly singular schemes, we use maps of the following form which arise in resolution of singularities

Definition 3.1 *Let X be a scheme of finite type over k and $Z \subset X$ be a closed subscheme in X which does not contain generic points of irreducible components of X. An abstract blow-up of X with center in Z is a proper surjective morphism $p : X' \to X$ such that $p^{-1}(X - Z)_{red} \to (X - Z)_{red}$ is an isomorphism.*

Denote by Sch/k the category of schemes of finite type over k. We recall the *Nisnevich topology* introduced by Y. Nisnevich in [16] and its modification the *cdh-topology* (the completely decomposed h-topology) introduced in [22].

Definition 3.2 *The Nisnevich topology is the minimal Grothendieck topology on Sch/k such that the following type of covering is a Nisnevich covering: etale coverings $\{U_i \overset{p_i}{\to} X\}$ such that for any point x of X there is a point x_i on one of the U_i such that $p_i(x_i) = x$ and the morphism $Spec(k(x_i)) \to Spec(k(x))$ is an isomorphism.*

The cdh-topology on Sch/k is the minimal Grothendieck topology on this category such that Nisnevich coverings and the following type of covering are cdh-covering: coverings of the form

$$Y \coprod X_1 \overset{p_Y \coprod i_{X_1}}{\Longrightarrow} X$$

such that p_Y is a proper morphism, i_{X_1} is a closed embedding and the morphism $p_Y^{-1}(X - X_1) \to X - X_1$ is an isomorphism.

Thus, the cdh-topology permits abstract blow-ups as coverings as well as the disjoint union of the embeddings of irreducible components of a reducible scheme.

Note in particular, that for any scheme X the closed embedding $X_{red} \to X$ where X_{red} is a maximal reduced subscheme of X is a cdh-covering. Thus, working with the cdh-topology, we do not see any difference between X and X_{red}. We will often use this fact below without additional reference.

The following elementary result about the cdh-topology will be used in the proof of Theorem 5.5 (which uses Theorem 5.1 and an investigation of the relationship of the cohomological behaviour of sheaves on the Nisnevich topology and sheaves on the cdh-topology).

Lemma 3.3 *Let F be a sheaf in Nisnevich topology on Sch/k such that $F_{cdh} = 0$, X be a scheme of finite type over k and $U \to X$ be a Nisnevich covering of X. Then for any element $a \in F(U)$, there is an abstract blow-up $X' \to X$ with center in $Z \subset X$ such that $dim(Z) < dim(X)$ and the restriction of a to $U \times_X X'$ equals zero.*

Proof: Using the definition of the cdh-topology and the fact that F is assumed to be a Nisnevich sheaf, we conclude for any $a \in F(U)$ that there is some abstract blow-up $p : U' \to U$ with $p^*(a) = 0$. We apply the platification theorem [17] to obtain an abstract blow-up $X' \to X$ with center in a closed subscheme $Z \subset X$ which does not contain generic points of X such that the proper transform $\tilde{U}' \to X'$ of $U' \to X$ with respect to this blow-up is flat over X'. One can easily see that we have $\tilde{U}' = U \times_X X'$, so that $U \times_X X'$ factors through U'.

To prove more about the cdh-topology we will often have to assume that the base field k "admits resolution of singularities". More precisely one says.

Definition 3.4 *Let k be a field. We say that k admits resolution of singularities if the following two conditions hold:*

1. *For any scheme of finite type X over k there is a proper surjective morphism $Y \to X$ such that U is a smooth scheme over k.*

2. *For any smooth scheme X over k and an abstract blow-up $q : X' \to X$ there exists a sequence of blow-ups $p : X_n \to \ldots \to X_1 = X$ with smooth centers such that p factors through q*

The following proposition follows immediately from standard results on resolution of singularities in characteristic zero ([12]).

Proposition 3.5 *Any field of characteristic zero admits resolution of singularities in the sense of Definition 3.4.*

Note that any field which admits resolution of singularities in the sense of Definition 3.4 is a perfect field.

Let $\pi : (Sch/k)_{cdh} \to (Sm/k)_{Zar}$ be the natural morphism of sites. Let further

$$\pi^* : Shv((Sm/k)_{Zar}) \to Shv((Sch/k)_{cdh})$$

be the corresponding functor of inverse image on the categories of sheaves. We will use the following simple lemma.

Lemma 3.6 *Let k be a field which admits resolution of singularities. Then the functor of inverse image*

$$\pi^* : Shv((Sm/k)_{Zar}) \to Shv((Sch/k)_{cdh})$$

is exact.

Proof: Consider the minimal Grothendieck topology t on the category Sm/k such that all cdh-coverings of the form $U' \to U$ with both U' and U being smooth are t-coverings. Clearly we have a morphism of sites

$$\pi_0 : (Sch/k)_{cdh} \to (Sm/k)_t$$

and a decomposition of π^* of the form:

$$Shv((Sm/k)_{Zar}) \to Shv((Sm/k)_t) \overset{\pi_0^*}{\to} Shv((Sch/k)_{cdh})$$

where the first arrow is the functor of associated t-sheaf. Note that this functor is exact. Resolution of singularities implies easily that the functor π_0^* is an equivalence. Thus π^* is exact.

Remark 3.7 The statement of Lemma 3.6 would be false if one considered the Zariski topology on Sch/k instead of the cdh-topology. The problem basically is that a fiber product of smooth schemes is not in general smooth and if we do not allow blow-ups to be coverings in our topology a non-smooth scheme can not be covered by smooth ones. This can also be reformulated as the observation that the inclusion functor $Sm/k \to Sch/k$ gives a morphism of cdh-sites but not a morphism of Zariski sites.

Note that Lemma 3.6 implies in particular that the functor π^* takes sheaves of abelian groups to sheaves of abelian groups. In an abuse of notation, for any presheaf F on Sm/k, we denote by F_{cdh} the sheaf $\pi^*(F_{Zar})$ where F_{Zar} is the sheaf in Zariski topology on Sm/k associated to F.

The results of the rest of this section are not used anywhere in this paper but in the definition of motivic cohomology with compact supports in Section 9. We are going to show that for *any* sheaf of abelian groups (and more generally any complex of abelian groups) F on $(Sch/k)_{cdh}$ one can define cdh-cohomology with compact supports with coefficients in F satisfying all the standard properties.

Let us recall that for any scheme X of finite type over k we denote by $\mathbf{Z}(X)$ the presheaf of abelian groups freely generated by the sheaf of sets represented by X on Sch/k. The universal property of freely generated sheaves implies immediately that for any cdh-sheaf of abelian groups F on Sch/k there are canonical isomorphisms:

$$H^i_{cdh}(X, F) = Ext^i(\mathbf{Z}_{cdh}(X), F)$$

where the groups on the right hand side are Ext-groups in the abelian category of cdh-sheaves on Sch/k. More generally for any complex of sheaves K one has:

$$\mathbf{H}^i_{cdh}(X, K) = Hom(\mathbf{Z}_{cdh}(X), K[i])$$

where groups on the left hand side are hypercohomology groups of X with coefficients in K and groups on the right hand side are morphisms in the derived category of cdh-sheaves on Sch/k.

Note that for any connected scheme U the group $\mathbf{Z}(X)(U)$ can be described as the free abelian group generated by closed subschemes Z in $X \times U$ such that the projection $Z \to U$ is an isomorphism. We denote for any U by $\mathbf{Z}^c(X)(U)$ the free abelian group generated by closed subschemes Z in $X \times U$ such that the projection $Z \to U$ is an open embedding. One can easily see that $\mathbf{Z}^c(X)(-)$ is a presheaf of abelian groups on Sch/k. We call it the presheaf of abelian groups freely generated by X with compact supports. If X is a proper scheme then for any topology t such that disjoint unions are t-coverings we have $\mathbf{Z}_t(X) = \mathbf{Z}_t^c(X)$. The following proposition summarizes elementary properties of sheaves $\mathbf{Z}_{cdh}^c(X)$ in cdh-topology.

Proposition 3.8 *Let k be a field. One has:*

1. *The sheaves $\mathbf{Z}_{cdh}^c(X)$ are covariantly functorial with respect to proper morphisms $X_1 \to X_2$.*

2. *The sheaves $\mathbf{Z}_{cdh}^c(X)$ are contravariantly functorial with respect to open embeddings $X_1 \to X_2$.*

3. *Let $j : U \to X$ be an open embedding and $i : X - U \to X$ be the corresponding closed embedding. Then the following sequence of cdh-sheaves is exact:*

$$0 \to \mathbf{Z}_{cdh}^c(X - U) \xrightarrow{i_*} \mathbf{Z}_{cdh}^c(X) \xrightarrow{j^*} \mathbf{Z}_{cdh}^c(U) \to 0.$$

Proof: The first two statements are obvious. To prove the last one note first that the sequence of presheaves

$$0 \to \mathbf{Z}^c(X - U) \xrightarrow{i_*} \mathbf{Z}^c(X) \xrightarrow{j^*} \mathbf{Z}^c(U)$$

is exact. Since the functor of associated sheaf is exact, we have only to show that j_* is a surjection of cdh-sheaves. Let Y be a scheme of finite type over k and Z be an element of $\mathbf{Z}^c(U)(Y)$. It is sufficient to show that there is a cdh-covering $p : Y' \to Y$ such that $Z \times_Y Y'$ as an element of $\mathbf{Z}^c(U)(Y')$ belongs to the image of the homomorphism $\mathbf{Z}^c(X)(Y') \to \mathbf{Z}^c(U)(Y')$. This follows trivially from the platification theorem [22, Th. 2.2.2], [17] and definition of cdh-topology.

Corollary 3.9 *Let $i : X \to \bar{X}$ be an open embedding of a scheme X over k to a proper scheme $\bar{X}$ over k. Then one has a short exact sequence of the form:*

$$0 \to \mathbf{Z}_{cdh}(\bar{X} - X) \to \mathbf{Z}_{cdh}(\bar{X}) \to \mathbf{Z}_{cdh}^c(X) \to 0.$$

For any complex of cdh-sheaves K we define the hypercohomology groups $\mathbf{H}_c^i(X, K)$ of X with compact supports with coefficients in K as the groups of morphisms $Hom(\mathbf{Z}_{cdh}^c(X), K[i])$ in the derived category of cdh-sheaves on Sch/k. Proposition 3.8 and Corollary 3.9 imply immediately that these groups satisfy standard properties of (hyper-)cohomology with compact supports.

4. Bivariant cycle cohomology

Denote by Δ^n the affine scheme

$$\Delta^n = Spec(k[x_0, \ldots, x_n]/\sum x_i = 1)$$

over k. We consider it as an algebro-geometrical analogue of the n-dimensional simplex. Proceeding exactly as in topological situation one can define boundary and degeneracy morphisms

$$\partial_i^n : \Delta^{n-1} \to \Delta^n$$

$$\sigma_i^n : \Delta^{n+1} \to \Delta^n$$

such that $\Delta^\bullet = (\Delta^n, \partial_i^n, \sigma_i^n)$ is a cosimplicial object in the category Sm/k.

For any presheaf F on Sch/k, the category of schemes of finite type over k, and any such k-scheme U, consider the simplicial set $F(\Delta^\bullet \times U)$. If F is a presheaf of abelian groups this is a simplicial abelian group and we denote by $\underline{C}_*(F)(U)$ the corresponding complex of abelian groups (i.e. $\underline{C}_n(F)(U) = F(\Delta^n \times U)$ and the differencial is given by alternating sums of homomorphisms $F(\partial_i^n \times Id_U)$). Clearly, $\underline{C}_*(F)$ is a complex of presheaves on Sch/k, which we call the *singular simplicial complex* of F. We denote by

$$\underline{h}_i(F) = H^{-i}(\underline{C}_*(F))$$

the cohomology presheaves of $\underline{C}_*(F)$.

If F is a presheaf defined on the full subcategory Sm/k of schemes smooth over k and if U is a smooth scheme, then we shall employ the same notation and terminology for the same constructions applied to F.

The following lemma shows that the presheaves of the form $\underline{h}_i(-)$ are homotopy invariant.

Lemma 4.1 *Let F be a presheaf on Sch/k (respectively, Sm/k). Then for any scheme U of finite type over k (resp., smooth of finite type over k), and any $i \in \mathbf{Z}$ the morphism $\underline{h}_i(F)(U) \to \underline{h}_i(F)(U \times \mathbf{A}^1)$ induced by the projection $U \times \mathbf{A}^1 \to U$ is an isomorphism.*

Proof: Denote by $i_0, i_1 : U \to U \times \mathbf{A}^1$ the closed embeddings $Id_U \times \{0\}$ and $Id_U \times \{1\}$ respectively. Let us show first that the morphisms

$$i_0^* : \underline{h}_i(F)(U \times \mathbf{A}^1) \to \underline{h}_i(F)(U)$$

$$i_1^* : \underline{h}_i(F)(U \times \mathbf{A}^1) \to \underline{h}_i(F)(U)$$

coincide. It is sufficient to prove that the corresponding morphisms of complexes of abelian groups $\underline{C}_*(F)(U \times \mathbf{A}^1) \to \underline{C}_*(F)(U)$ are homotopic.

Define a homomorphism

$$s_n : F(U \times \mathbf{A}^1 \times \Delta^n) \to F(U \times \Delta^{n+1})$$

by the formula

$$s_n = \sum_{i=0}^{n} (-1)^i (Id_U \times \psi_i)^*$$

where $\psi_i : \Delta^{n+1} \to \Delta^n \times \mathbf{A}^1$ is the linear isomorphism taking v_j to $v_j \times 0$ if $j \leq i$ or to $v_{j-1} \times 1$ if $j > i$ (here $v_j = (0, \ldots, 1, \ldots, 0)$ is the j-th vertex of Δ^{n+1} (resp. Δ^n)). A straightforward computation shows that $sd + ds = i_1^* - i_0^*$.

Consider now the morphism

$$Id_U \times \mu : U \times \mathbf{A}^1 \times \mathbf{A}^1 \to U \times \mathbf{A}^1$$

where $\mu : \mathbf{A}^1 \times \mathbf{A}^1 \to \mathbf{A}^1$ is given by multiplication of functions. Applying the previous result to the embeddings

$$i_0, i_1 : U \times \mathbf{A}^1 \to (U \times \mathbf{A}^1) \times \mathbf{A}^1$$

we conclude that the homomorphism

$$\underline{h}_i(F)(U \times \mathbf{A}^1) \to \underline{h}_i(F)(U \times \mathbf{A}^1)$$

induced by the composition

$$U \times \mathbf{A}^1 \overset{pr_1}{\to} U \overset{i_0}{\to} U \times \mathbf{A}^1$$

is the identity homomorphism which implies immediately the assertion of the lemma.

We will use frequently (without explicit mention) the following elementary fact. Let

$$0 \to F \to G \to H \to 0$$

be a short exact sequence of presheaves of abelian groups on either Sch/k or Sm/k. Then the corresponding sequence of complexes of presheaves

$$0 \to \underline{C}_*(F) \to \underline{C}_*(G) \to \underline{C}_*(H) \to 0$$

is exact. In particular there is a long exact sequence of presheaves of the form

$$\ldots \to \underline{h}_i(F) \to \underline{h}_i(G) \to \underline{h}_i(H) \to \underline{h}_{i-1}(F) \to \ldots.$$

For any smooth scheme U over k and a scheme of finite type X over k consider the abelian groups $\underline{h}_i(z_{equi}(X,r))(U)$. These groups are contravariantly functorial with respect to U and covariantly (resp. contravariantly) functorial with respect to proper morphisms (resp. flat equidimensional morphisms) in X. The following theorem summarizes most of the known results which relate the groups $\underline{h}_i(z_{equi}(X,r))(Spec(k))$ to other theories.

Theorem 4.2 *Let k be a field.*

1. *For any scheme X over k the group $\underline{h}_0(z_{equi}(X,r)(Spec(k))$ is canonically isomorphic to the group $A_r(X)$ of cycles of dimension r on X modulo rational equivalence.*

2. *For any equidimensional affine scheme X over k and any $r \geq 0$ there are canonical isomorphisms*

$$\underline{h}_i(z_{equi}(X,r))(Spec(k)) \to CH^{dim(X)-r}(X,i)$$

    where the groups on the right hand side are higher Chow groups of X defined in [1].

3. *Assume k is of characteristic 0 and X is a normal equidimensional scheme of pure dimension n or that k is a perfect field and that X is a normal affine scheme of pure dimension n. Then the groups $\underline{h}_i(z_{equi}(X,n-1))(Spec(k))$ are of the form:*

$$\underline{h}_i(z_{equi}(X,n-1))(Spec(k)) = \begin{cases} A_{n-1}(X) & \text{for } i=0 \\ \mathcal{O}^*(X) & \text{for } i=1 \\ 0 & \text{for } i \neq 0,1 \end{cases}$$

4. *The groups $\underline{h}_n(z_{equi}(\mathbf{A}^n,0))(Spec(k))$ are isomorphic to the Milnor K-groups $K_n^M(k)$.*

5. *If k is an algebraically closed field which admits resolution of singularities, X is a smooth scheme over k of dimension m, and $n \neq 0$ is an integer prime to $char(k)$, the groups*

$$\underline{h}_i(z_{equi}(X,0) \otimes \mathbf{Z}/n\mathbf{Z})(Spec(k))$$

    are isomorphic to the etale cohomology groups $H_{et}^{2m-i}(X,\mathbf{Z}/n\mathbf{Z})$.

6. If $k = \mathbf{C}$, then for any quasi-projective variety X (i.e., reduced, irreducible $\mathbf{C}$-scheme) the groups

$$\underline{h}_i(z_{equi}(X,r) \otimes \mathbf{Z}/n\mathbf{Z})(Spec(k))$$

are isomorphic to the corresponding Lawson homology with finite coefficients.

Proof:

1. Elementary (see for instance [4]).

2. Note that for any equidimensional scheme X over k and any r the complex $\underline{C}_*(z_{equi}(X,r))(Spec(k))$ which computes the groups $\underline{h}_i(z_{equi}(X,r)(Spec(k))$ can be considered as a subcomplex in the Bloch complex $Z^{dim(X)-r}(X,*)$ which computes the higher Chow groups of X (see [1],[6]). The fact that this morphism is a quasi-isomorphism for affine schemes X was proven by A. Suslin (see [20])[1].

3. In [15], $CH^1(X,i)$ was computed for X a normal scheme of pure dimension n over a perfect field thus, the assertion for X affine and normal follows from (2). In [4], $z_{equi}(X,n-1)_h$ was computed for X arbitrary and the computation agrees with that of assertion (3) for X normal; by Remarks 4.6, 5.10, this implies assertion (3) provided k has characteristic 0.

4. This follows from part (2), the homotopy invariance property for higher Chow groups, and [1][2].

5. This was proven in [21].

6. This was proven in [21] for X projective and the extended to quasi-projective X in [4].

One would like to consider $\underline{h}_i(Z(X,r))(U)$ as a natural bivariant generalization of groups $A_r(X)$ (and more generally of the higher Chow groups of X). In fact, we shall consider a somewhat more sophisticated construction (cf. Definition 4.3) in order to enable the Mayer-Vietoris property with respect to the first argument of our bivariant theory.

---

[1] The proof is based on an elementary moving technique and does not use any of the advanced properties of either higher Chow groups or groups $\underline{h}_i(z_{equi}(X,r))(-)$.

[2] Another way to prove this (in the case of a perfect field k) is to use Mayer-Vietoris sequence in Suslin homology proven in [23] together with a direct computation of Suslin homology of $(\mathbf{A}^1 - \{0\})^n$.

Note that functoriality of the groups $\underline{h}_i(Z(X,r))(U)$ shows that they behave as a cohomology theory with respect to the first argument (U) and as a Borel-Moore homology theory with respect to the second (X). Thus, one would expect that other properties of these classes of theories should hold for the groups $\underline{h}_i(z_{equi}(X,r))(U)$. In particular, there should exist a localization long exact sequence with respect to X and a Mayer-Vietoris long exact sequence with respect to U. We will show below that this is indeed true for quasi-projective schemes U if k admits resolution of singularities. It turns out though that it is inconvenient to work with the groups $\underline{h}_i(z_{equi}(X,r))(U)$ directly. Instead we define for all schemes of finite type Y, X over k the *bivariant cycle cohomology* groups $A_{r,i}(Y,X)$ of Y with coefficients in r-cycles on X as certain hypercohomology groups of Y which automatically gives us most of the good properties with respect to Y. We will show then that for a field k which admits resolution of singularities and for a smooth quasi-projective k-scheme U the groups $A_{r,i}(U,X)$ are canonically isomorphic to the groups $\underline{h}_i(z_{equi}(X,r)(U))$.

Definition 4.3 *Let X, Y be schemes of finite type over a field k and $r \geq 0$ be an integer. The bivariant cycle cohomology groups of Y with coefficients in cycles on X are the groups*

$$A_{r,i}(Y,X) = \mathbf{H}_{cdh}^{-i}(Y, (\underline{C}_*(z_{equi}(X,r))_{cdh}).$$

We will also use the notation $A_{r,i}(X)$ for the groups $A_{r,i}(Spec(k), X)$.

It follows immediately from this definition that the groups $A_{r,i}(Y,X)$ are contravariantly functorial with respect to Y and covariantly (resp. contravariantly) functorial with respect to proper morphisms (resp. flat equidimensional morphisms) in X. It is also clear that the groups $A_{r,i}(X)$ coincide with the groups $\underline{h}_i(z_{equi}(X,r))(Spec(k))$. In particular, Theorem 4.2(2) implies that for affine equidimensional schemes X the groups $A_{r,i}(X)$ are isomorphic to the corresponding higher Chow groups.

Since open coverings are cdh-coverings the following proposition is a trivial corollary of our definition.

Proposition 4.4 *Let X, Y be schemes of finite type over a field k. Let further $Y = U_1 \cup U_2$ be a Zariski open covering of Y. Then for any $r \geq 0$ there is a canonical long exact sequence of the form:*

$$\ldots \to A_{r,i}(Y,X) \to A_{r,i}(U_1,X) \oplus A_{r,i}(U_2,X) \to A_{r,i}(U_1 \cap U_2, X) \to$$

$$\to A_{r,i-1}(Y,X) \to \ldots .$$

Since abstract blow-ups are cdh-coverings (Definition 3.2) we immediately obtain the following blow-up exact sequence with respect to the first argument.

Proposition 4.5 *Let $p : Y' \to Y$ be a proper morphism of schemes of finite type over k and $Z \subset Y$ be a closed subscheme of Y such that the morphism $p^{-1}(Y - Z) \to Y - Z$ is an isomorphism. Then for any scheme of finite type X over k and any $r \geq 0$ there is a canonical long exact sequence of the form*

$$\dots \to A_{r,i}(Y, X) \to A_{r,i}(Y', X) \oplus A_{r,i}(Z, X) \to A_{r,i}(p^{-1}(Z), X) \to$$

$$\to A_{r,i-1}(Y, X) \to \dots.$$

Remark 4.6 The first definition of sheaves of relative cycles similar to $z_{equi}(X, r)_{cdh}$ was given in [6] together with a definition of the corresponding "naive" bivariant cycle cohomology groups. As was shown in [4] the presheaves defined in [6] are isomorphic to the sheaves $z_{equi}(X, r)_h$ in the h-topology (see [25] or [21]) on Sch/k associated with $z_{equi}(X, r)_{cdh}$.

The cdh-sheaves $z_{equi}(X, r)_{cdh}$ were considered in [22] where they are denoted by $z(X, r)$ (see [22, Th. 4.2.9(2)]). As was shown in [22], the canonical morphism

$$z_{equi}(X, r)_{cdh} \to z_{equi}(X, r)_h$$

is a monomorphism and it becomes an isomorphism after tensoring with $\mathbb{Z}[1/p]$ where p is the exponential characteristic of k. In particular for a quasi-projective scheme X over a field of characteristic zero, the sheaves $z_{equi}(X, r)_{cdh}$ are isomorphic to the presheaves constructed in [6].

One can show easily (we will not use this fact in the paper) that for any scheme of finite type X over k the restriction of the cdh-sheaf $z_{equi}^{eff}(X, r)_{cdh}$ to Sm/k coincides with $z_{equi}^{eff}(X, r)$. On the other hand, the corresponding statement for $z_{equi}(X, r)$ is false: the natural homomorphism $z_{equi}(X, r)(V) \to z_{equi}(X, r)_{cdh}(V)$ is a monomorphism but not in general an epimorphism. Nevertheless, it turns out that this difference is insignificant from the point of view of bivariant cycle cohomology. More precisely, it will be shown in the next section (see remark 5.10) that for any scheme of finite type X over k the groups $\underline{h}_i(z_{equi}(X, r))(Spec(k))$ and the groups $\underline{h}_i(z_{equi}(X, r)_{cdh})(Spec(k))$ are isomorphic.

5. Pretheories

One of our main technical tools is the "theory of pretheories" developed in [23]. Informally speaking, a pretheory is a presheaf of abelian groups on Sm/k which has transfers with respect to finite coverings (the precise definition is given below). We will show below how to define a natural structure of pretheory on presheaves $z_{equi}(X, r)$. The main reason why this class of presheaves is important for us is that the singular simplicial complexes $\underline{C}_*(F)$ of pretheories F behave "nicely"; in particular we obtain the localization theorem (5.11) in our theory.

Let U be a smooth scheme over k and $C \to U$ be a smooth curve over U. Denote by $c_{equi}(C/U, 0)$ the free abelian group generated by integral closed subschemes in C which are finite over U and dominant over an irreducible component of U. Any such subscheme is flat over U. In particular for a morphism $f : U' \to U$ there is defined a homomorphism $cycl(f) : c_{equi}(C/U, 0) \to c_{equi}(C \times_U U'/U', 0)$. For any section $s : U \to C$ of the projection $C \to U$ its image is an element in $c_{equi}(C/U, 0)$ which we denote by $[s]$. A *pretheory* (with values in the category of abelian groups) on the category Sm/k is a presheaf F of abelian groups together with a family of homomorphisms $\phi_{C/U} : c_{equi}(C/U, 0) \to Hom(F(C), F(U))$ given for all smooth curves $C \to U$ over smooth schemes over k and satisfying the following conditions:

1. For any smooth schemes U_1, U_2 over k the canonical homomorphism $F(U_1 \coprod U_2) \to F(U_1) \oplus F(U_2)$ is an isomorphism.

2. For a section $s : U \to C$ one has $\phi_{C/U}([s]) = F(s)$.

3. For a morphism $f : U' \to U$, an element $a \in F(C)$ and an element $\mathcal{Z}$ in $c_{equi}(C/U, 0)$ one has

$$F(f)(\phi_{C/U}(\mathcal{Z})(a)) = \phi_{C \times_U U'/U'}(cycl(f))(\mathcal{Z})(F(f \times_U C)(a)).$$

A morphism of pretheories is a morphism of presheaves which is consistent with the structures of pretheories in the obvious way.

Note that the category of pretheories is abelian and the forgetful functor from this category to the category of presheaves of abelian groups is exact. In particular for any pretheory F on Sm/k the cohomology presheaves $\underline{h}_*(F)$ are pretheories.

A pretheory F is called *homotopy invariant* if for any smooth scheme U over k the homomorphism $F(U) \to F(U \times \mathbf{A}^1)$ induced by the projection $U \times \mathbf{A}^1 \to U$ is an isomorphism. Note that by Lemma 4.1 for any pretheory F the pretheories $\underline{h}_i(F)$ are homotopy invariant.

Theorem 5.1 *Let k be a perfect field and F be a homotopy invariant pretheory on Sm/k. Then one has:*

1. *For any $i \geq 0$ the presheaf $H^i_{Zar}(-, F_{Zar})$ has a canonical structure of a homotopy invariant pretheory. In particular F_{Zar} is a homotopy invariant pretheory and for any smooth scheme U over k the homomorphisms*

$$H^i_{Zar}(U, F_{Zar}) \to H^i_{Zar}(U \times \mathbf{A}^1, F_{Zar})$$

are isomorphisms.

2. *For any smooth scheme U over k and any $i \geq 0$ the canonical homomorphism*

$$H^i_{Zar}(U, F_{Zar}) \to H^i_{Nis}(U, F_{Nis})$$

is an isomorphism. In particular $F_{Zar} = F_{Nis}$.

Proof: See [23, Th. 4.27] for the first part and [23, Th. 5.7] for the second.

This theorem implies easily the following important criterion of vanishing for sheaves $\underline{h}_i(F)_{Zar}$ if F is a pretheory.

Proposition 5.2 *Let k be a perfect field and F be a pretheory on Sm/k. Then the following conditions are equivalent.*

1. *The sheaves $\underline{h}_i(F)_{Zar}$ are zero for all $i \leq n$.*

2. *For any homotopy invariant pretheory G and any $i < n$, one has*

$$Ext^i(F_{Nis}, G_{Nis}) = 0$$

(here $Ext^i(-, -)$ is the Ext-group in the category of Nisnevich sheaves on Sm/k).

Proof: See [23, Th. 5.9].

The key to our localization theorem for bivariant cycle homology is the following statement (cf Theorem 5.5): if F is a pretheory such that $F_{cdh} = 0$ and if k admits resolution of singularities, then $\underline{h}_i(F)_{Zar} = 0$. We observe that if we the weaken the condition $F_{cdh} = 0$ to $F_{Nis} = 0$, then this follows immediately from Proposition 5.2. To obtain the stronger result we need, we first require the following two lemmas.

Let us recall that for a smooth scheme U over k we denote by $\mathbf{Z}(U)$ the presheaf of abelian groups on Sm/k freely generated by the presheaf of sets represented by U. We will use the following lemma.

Lemma 5.3 *Let k be a perfect field, U be a smooth scheme over k and $Z \subset U$ be a smooth closed subscheme of U. Denote by $p : U_Z \to U$ the blow-up of U with center in Z. Then for any homotopy invariant pretheory G on Sm/k and any $i \geq 0$ one has:*

$$Ext^i(coker(\mathbf{Z}(U_Z) \to \mathbf{Z}(U))_{Nis}, G_{Nis}) = 0.$$

Proof: See [23, Prop. 5.21].

The main technical ingredient for the proof of Theorem 5.5 is isolated in the following lemma.

Lemma 5.4 *Let k be a field which admits resolution of singularities and let F be a Nisnevich sheaf of abelian groups on Sm/k such that $F_{cdh} = 0$. Then for any homotopy invariant pretheory G and any $i \geq 0$ one has*

$$Ext^i(F, G_{Nis}) = 0.$$

Proof: In view of Theorem 5.1 we may assume that $G_{Nis} = G$, i.e. that G is a sheaf in the Nisnevich topology. We use induction on i. Since our statement is trivial for $i < 0$ we may assume that for any sheaf F satisfying the condition of the proposition and any $j < i$ one has $Ext^j(F, G) = 0$.

Let U be a smooth scheme over k and $p : U' \to U$ be a morphism which is a composition of n blow-ups with smooth centers. Let us show that $Ext^i(coker(\mathbf{Z}(p))_{Nis}, G) = 0$. Let $p = p_0 \circ p_1$ where p_0 is a blow-up with a smooth center and p_1 is a composition of $n - 1$ blow-ups with smooth centers. By induction on n and Lemma 5.3 we may assume that $Ext^i(coker(\mathbf{Z}(p_\epsilon))_{Nis}, G) = 0$ for $\epsilon = 0, 1$. We have an exact sequence of presheaves

$$0 \to \Psi \to coker(\mathbf{Z}(p_1)) \to coker(\mathbf{Z}(p)) \to coker(\mathbf{Z}(p_0)) \to 0.$$

We rewrite this exact sequence as two short exact sequences

$$0 \to \Psi \to coker(\mathbf{Z}(p_1)) \to \Psi' \to 0$$

$$0 \to \Psi' \to coker(\mathbf{Z}(p)) \to coker(\mathbf{Z}(p_0)) \to 0$$

Since $(\Psi)_{cdh} = 0$ the long exact sequences of Ext-groups associated with two short exact sequences above together with the induction assumption and Lemma 5.3 imply now that $Ext^i(coker(\mathbf{Z}(p))_{Nis}, G) = 0$.

Let now F be any Nisnevich sheaf such that $F_{cdh} = 0$. Consider the epimorphism

$$\oplus \phi_\alpha : \oplus \mathbf{Z}(U_\alpha) \to F$$

where the sum is taken over all pairs of the form (U_α, ϕ_α) where U_α is a smooth scheme over k and $\phi_\alpha \in F(U_\alpha)$. It follows from resolution of singularities and Lemma 3.3 that for any smooth scheme U_α over k and any section $\phi_\alpha \in F(U_\alpha)$ there is a sequenece of blow-ups with smooth centers $p_\alpha : U_\alpha' \to U_\alpha$ such that $p_\alpha^*(\phi_\alpha) = 0$. Thus our epimorphism can be factorized through an epimorphism $\Psi \to F$ where $\Psi = \oplus_\alpha coker(\mathbf{Z}(p_\alpha))$. Let Ψ_0 be the kernel of this epimorphism. Then $(\Psi_0)_{cdh} = 0$. Since we have already proven that $Ext^i(\Psi_{Nis}, G) = 0$ we conclude that $Ext^i(F_{Nis}, G) = 0$ from the induction assumption and the long exact sequence of Ext-groups associated with the short exact sequence of presheaves

$$0 \to \Psi_0 \to \Psi \to F \to 0.$$

Theorem 5.5 *Let k be a field which admits resolution of singularities and F be a pretheory over k.*

1. *For any smooth scheme U over k, one has canonical isomorphisms*

$$\mathbf{H}^i_{cdh}(U, \underline{C}_*(F)_{cdh}) = \mathbf{H}^i_{Zar}(U, \underline{C}_*(F)_{Zar}).$$

2. *If $F_{cdh} = 0$, then $\underline{C}_*(F)_{Zar}$ is acyclic.*

3. *For any scheme of finite type X over k, the projection $X \times \mathbf{A}^1 \to X$ induces isomorphisms*

$$\mathbf{H}^i_{cdh}(X, \underline{C}_*(F)_{cdh}) \to \mathbf{H}^i_{cdh}(X \times \mathbf{A}^1, \underline{C}_*(F)_{cdh}).$$

Proof: To prove part (1) of the theorem it is sufficient in view of hyper-cohomology spectral sequence and Theorem 5.1(2) to show that for any homotopy invariant pretheory G one has:

$$H^i_{cdh}(U, G_{cdh}) = H^i_{Nis}(U, G_{Nis})$$

(we then apply it to $G = \underline{h}_n(F)$).

Let $\mathcal{U}$ be a cdh-hypercovering of U. We say that it is smooth if all schemes U_i are smooth. Resolution of singularities guarantees that any cdh-hypercovering of U has a smooth refiniment. For any smooth hyper-covering $\mathcal{U}$ denote by $\mathbf{Z}(\mathcal{U})$ the complex of presheaves on Sm/k with terms $\mathbf{Z}(U_i)$ and differentials given by alternating sums of morphisms induced by boundary morphisms of the simplicial scheme $\mathcal{U}$. The standard description of cohomology in terms of hypercoverings gives us for any presheaf G a canonical isomorphism

$$H^i_{cdh}(U, G_{cdh}) = \varinjlim_{\mathcal{U}} (Hom(\mathbf{Z}(\mathcal{U})_{Nis}, G_{Nis}[i]))$$

where the limit on the right hand side is taken over all smooth cdh-hyper-coverings of U and $Hom(-,-)$ refers to morphisms in the derived category of the category of sheaves of abelian groups on $(Sm/k)_{Nis}$ (note that we could replace in this isomorphism Nis by any topology which is weaker then the cdh-toplogy). There is a canonical morphism of complexes $\mathbf{Z}(\mathcal{U}) \to \mathbf{Z}(U)$. Denote its cone by $K_{\mathcal{U}}$. By definition of a hypecovering the complex of cdh-sheaves $(K_{\mathcal{U}})_{cdh}$ is quasi-isomorphic to zero, i.e. for any $j \in \mathbf{Z}$ one has $\underline{H}^j(K_{\mathcal{U}})_{cdh} = 0$ where $\underline{H}^j(K_{\mathcal{U}})$ are cohomology presheaves of $K_{\mathcal{U}}$. Thus by Lemma 5.4 for a homotopy invariant pretheory G the homomorphisms

$$Hom(\mathbf{Z}(U)_{Nis}, G_{Nis}[i]) \to Hom(\mathbf{Z}(\mathcal{U})_{Nis}, G_{Nis}[i])$$

are isomorphisms for all $i \in \mathbf{Z}$. Since the left hand side groups are canonically isomorphic to $H^i_{Nis}(U, G_{Nis})$ this proves part (1) of the theorem.

Let now F be a pretheory such that $F_{cdh} = 0$. We want to show that $\underline{C}_*(F)_{Zar}$ is quasi-isomorphic to zero. By Theorem 5.1(2) it is sufficient to show that $\underline{C}_*(F)_{Nis}$ is quasi-isomiorphic to zero. Suppose that $\underline{h}_n(F)_{Nis} \neq 0$ for some n. We may assume that $\underline{h}_i(F)_{Nis} = 0$ for all $i < n$. Then there is a non trivial morphism in the derived category $\mathcal{D}(Sm/k)_{Nis})$ of sheaves on $(Sm/k)_{Nis}$ of the form $\underline{C}_*(F)_{Nis} \to \underline{h}_n(F)_{Nis}[n]$. The second part of our theorem follows now from 5.4 and lemma below.

Lemma 5.6 *Let F be a presheaf on Sm/k and G be a homotopy invariant pretheory on Sm/k. Then for any $n \in \mathbf{Z}$ one has a canonical isomorphism*

$$Hom_{\mathcal{D}(Sm/k)_{Nis}}(\underline{C}_*(F)_{Nis}, G_{Nis}[n]) = Ext^n(F_{Nis}, G_{Nis}).$$

Proof: See proof of [23, Th. 5.9]. $\blacksquare$

To prove the third part of our theorem, we use smooth hypercoverings of X as in the proof of part (1) and apply part (1) and Theorem 5.1(1).

Let X be a scheme of finite type over k. We define a structure of pretheory on the presheaf $z_{equi}(X, r)$ as follows. Let $C \to U$ be a smooth curve over a smooth scheme over k, Z be an integral closed subscheme of C which belongs to $c_{equi}(C/U, 0)$ and $\mathcal{W}$ be an element of $z_{equi}(X, r)(C)$. Denote by z the generic point of Z and consider the specialization $\mathcal{W}_z$. Namely, let $\mathcal{O}_{C,z}$ denote the discrete valuation ring obtained by localizing C at z and let $k(z)$ denote its residue field, equal to the field of fractions of Z. Then the restriction $W \times_C Spec(\mathcal{O}_{C,z})$ of any irreducible component $W \subset X \times C$ of $\mathcal{W}$ is flat over $Spec(\mathcal{O}_{C,z})$ and $\mathcal{W}_z$ is defined to be the sum over the irreducible components W of the cycles associated to $W \times_C Spec(k(z))$.

We set $\phi_{C/U}(Z)(W)$ to be the push-forward of the closure of W_z in $Z \times X$ to $U \times X$.

Proposition 5.7 *Let X be a scheme of finite type over k, U be a smooth scheme over k and C/U be a smooth curve over U. Then one has:*

1. *For any element W in $z_{equi}(X,r)(C)$ the cycle $\phi_{C/U}(Z)(W)$ defined above belongs to $z_{equi}(X,r)(U)$.*

2. *The presheaf $z_{equi}(X,r)$ together with homomorphisms $\phi_{C/U}$ is a pretheory.*

Proof: Our homomorphism

$$\phi_{C/U} : z_{equi}(X,r)(C) \otimes c_{equi}(C/U,0) \to Cycl(U \times X)$$

is a particular case of correspondence homomorphisms considered in [22, Section 3.7]. The first statement of the proposition is a particular case of the first statement of [22, Th. 3.7.3].

First two conditions of the definition of pretheory are hold for trivial reasons. The third one is a particular case of the second statement of [22, Th. 3.7.3].

Proposition 5.8 1. *For any proper morphism $p : X' \to X$ the push-forward homomorphism of presheaves*

$$p_* : z_{equi}(X',r) \to z_{equi}(X,r)$$

is a morphism of pretheories.

2. *For any flat equidimensional morphism $f : X' \to X$ the pull-back homomorphism*

$$f^* : z_{equi}(X,r) \to z_{equi}(X',r+dim(X'/X))$$

is a homomorphism of pretheories.

3. *For any smooth equidimensional scheme U and any scheme of finite type X over k the duality homomorphisms*

$$\mathcal{D} : z_{equi}(U,X,r) \to z_{equi}(U \times X, r + dim(U))$$

are morphisms of pretheories.

Proof: The second and the third assertions are trivial. The first one is a particular case of [22, Prop. 3.7.6].

Proposition 5.9 *Let k be a field which admits resolution of singularities and let X be a scheme of finite type over k. Then for any scheme Y over k the homomorphisms*

$$A_{r,i}(Y, X) \to A_{r,i}(Y \times \mathbf{A}^1, X)$$

induced by the projection are isomorphisms.

Proof: This is a particular case of Theorem 5.5(3) with $F = z_{equi}(X, r)$.

Remark 5.10 It is easy to show that for any scheme of finite type X over k the restriction of the sheaf $z_{equi}(X, r)_{cdh}$ to Sm/k has a unique structure of pretheory such that the morphism of presheaves on Sm/k

$$z_{equi}(X, r) \to z_{equi}(X, r)_{cdh}$$

is a morphism of pretheories. Applying Theorem 5.5(2) to kernel and cokernel of this morphism we conclude that

$$\underline{h}_i(z_{equi}(X, r))(Spec(k)) = \underline{h}_i(z_{equi}(X, r)_{cdh})(Spec(k)).$$

Observe that Theorem 5.5(2) implies that any sequence of pretheories

$$0 \to F \to G \to H \to 0$$

such that the corresponding sequence of cdh-sheaves

$$0 \to F_{cdh} \to G_{cdh} \to H_{cdh} \to 0$$

is exact gives us an exact triangle of complexes of sheaves in Zariski topology of the form

$$\underline{C}_*(F)_{Zar} \to \underline{C}_*(G)_{Zar} \to \underline{C}_*(H)_{Zar} \to \underline{C}_*(F)_{Zar}[1]$$

and in particular a long exact sequence of the corresponding groups $\underline{h}_i(-)(Spec(k))$.

By Lemma 3.6 we also have in this case an exact triangle

$$\underline{C}_*(F)_{cdh} \to \underline{C}_*(G)_{cdh} \to \underline{C}_*(H)_{cdh} \to \underline{C}_*(F)_{cdh}[1]$$

and hence a long exact sequence of the corresponding hypercohomology groups.

We now prove localization and Mayer-Vietoris in our theory.

Theorem 5.11 *Let k be a field which admits resolution of singularities, let X be a scheme of finite type over k, let $Y \subset X$ be a closed subscheme of X, and let $U_1, U_2 \subset X$ be Zariski open subsets with $X = U_1 \cup U_2$. Then there are canonical exact triangles (in the derived category of complexes of sheaves on $(Sm/k)_{Zar}$) of the form*

$$\underline{C}_*(z_{equi}(Y,r))_{Zar} \to \underline{C}_*(z_{equi}(X,r))_{Zar} \to \underline{C}_*(z_{equi}(X-Y,r))_{Zar} \to$$

$$\to \underline{C}_*(z_{equi}(Y,r))_{Zar}[1] \qquad (4.10.1)$$

and

$$\underline{C}_*(z_{equi}(X,r))_{Zar} \to \underline{C}_*(z_{equi}(U_1,r))_{Zar} \oplus \underline{C}_*(z_{equi}(U_2,r))_{Zar} \to$$

$$\to \underline{C}_*(z_{equi}(U_1 \cap U_2,r))_{Zar} \to \underline{C}_*(z_{equi}(X,r))_{Zar}[1]. \qquad (4.10.2)$$

Proof: The sequences of presheaves

$$0 \to z_{equi}(Y,r) \to z_{equi}(X,r) \to z_{equi}(X-Y,r)$$

$$0 \to z_{equi}(X,r) \to z_{equi}(U_1,r) \oplus z_{equi}(U_2,r) \to z_{equi}(U_1 \cap U_2,r)$$

on Sm/k are exact for obvious reasons. Hence by Theorem 5.5(2), it suffices to show that

$$coker(i^* : z_{equi}(X,r) \to z_{equi}(X-Y,r))_{cdh} = 0$$

$$coker(j_1^* - j_2^* : z_{equi}(U_1,r) \oplus z_{equi}(U_2,r) \to z_{equi}(U_1 \cap U_2,r))_{cdh} = 0.$$

In view of definition of cdh-topology, to verify the first asserted vanishing it suffices to show for any smooth scheme U over k and any element $\mathcal{Z}$ in $z_{equi}(X-Y,r)(U)$ that there is a blow-up $p : U' \to U$ such that U' is smooth and $cycl(p)(\mathcal{Z})$ lies in the image of $z_{equi}(X,r)(U')$. We may clearly assume that $\mathcal{Z} = Z$ for a closed integral subscheme Z of $(X-Y) \times U$. Let $\bar{Z}$ be the closure of Z in $X \times U$. By the platification theorem [17], there is a blow-up $p : U' \to U$ such that the proper transform $\tilde{Z}$ of $\bar{Z}$ with respect to p is flat over U'. Since k admits resolution of singularities we may choose U' to be smooth. Then $\tilde{Z} \in z_{equi}(X,r)(U')$ and clearly its image in $z_{equi}(X-Y,r)(U')$ coincides with $cycl(p)(Z)$.

The proof of the second asserted vanishing differs only in notation.

The following long exact sequence of bivariant cycle cohomology groups is an immediate corollary of Theorem 5.11 together with Lemma 3.6.

Corollary 5.12 *Let k be a field which admits resolution of singularities, let X be a scheme of finite type over k, let $Y \subset X$ be a closed subscheme of X, and let $U_1, U_2 \subset X$ be Zariski open subsets with $X = U_1 \cup U_2$. Then for any scheme U over k, there are canonical long exact sequences*

$$\ldots \to A_{r,i}(U,Y) \to A_{r,i}(U,X) \to A_{r,i}(U, X - Y) \to A_{r,i-1}(U,Y) \to \ldots$$

$$\ldots \to A_{r,i}(Y,X) \to A_{r,i}(Y,U_1) \oplus A_{r,i}(Y,U_2) \to A_{r,i}(Y, U_1 \cap U_2) \to$$

$$\to A_{r,i-1}(Y,X) \to \ldots.$$

The following theorem provides us with another class of long exact sequences in our theory.

Theorem 5.13 *Let $p: X' \to X$ be a proper morphism of schemes of finite type over field k which admits resolution of singularities and let and $Z \subset X$ be a closed subscheme of X such that the morphism $p^{-1}(X - Z) \to X - Z$ is an isomorphism. Then there is a canonical exact triangle (in the derived category of complexes of sheaves on $(Sm/k)_{Zar}$):*

$$\underline{C}_*(z_{equi}(p^{-1}(Z),r))_{Zar} \to \underline{C}_*(z_{equi}(Z,r))_{Zar} \oplus \underline{C}_*(z_{equi}(X',r))_{Zar} \to$$

$$\to \underline{C}_*(z_{equi}(X,r))_{Zar} \to \underline{C}_*(z_{equi}(p^{-1}(Z),r))_{Zar}[1].$$

Proof: The proof is exactly the same as for Theorem 5.11, except that we need the vanishing assertion

$$coker(z_{equi}(Z,r) \oplus z_{equi}(X',r) \to z_{equi}(X,r))_{cdh} = 0.$$

This is proved exactly as in the proof of the vanishing assertions of the proof of Theorem 5.11.

Corollary 5.14 *Let $p: X' \to X$ be a proper morphism of schemes of finite type over a field k which admits resolution of singularities and let $Z \subset X$ be a closed subscheme of X such that the morphism $p^{-1}(X - Z) \to X - Z$ is an isomorphism. Then for any scheme U over k there is a canonical long exact sequence of the form:*

$$\ldots \to A_{r,i}(U, p^{-1}(Z)) \to A_{r,i}(U,Z) \oplus A_{r,i}(U,X') \to A_{r,i}(U,X) \to$$

$$\to A_{r,i-1}(U, p^{-1}(Z)) \to \ldots$$

6. The moving Lemma

In this section we return to the "naive" groups $\underline{h}_i(z_{equi}(X,r))(U)$. We study them in the case of smooth varieties X, U using the "moving lemma" techniques developed in [9]. These techniques are summarized in Theorems 6.1 and 6.3 below. It is worthy of note that the results in this section apply to varieties over an arbitrary field k.

Let X be an equidimensional smooth scheme of finite type over k. Let further $\mathcal{Z} = \sum n_i Z_i$, $\mathcal{W} = \sum m_j W_j$ be two effective cycles on X of dimensions r and s respectively. We say that $\mathcal{Z}$ and $\mathcal{W}$ intersect properly if the schemes $Z_i \cap W_j$ are equidimensional of dimension $r + s - dim(X)$.

We begin with a "presheafication" of the "Moving Lemma for Bounded Families" of cycles on projective space proved in [9].

Theorem 6.1 *Let k be a field and $m, r, s, d, e \geq 0$ be integers such that $r + s \geq m$. Then there are homomorphisms of abelian monoids*

$$H_U^+ : z_{equi}^{eff}(\mathbf{P}^m, r)(U) \to z_{equi}^{eff}(\mathbf{P}^m, r)(U \times \mathbf{A}^1)$$

$$H_U^- : z_{equi}^{eff}(\mathbf{P}^m, r)(U) \to z_{equi}^{eff}(\mathbf{P}^m, r)(U \times \mathbf{A}^1)$$

defined for all smooth schemes U over k and satisfying the following conditions:

1. *For any morphism $f : U' \to U$ of smooth schemes over k, one has*

$$H_{U'}^+ cycl(f) = cycl(f \times Id_{\mathbf{A}^1})H_U^+$$

$$H_{U'}^- cycl(f) = cycl(f \times Id_{\mathbf{A}^1})H_U^-$$

2. *One has*

$$cycl_{i_U} H_U^+ = (n+1)Id_{z_{equi}^{eff}(\mathbf{P}^m,r)(U)}$$

$$cycl_{i_U} H_U^- = nId_{z_{equi}^{eff}(\mathbf{P}^m,r)(U)}$$

where i_U is the closed embedding $Id_U \times \{0\} : U \to U \times \mathbf{A}^1$ and $n \geq 0$ is an integer.

3. *For any geometric point*

$$\bar{x} : Spec(\bar{k}) \to U \times (\mathbf{A}^1 - \{0\})$$

any effective cycle $\mathcal{W}$ of degree $\leq e$ and dimension s on $\mathbf{P}_k^m$ and any cycle $\mathcal{Z} \in z_{equi}^{eff}(\mathbf{P}^m, r, \leq d)(U)$ the cycle $\mathcal{W}$ intersects properly in $\mathbf{P}_{\bar{k}}^m$ both $cycl(\bar{x})(H_U^+(\mathcal{Z}))$ and $cycl(\bar{x})(H_U^-(\mathcal{Z}))$.

Proof: As shown in [9, 3.1], there is a continuous algebraic map satisfying the analogues of the above properties (1), (2), (3) of the following form

$$\tilde{\Theta} : \mathcal{C}_r(\mathbf{P}^m) \times V \to \mathcal{C}_r(\mathbf{P}^m)^2,$$

where $\mathcal{C}_r(\mathbf{P}^m)$ is the Chow monoid of effective r-cycles on $\mathbf{P}^m$ and $V \subset \mathbf{A}^1$ is a Zariski open neighborhood of $0 \in \mathbf{A}^1$. This map induces maps natural in U:

$$\mathcal{H}_{U,V}^+ : Hom(U, \mathcal{C}_r(\mathbf{P}^m)) \to Hom(U \times V, \mathcal{C}_r(\mathbf{P}^m))$$

$$\mathcal{H}_{U,V}^- : Hom(U, \mathcal{C}_r(\mathbf{P}^m)) \to Hom(U \times V, \mathcal{C}_r(\mathbf{P}^m)).$$

By Proposition 2.3, $z_{equi}^{eff}(\mathbf{P}^m, r)(U) \subset Hom(U, \mathcal{C}_r(\mathbf{P}^m))$ consists of those "graphs" of maps $f : U \to \mathcal{C}_r(\mathbf{P}^m)$ defined over k, so that the above maps restrict to maps

$$H_{U,V}^+ : z_{equi}^{eff}(\mathbf{P}^m, r)(U) \to z_{equi}^{eff}(\mathbf{P}^m, r)(U \times V)$$

$$H_{U,V}^- : z_{equi}^{eff}(\mathbf{P}^m, r)(U) \to z_{equi}^{eff}(\mathbf{P}^m, r)(U \times V)$$

satisfying (1), (2), (3) with $\mathbf{A}^1$ replaced by V.

To extend these maps to $U \times \mathbf{A}^1$, we let u_0 be a regular function on $\mathbf{A}^1$ whose divisor equals $W \equiv \mathbf{A}^1 - V$ (as a reduced closed subscheme) and we let w_0 be another regular function on $\mathbf{A}^1$ whose divisor misses $W \cup \{0\}$ and whose degree is prime to $1 + deg(u_0)$. We set

$$u = \frac{t^{deg(u_0)+1}}{u_0} \quad , \quad w = \frac{t^{deg(u_0)+deg(w_0)+1}}{u_0 w_0},$$

where t is the tautological regular function on $\mathbf{A}^1$. Let Γ_u denote the graph of $u : V \to \mathbf{A}^1$, and let Γ_w denote the graph of the rational map $w : \mathbf{A}^1 \to \mathbf{A}^1$. These graphs are finite over $\mathbf{A}^1$, in view of the fact that they equal the restrictions to $\mathbf{P}^1 \times \mathbf{A}^1$ of graphs of morphisms from $\mathbf{P}^1$ to itself.

Let m, n be chosen so that $m\,deg(u) + n\,deg(w) = 1$ and set $\Gamma = m\Gamma_u + n\Gamma_w$. Then, $\Gamma \in c_{equi}(V \times \mathbf{A}^1/\mathbf{A}^1, 0)$ and $V \times \mathbf{A}^1$ is a smooth relative curve over $\mathbf{A}^1$. We consider the composition

$$\Psi = \phi(\Gamma) \circ \pi^* : z_{equi}^{eff}(\mathbf{P}^m, r)(U \times V) \to z_{equi}^{eff}(\mathbf{P}^m, r)(U \times V \times \mathbf{A}^1)$$

$$\to z_{equi}^{eff}(\mathbf{P}^m, r)(U \times \mathbf{A}^1),$$

the composition of flat pull-back followed by the "transfer" with respect to Γ for the pre-theoretical structure of $z_{equi}(\mathbf{P}^m, r)$. The latter map is described more explicitly as follows. For each

$$\mathcal{Z} \in z_{equi}^{eff}(Y, r)(\mathbf{P}^m, r)(U \times V \times \mathbf{A}^1)$$

and each generic point γ of an irreducible component G_γ of $U \times \Gamma$ in $U \times V \times \mathbf{A}^1$, we consider the push-forward to $\mathbf{P}^m \times \mathbf{A}^1$ of the closure in $\mathbf{P}^m \times G_\gamma$ of the specializaton $\mathcal{Z}_\gamma$ on $\mathbf{P}^m \times Spec(k(\gamma))$. Then,

$$\phi(\Gamma)(\mathcal{Z}) = \sum_\gamma (1 \times p_\gamma)_* (\overline{\mathcal{Z}}_\gamma),$$

where $p_\gamma : G_\gamma \to U \times \mathbf{A}^1$.

To verify that

$$H_U^+ \equiv \Psi \circ \mathcal{H}_{U,V}^+ \ , \ H_U^- \equiv \Psi \circ \mathcal{H}_{U,V}^-$$

satisfy (1), (2), and (3), we consider for each $x \in U$ the following diagram

$$
\begin{array}{ccccc}
\Gamma_x \ \subset & \{x\} \times V \times \mathbf{A}^1 & \subset & U \times V \times \mathbf{A}^1 & \supset \ \Gamma \\
& \downarrow & & \downarrow & \\
& \{x\} \times \mathbf{A}^1 & \subset & U \times \mathbf{A}^1 &
\end{array}
$$

By property (3) of the pretheoretical structure for $z_{equi}(\mathbf{P}^m, r)$, we obtain the commutative square

$$
\begin{array}{ccc}
z_{equi}^{eff}(\mathbf{P}^m, r)(U \times V \times \mathbf{A}^1) & \to & z_{equi}^{eff}(\mathbf{P}^m, r)(\{x\} \times V \times \mathbf{A}^1) \\
\phi(\Gamma) \downarrow & & \downarrow \phi(\Gamma_x) \\
z_{equi}^{eff}(\mathbf{P}^m, r)(U \times \mathbf{A}^1) & \to & z_{equi}^{eff}(\mathbf{P}^m, r)(\{x\} \times \mathbf{A}^1)
\end{array}
\qquad (*)
$$

Property (1) is immediate. Since the restriction of Γ to $\{0\} \times \mathbf{A}^1$ is the graph of the embedding $\{0\} \subset \mathbf{A}^1$, we conclude that $\phi(\Gamma) \circ \pi^*$ restricted to $U \times \{0\}$ is the identity, so that property (2) follows. Finally, property (3) follows easily from commutative square (*) and the identification of $\Gamma \cap (V - \{0\} \times \mathbf{A}^1)$ as the sum of graphs of embeddings of $V - \{0\}$ into $\mathbf{A}^1 - \{0\}$.

This theorem provides us with a method to move families of cycles on projective space. To move families of cycles on more general projective varieties, one uses a version of classical projective cones technique (see [18]) as developed in [9].

Let $X \subset \mathbf{P}^n$ be a smooth equidimensional projective variety of dimension m over k. For a positive integer D, we consider the projective space $\mathbf{P}(H^0(\mathcal{O}(D))^{m+1})$ which parametrize projections $\mathbf{P}^n \to \mathbf{P}^m$ given by families $\underline{F} = (F_0, \ldots, F_m)$ of homogeneous polynomials of degree D. Let further $\mathcal{U}_{X,D}$ be the open subscheme in this projective space which consists of families $\underline{F}$ such that the corresponding rational map $\pi_{\underline{F}} : \mathbf{P}^n \to \mathbf{P}^m$ restricts to a finite flat morphism $p_{\underline{F}} : X \to \mathbf{P}^m$. Note that this open subscheme is always nonempty.

For an effective cycle $\mathcal{Z}$ on X and a point $\underline{F}$ of $\mathcal{U}_{X,D}$ denote by $R_{\underline{F}}(\mathcal{Z})$ the effective cycle $p_{\underline{F}}^*(p_{\underline{F}})_*(\mathcal{Z}) - \mathcal{Z}$. Since this construction only uses flat pull-backs and proper push-forwards of cycles it has a relative analog. More precisely, Proposition 2.1 implies that any point $\underline{F}$ of $\mathcal{U}_{X,D}$ defines an endomorphism of presheaves

$$R_{\underline{F}} : z_{equi}^{eff}(X, r) \to z_{equi}^{eff}(X, r).$$

For a point $\underline{F}$ in $\mathcal{U}_{X,D}$ denote by $Ram_{\underline{F}}$ the closed subset in X which consists of ramification points of the projection $p_{\underline{F}} : X \to \mathbf{P}^m$. The key result of the projective cones technique is the following proposition.

Proposition 6.2 *Let k be an field, $X \subset \mathbf{P}^n$ be a smooth equidimensional projective scheme of dimension m and $d, e, r, s \geq 0$ be integers such that $r + s \geq m$. Then there exists $D > 0$ and a nonempty open subscheme $\mathcal{V} = \mathcal{V}_{d,e,r,s}$ in $\mathcal{U}_{X,D}$ such that for any point $\underline{F}$ in $\mathcal{V}$ and any pair of effective cycles $\mathcal{Z}, \mathcal{W}$ on X of degrees d, e and dimensions r, s respectively all components of dimension $> r + s - m$ in $Supp(\mathcal{W}) \cap Supp(R_{\underline{F}}(\mathcal{Z}))$ belong to $Supp(\mathcal{W}) \cap Supp(\mathcal{Z}) \cap Ram_{\underline{F}}$.*

Proof: See [9, 1.7].

We apply Proposition 6.2 several times to obtain a sequence $\underline{F}_0, \dots, \underline{F}_m$ of points in $\mathcal{U}_{X,D_i}$ for some $D_0, \dots, D_m > 0$ such that for any effective cycle $\mathcal{Z}$ of dimension r and degree d the (effective) cycle $R_{\underline{F}_m} \dots R_{\underline{F}_0}(\mathcal{Z})$ intersects properly all effective cycles of dimension s and degree e (provided $r + s \geq m$). Since the cycle $R_{\underline{F}_m} \dots R_{\underline{F}_0}(\mathcal{Z})$ differs from the cycle $\mathcal{Z}$ by a linear combination of pull-backs of cycles on $\mathbf{P}^m$ we may further apply Theorem 6.1 to obtain a "move" of any relative cycle which will intersect given relative cycles properly. The precise formulation of the corresponding "moving lemma for families of cycles" is given by the following theorem.

Theorem 6.3 *Let k be a field, $X \subset \mathbf{P}^n$ be a smooth projective equidimensional scheme of dimension m over k and $d, e, r, s \geq 0$ be integers such that $r + s \geq m$. Then there are homomorphisms of abelian monoids*

$$H_U^+ : z_{equi}^{eff}(X, r)(U) \to z_{equi}^{eff}(X, r)(U \times \mathbf{A}^1)$$

$$H_U^- : z_{equi}^{eff}(X, r)(U) \to z_{equi}^{eff}(X, r)(U \times \mathbf{A}^1)$$

defined for all smooth schemes U over k and satisfying the following conditions:

1. *For any morphism $f : U' \to U$ of smooth schemes over k one has*

$$H_{U'}^+ cycl(f) = cycl(f \times Id_{\mathbf{A}^1}) H_U^+$$

$$H_{U'}^- cycl(f) = cycl(f \times Id_{\mathbf{A}^1}) H_U^-$$

2. *Let $\mathcal{W}$ be an effective cycle of degree $\leq e$ and dimension s on $X_{\bar{k}}$ and $\mathcal{Z}$ be an element of $z_{equi}^{eff}(X, r, \leq d)(U)$. Then for any geometric point $\bar{x} : \bar{k} \to U \times \{0\} \subset U \times \mathbf{A}^1$ of U one has:*

   (a) *The components of the closed subschemes*

$$Supp(cycl(\bar{x})(H_U^+(\mathcal{Z}))) \cap Supp(\mathcal{W})$$

$$Supp(cycl(\bar{x})(H_U^-(\mathcal{Z}))) \cap Supp(\mathcal{W})$$

   of dimension $> r + s - m$ belong to $Supp(cycl(\bar{x})(\mathcal{Z})) \cap Supp(\mathcal{W})$.

   (b)
$$cycl(\bar{x})(H_U^+(\mathcal{Z})) = cycl(\bar{x})(H_U^-(\mathcal{Z})) + \mathcal{Z}.$$

3. *For any geometric point $\bar{x} : Spec(\bar{k}) \to U \times (\mathbf{A}^1 - \{0\})$, any effective cycle $\mathcal{W}$ of degree $\leq e$ and dimension s on $X_{\bar{k}}$ and any cycle $\mathcal{Z} \in z_{equi}^{eff}(X, r, \leq d)(U)$ the cycles $\mathcal{W}$ and $cycl(\bar{x})(H_U^+(\mathcal{Z}))$ (resp. $cycl(\bar{x})(H_U^-(\mathcal{Z}))$) on $X_{\bar{k}}$ intersect properly.*

Proof: See [9, 3.2]. $\quad\blacksquare$

Let $T = X \times Y$ be a product of projective, smooth schemes over k and let $e \geq 0$ be such that $\{x\} \times Y \subset T$ has degree $\leq e$ for all points $x \in X$. Then Theorem 6.3 asserts that $\mathcal{Z} \in z_{equi}^{eff}(T, r)(U)$ can be moved to $\mathcal{Z}'$ on $T \times U$ with the property that $cycl(\bar{u})(\mathcal{Z}')$ meets $\{x\} \times Y \times Spec(\bar{k})$ properly for all geometric points $\bar{u} : Spec(\bar{k}) \to U$, all points $x \in X$. The following proposition enables us to conclude that such a $\mathcal{Z}'$ lies in the image of $z_{equi}^{eff}(X, r)(U \times Y)$.

Proposition 6.4 *Let X be a smooth scheme of pure dimension n over a field k, let Y be a projective scheme over k, and let U be a smooth scheme over k. Then $\mathcal{Z} \in z_{equi}^{eff}(X \times Y, r + n)(U)$ lies in the image of*

$$\mathcal{D}^{eff} : z_{equi}(Y, r)^{eff}(X \times U) \to z_{equi}(X \times Y, r + n)^{eff}(U)$$

if and only if for every geometric point $\bar{u} : Spec(\bar{k}) \to U$ the cycle $cycl(\bar{u})(\mathcal{Z})$ belongs to the image of the homomorphism

$$\mathcal{D}^{eff} : z_{equi}(Y, r)^{eff}(X \times Spec(\bar{k})) \to z_{equi}(X \times Y, r + n)^{eff}(Spec(\bar{k})).$$

Proof: We may assume that $\mathcal{Z}$ is a closed integral subscheme of $X \times Y \times U$ satisfying the condition that for every geometric point $\bar{u} : Spec(\bar{k}) \to U$ the subscheme $Z_{\bar{u}} \subset X \times Y \times Spec(\bar{k})$ is equidimensional over $X \times Spec(\bar{k})$. The generic fibre of the projection $\mathcal{Z} \to X \times U$ defines a rational map $\phi_{\mathcal{Z}} : X \times U ---> \mathcal{C}_r(Y)$. It suffices to prove that the graph $\Gamma(\phi_{\mathcal{Z}}) \subset X \times U \times \mathcal{C}_r(Y)$ projects bijectively onto $X \times U$, since the resulting continuous algebraic map is a morphism by the normality of $X \times U$.

To prove the required bijectivity, it suffices to show for any specialization $\eta \searrow \bar{x} \times \bar{u}$ in $X \times U$ and any point $\mathcal{Z}_\eta \in \Gamma(\phi_{\mathcal{Z}})$ that the only specialization of $\mathcal{Z}_\eta$ extending $\eta \searrow \bar{x} \times \bar{u}$ is $\mathcal{Z}_\eta \searrow f_{\bar{u}}(\bar{x})$, where $f_{\bar{u}} : X \times Spec(\bar{k}) \to \mathcal{C}_r(Y)$ is associated to $cycl(\bar{u})(\mathcal{Z}) \in z_{equi}(Y, r)^{eff}(X \times Spec(\bar{k}))$. This is a consequence of the observation that if $\mathcal{Z}_C \subset C \times Y$ is a cycle assoicated to a map $f : C \to \mathcal{C}_r(Y)$ for some smooth curve C, then for any geometric point $\bar{c} : Spec(\bar{k}) \to C$ the Chow point of the cycle $\mathcal{Z}_{\bar{x}} \subset Y \times Spec(\bar{k})$ equals $f(\bar{c})$.

The main reason we have worked explicitly with differences of effective cycles instead of all cycles is the fact that Proposition 6.4 becomes false without effectivity assumption on $\mathcal{Z}$ as shown in the following example.

Example 6.5 *Let $V = \mathbf{A}^1$, $U = X = \mathbf{P}^1$. Consider the cycle $W = W_+ - W_-$ on $V \times U \times X$ where W_+ (resp. W_-) is the graph of the rational map $U \times V \to X$ of the form x/y (resp. $2x/y$). Then both W_+ and W_- are relative cycles over V. Moreover the specialization of W (but not of W_+ or W_-!) to any point v of V is a relative cycle on $(U \times X)_v$ over U_v, but W is not relative over $U \times V$.*

In order to apply Theorem 6.3 in the next section to prove that the "duality map" is a quasi-isomorphism, we shall require the following simple result.

Lemma 6.6 *Let k be a field, F be a presheaf of abelian groups on the category Sm/k and $G \subset F$ be a subpresheaf (of abelian groups) of F. Let further*

$$F_0 \subset F_1 \subset \ldots \subset F_n \subset \ldots$$

be an increasing sequence of subpresheaves of sets in F such that $F = \cup_{d \geq 0} F_d$.

Assume that for any $d \geq 0$ there exist a family of homomorphisms

$$H_U : F(U) \to F(U \times \mathbf{A}^1)$$

given for all smooth schemes U over k and satisfying the following conditions:

1. *For any morphism $f : U' \to U$ one has $H_{U'}F(f) = F(f \times Id_{\mathbf{A}^1})H_U$.*

2. *For any smooth scheme U one has*

$$F(i_1)H_U(F_d(U)) \subset G(U)$$

$$F(i_0)H_U = Id_{F(U)}$$

   where i_0, i_1 are the closed embeddings $Id_U \times \{0\}$ and $Id_U \times \{1\}$ respectively.

3. *For any smooth scheme U one has*

$$H_U(G \cap F_d) \subset G.$$

Then the morphism of complexes of presheaves $\underline{C}_*(G) \to \underline{C}_*(F)$ is a quasi-isomorphism.

Proof: As in the proof of Lemma 4.1, the natural (with respect to U) homomorphism H_U determines a natural chain homotopy

$$s_* : \underline{C}_*(F)(U) \to \underline{C}_{*+1}(F)(U)$$

whose restriction to $\underline{C}(G)_*(U)$ lies in $\underline{C}_{*+1}(G)(U)$ and which relates the identity to a map $h_U = F(i_1) \circ H_U$ satisfying $h_U(\underline{C}(F_d)_*(U)) \subset \underline{C}(G)_*(U)$. Since $F = \cup_{d \geq 0} F_d$, this easily implies that $\underline{C}_*(G)(U) \to \underline{C}_*(F)(U)$ is a quasi-isomorphism.

7. Duality

In this section, we prove duality theorems relating $z_{equi}(U, X, r)$ to $z_{equi}(X \times U, r + dim(U))$ for a smooth scheme U. The proofs of these theorems use techniques which were originally developed for the duality theorems of [8]. In the next section, we shall apply duality to conclude the basic properties of bivariant cycle cohomology groups $A_{r,i}(Y, X)$ for all schemes of finite type Y, X over a field k which admit resolution of singularities.

We begin with the following duality theorem for projective, smooth varieties over an arbitrary field k.

Theorem 7.1 *Let* X, Y *be smooth projective equidimensional schemes over a field* k. *Then the embedding of presheaves*

$$\mathcal{D} : z_{equi}(Y, X, r) \to z_{cqui}(X \times Y, r + dim(Y))$$

induces isomorphisms

$$\underline{h}_i(z_{equi}(Y, X, r)) \to \underline{h}_i(z_{equi}(X \times Y, r + dim(Y)))$$

for all $i \in \mathbf{Z}$.

Proof: We apply Theorem 6.3 with X replaced by $X \times Y \subset \mathbf{P}^N$ and e the maximum of the degrees of $x \times Y \subset X \times Y$. Lemma 6.6 enables us to obtain from Theorem 6.3 a quasi-isomorphism of complexes of presheaves. We interpret this as the required quasi-isomorphism

$$\underline{C}_*(z_{equi}(X, Y, r)) \to \underline{C}_*(z_{equi}(X \times Y, r + dim(Y)))$$

by applying Proposition 6.4

We now proceed to remove the hypotheses that X, Y be projective and smooth but add the hypothesis that k admit resolution of singularities. In so doing, we shall obtain a quasi-isomorphism of chain complexes obtained by evaluating the appropriate complexes of presheaves at $Spec(\bar{k})$. The key reason why we do not conclude a quasi-isomorphism of presheaves is that the conclusion of Theorem 5.5(2) concerns the associated sheaf of a presheaf.

For the remainder of this section, let k be a field which admits resolution of singularities, let U be a smooth scheme over k of pure dimension n, let $i_U : U \subset \bar{U}$ be a smooth compactification, let X be a scheme of finite type over k, and let $i_X : X \subset \bar{X}$ be an embedding of X in a proper scheme of finite type over k (see [14]).

For any proper scheme $q : \bar{Y} \to \bar{X}$ of finite type over $\bar{X}$, consider the morphism of presheaves of abelian monoids

$$\alpha_{\bar{Y}}^{eff} : z_{equi}^{eff}(\bar{U} \times \bar{Y}, n + r) \to z_{equi}^{eff}(U \times X, n + r)$$

which is the composition of proper push-forward morphism

$$(Id_{\bar{U}} \times q)_* : z_{equi}^{eff}(\bar{U} \times \bar{Y}, n + r) \to z_{equi}^{eff}(\bar{U} \times \bar{X}, n + r)$$

with the flat pull-back

$$(i_U \times i_X)^* : z_{equi}^{eff}(\bar{U} \times \bar{X}, n+r) \to z_{equi}^{eff}(U \times X, n+r).$$

Let $\alpha_{\bar{Y}}$ be the corresponding morphism of presheaves of abelian groups.

We denote by $\Phi_{\bar{Y}}$ the subpresheaf of abelian groups in $z_{equi}(\bar{U} \times \bar{Y}, n+r)$ generated by the subpresheaf of abelian monoids $(\alpha_{\bar{Y}}^{eff})^{-1}(z_{equi}^{eff}(U, X, r))$ (where we identify $z_{equi}^{eff}(U, X, r)$ with its image in $z_{equi}^{eff}(U \times X, r+n)$). Thus, $\Phi_{\bar{Y}}$ fits in the following commutative square

$$\Phi_{\bar{Y}} = [(\alpha_{\bar{Y}}^{eff})^{-1}(z_{equi}^{eff}(U, X, r))]^+ \quad \to \quad z_{equi}(U, X, r)$$

$$\downarrow \qquad\qquad\qquad\qquad\qquad\qquad \downarrow$$

$$z_{equi}(\bar{U} \times \bar{Y}, n+r) \quad \overset{\alpha_{\bar{Y}}}{\to} \quad z_{equi}(U \times X, r+n)$$

where $[-]^+$ denotes the abelian group associated with the abelian monoid $-$.

We should be wary that for general $q : \bar{Y} \to \bar{X}$,

$$\Phi_{\bar{Y}} \neq (\alpha_{\bar{Y}})^{-1}(z_{equi}(U, X, r));$$

for example, distinct effective cycles on $\bar{U} \times \bar{Y}$ might map to the same, non-equidimensional cycle on $\bar{U} \times \bar{X}$). Of course, $\Phi_{\bar{X}} \neq (\alpha_{\bar{X}})^{-1}(z_{equi}(U, X, r))$.

We require the following lemma, analogous to but considerably more elementary than Proposition 6.4.

Lemma 7.2 *Let V be a smooth scheme over k, $q : \bar{Y} \to \bar{X}$ be a smooth projective scheme over $\bar{X}$ and $\mathcal{Z} = \sum n_i Z_i$ ($n_i \neq 0$) be an element of $z_{equi}(\bar{U} \times \bar{Y}, n+r)(V)$. $\mathcal{Z}$ belongs to $\Phi_{\bar{Y}}$ if and only if for any geometric point $\bar{x} : Spec(\bar{k}) \to V \times U$ of $V \times U$ and any i we have*

$$dim(q_{\bar{k}}(Z_i \times_{V \times U} Spec(\bar{k})) \cap X_{\bar{k}}) \leq r$$

where

$$q_{\bar{k}} = q \times_{Spec(k)} Spec(\bar{k})$$

$$X_{\bar{k}} = X \times_{Spec(k)} Spec(\bar{k}).$$

Proof: Note first that due to our definition of $\Phi_{\bar{Y}}$ we may assume that $\mathcal{Z} = Z$ for a closed integral subscheme Z in $V \times \bar{U} \times \bar{Y}$ which is equidimensional of relative dimension $r + n$ over V. Obviously $Z \in \Phi_{\bar{Y}}(V)$ if and only if

$$(Id_V \times Id_{\bar{U}} \times q)(Z) \cap V \times U \times X$$

is equidimensional of relative dimension r over $V \times U$ or is of dimension everywhere less then r over $U \times V$ (the last case corresponds to $\alpha_{\bar{Y}}(\mathcal{Z}) = 0$). Equivalently, this means that for any geometric point $\bar{x}$ of $V \times U$ we have

$$dim((Id_V \times Id_{\bar{U}} \times q)(Z) \times_{V \times U} Spec(\bar{k}) \cap X_{\bar{k}}) \leq r.$$

Our statement follows now from the obvious equality:

$$(Id_V \times Id_{\bar{U}} \times q)(Z) \times_{V \times U} Spec(\bar{k}) = q_{\bar{k}}(Z \times_{V \times U} Spec(\bar{k})).$$

The following somewhat technical proposition generalizes Theorem 7.1 in the following sense: if X is projective but not necessarily smooth and if U is both projective and smooth, then Proposition 7.3 immediately implies the quasi-isomorphism of chain complexes

$$\underline{C}_*(z_{equi}(U, X, r))(Spec(k)) \to \underline{C}_*(z_{equi}(X \times U, r + n))(Spec(k)).$$

Proposition 7.3 Let $q : \bar{Y} \to \bar{X}$ be a proper scheme over $\bar{X}$. Then for any smooth scheme U of pure dimension n and any $i \in \mathbf{Z}$, the morphisms

$$\underline{h}_i(\Phi_{\bar{Y}})(Spec(k)) \to \underline{h}_i(z_{equi}(\bar{U} \times \bar{Y}, r + n))(Spec(k))$$

are isomorphisms.

Proof: We first assume that $\bar{Y}$ is smooth; then we may clearly also assume that $\bar{Y}$ is connected (and thus equidimensional). Choose a projective embedding $\bar{U} \times \bar{Y} \subset \mathbf{P}^N$. Let e be the degree of the closed subschemes $(\bar{Y})_{\bar{u}}$, $\bar{u} \in U(\bar{k})$ of $\mathbf{P}^N_{\bar{k}}$ and let $d \geq 0$ be an integer. Applying Theorem 6.3 to the scheme $\bar{U} \times \bar{Y}$ (with numbers being $d, e, r + n, dim(\bar{Y})$) we get a family of natural homomorphisms

$$H_V = H_V^+ - H_V^- : z_{equi}(\bar{U} \times \bar{Y}, n + r)(V) \to z_{equi}(\bar{U} \times \bar{Y}, n + r)(V \times \mathbf{A}^1)$$

given for all smooth schemes V over k. It is sufficient to show now that they satisfy conditions of Lemma 6.6 for $F = z_{equi}(\bar{U} \times \bar{Y}, n + r)$,

$$F_d = (z_{equi, \leq d}^{eff}(\bar{U} \times \bar{Y}, n + r))_+$$

where $(-)_+$ denotes the abelian group associated with an abelian monoid $-$ and $G = \Phi_{\bar{Y}}$. Property (1) follows from Theorem 6.3(1). Property (2) follows from our choice of e, Lemma 7.2 and Theorem 6.3(2b, 3). Finally the property (3) follows from Lemma 7.2 and Theorem 6.3(2a,3).

We now consider the general case in which $q : \bar{Y} \to \bar{X}$ is a proper scheme over $\bar{X}$. By the above proof for smooth $\bar{Y}$, we may assume that our quasi-isomorphism is proven for all smooth projective schemes as well as for all schemes of dimension $< dim(\bar{Y})$.

Since k admits resolution of singularities there is a proper surjective morphism $p : \bar{Y}' \to \bar{Y}$ such that $\bar{Y}'$ is smooth and projective and there exits a closed subscheme $j : \bar{Z} \to \bar{Y}$ in $\bar{Y}$ such that $p^{-1}(\bar{Y} - \bar{Z}) \to \bar{Y} - \bar{Z}$ is an isomorphism and $dim(\bar{Z}) < dim(\bar{Y})$. Consider the following diagram of presheaves (we write $z(-)$ instead of $z(-, r+n)$):

$$
\begin{array}{ccccccccc}
0 \to & \Phi_{p^{-1}(\bar{Z})} & \to & \Phi_{\bar{Y}'} \oplus \Phi_{\bar{Z}} & \to & \Phi_{\bar{Y}} & \to & coker_1 & \to 0 \\
& \downarrow & & \downarrow & & \downarrow & & \downarrow & \\
0 \to & z(\bar{U} \times p^{-1}(\bar{Z})) & \to & z(\bar{U} \times \bar{Y}') \oplus z(\bar{U} \times \bar{Z}) & \to & z(\bar{U} \times \bar{Y}) & \to & coker_2 & \to 0.
\end{array}
$$

Since $dim(p^{-1}(\bar{Z})) < dim(\bar{Y})$, the induction assumption together with the isomorphism for the smooth, projective $\bar{Y}$ imply that the first two vertical arrows induces isomorphisms on the corresponding groups $\underline{h}_i(-)(Spec(k))$. One can verify easily that both horizontal sequences are exact and the last vertical arrow is a monomorphism. Proposition 5.8 implies that all horizontal and vertical arrows are morphisms of pretheories. Applying Lemma 3.6 and Theorem 5.5(2) to $coker_j$ we see that it is sufficient to show that $(coker_2)_{cdh} = 0$. This follows exactly as in the proof of Theorem 5.11 (cf. Theorem 5.13).

We now can easily prove our main duality theorem.

Theorem 7.4 *Let k be a field which admits resolution of singularities, U be a smooth quasi-projective equidimensional scheme of dimension n over k and X be a scheme of finite type over k. Then for any $r \geq 0$ the embedding*

$$
\mathcal{D} : z_{equi}(U, X, r) \to z_{equi}(X \times U, r+n)
$$

induce quasi-isomorphisms of complexes of abelian groups

$$
\underline{C}_*(z_{equi}(U, X, r))(Spec(k)) \to \underline{C}_*(z_{equi}(X \times U, r+n))(Spec(k)).
$$

Proof: We have the following diagram of morphisms of presheaves

$$
\begin{array}{ccccccccc}
0 \to & ker' & \to & \Phi_{\bar{X}} & \to & z_{equi}(U, X) & \to & coker_1 & \to 0 \\
& a \downarrow & & b \downarrow & & \mathcal{D} \downarrow & & c \downarrow & \\
0 \to & ker(\alpha_{\bar{X}}) & \to & z_{equi}(\bar{U} \times \bar{X}) & \to & z_{equi}(U \times X) & \to & coker_2 & \to 0.
\end{array}
$$

(where

$$z_{equi}(U, X) = z_{equi}(U, X, r)$$

$$z_{equi}(\bar{U} \times \bar{X}) = z_{equi}(\bar{U} \times \bar{X}, r + n)$$

$$z_{equi}(U \times X) = z_{equi}(U \times X, r + n))$$

Since $\Phi_{\bar{X}} = (\alpha_{\bar{X}})^{-1}(z_{equi}(U, X, r))$, the morphism a is an isomorphism and all other vertical morphisms are monomorphisms. Moreover, Proposition 7.3 asserts that $b(Spec(\bar{k}))$ is a quasi-isomorphism. Thus, it suffices to show that

$$\underline{h}_i(coker_j)(Spec(k)) = 0$$

for $j = 1, 2$ and all $i \in \mathbf{Z}$. By Proposition 5.8(2) (resp. 5.8(3)) the morphism α (resp. $\mathcal{D}$) is a morphism of pretheories. In particular $coker_j$ are pretheories. In view of Theorem 5.5(2) it is sufficient to show that $(coker_j)_{cdh} = 0$. Since $coker_1 \subset coker_2$, Lemma 3.6 implies we have only to consider the case of $coker_2$ whose vanishing is given in the proof of Theorem 5.11.

8. Properties and pairings

We can prove now the comparison theorem for "naive" $(\underline{h}_i(z_{equi}(X, r)(U))$ and "fancy" $(A_{r,i}(U, X))$ definitions of bivariant cycle cohomology discussed in Section 4.

Theorem 8.1 *Let k be a field which admits a resolution of singularities, let X be a scheme of finite type over k, and U be a smooth quasi-projective scheme over k. Then the natural homomorphisms of abelian groups*

$$\underline{h}_i(z_{equi}(X, r))(U) \to A_{r,i}(U, X)$$

are isomorphisms for all $i \in \mathbf{Z}$.

Proof: For any smooth scheme V of pure dimension n and Zariski open subsets $V_1, V_2 \subset V$ with $V = V_1 \cup V_2$, we have the following exact sequences of complexes

$$0 \to \underline{C}_*(z_{equi}(X, r))(V) \to \underline{C}_*(z_{equi}(X, r))(V_1) \oplus \underline{C}_*(z_{equi}(X, r))(V_2) \to$$

$$\to \underline{C}_*(z_{equi}(X, r))(V_3)$$

where $V_3 = V_1 \cap V_2$. By Theorem 7.4, this sequence is quasi-isomorphic in the derived category to the distinguished triangle obtained by evaluating the following distinguished triangle provided by Theorem 5.11

$$\underline{C}_*(z_{equi}(X \times V, r+n) \to \underline{C}_*(z_{equi}(X \times V_1, r+n)) \oplus \underline{C}_*(z_{equi}(X \times V_2, r+n)) \to$$

$$\underline{C}_*(z_{equi}(X \times V_3, r + n)) \to \underline{C}_*(z_{equi}(X \times V, r + n))[1]$$

at $Spec(k)$. Hence, in the sense of [2], the presheaf $\underline{C}_*(z_{equi}(X, r))$ is "pseudo-flasque". Thus, we may apply [2, Th. 4] to conclude that

$$\mathbf{H}^{-i}(U, \underline{C}_*(z_{equi}(X, r))_{Zar}) = h_i(\underline{C}_*(z_{equi}(X, r))(U)).$$

On the other hand, Theorem 5.5 asserts that the left hand side of the above equality equals $A_{r,i}(U, X)$ whereas the right hand side equals $\underline{h}_i(z_{equi}(X, r))(U)$ by definition.

The following theorem provides our strongest duality theorem, in which both Y and X are permited to be arbitrary schemes of finite type over k.

Theorem 8.2 *Let k be a field which admits resolution of singularities, let X, Y be schemes of finite type over k, and let U be a smooth scheme of pure dimension n over k. There are canonical isomorphisms*

$$A_{r,i}(Y \times U, X) \to A_{r+n,i}(Y, X \times U).$$

Proof: Let $\mathcal{U} = \{U_i \to U\}$ be a finite open covering of U by quasi-projective schemes. The complexes $C_*(z(U_i, X, r))$ form a bicomplex and we denote by $\underline{C}_*(z(\mathcal{U}, X, r))$ its total complex. It follows immediately from definition of bivariant cycle cohomology groups that there are canonical homomorphisms

$$\mathbf{H}^{-i}_{cdh}(Y, \underline{C}_*(z(\mathcal{U}, X, r))_{cdh}) \to A_{r,i}(Y \times U, X).$$

On the other hand the duality embeddings $z(U_i, X, r) \to z(X \times U_i, r + dim(U))$ give us together with Theorem 7.4 a canonical quasi-isomorphism of complexes

$$\underline{C}_*(z(\mathcal{U}, X, r))_{cdh} \to \underline{C}_*(z(X \times U, r + dim(U))).$$

It is sufficient therefore to show that the homomorphisms

$$\mathbf{H}^{-i}_{cdh}(Y, \underline{C}_*(z(\mathcal{U}, X, r))_{cdh}) \to A_{r,i}(Y \times U, X)$$

are isomorphisms. Since these homomorphisms are canonical and both sides are cohomology groups in cdh-topology the problem is cdh-local. In particular since k admits resolution of singularities we may assume that Y is smooth and quasi-projective. In this case our statement follows easily from Theorem 8.1 and Theorem 7.4.

We provide additional good properties of the bivariant theory $A_{*,*}(-,-)$ in the next theorem.

Theorem 8.3 *Let k be a field which admits resolution of singularities and let X, Y be schemes of finite type over k.*

1. *(Homotopy invariance) The pull-back homomorphism $z_{equi}(X, r) \to z_{equi}(X \times \mathbf{A}^1, r+1)$ induces for any $i \in \mathbf{Z}$ an isomorphism*

$$A_{r,i}(Y, X) \to A_{r+1,i}(Y, X \times \mathbf{A}^1).$$

2. *(Suspension) Let*
$$p : X \times \mathbf{P}^1 \to X$$
$$i : X \to X \times \mathbf{P}^1$$

   be the natural projection and closed embedding. Then the morphism

$$i_* \oplus p^* : z_{equi}(X, r+1) \oplus z_{equi}(X, r) \to z_{equi}(X \times \mathbf{P}^1, r+1)$$

   induces an isomorphism

$$A_{r+1,i}(Y, X) \oplus A_{r,i}(Y, X) \to A_{r+1,i}(Y, X \times \mathbf{P}^1)$$

3. *(Cosuspension) There are canonical isomorphisms:*

$$A_{r,i}(Y \times \mathbf{P}^1, X) \to A_{r+1,i}(Y, X) \oplus A_{r,i}(Y, X).$$

4. *(Gysin) Let $Z \subset U$ be a closed immersion of smooth schemes everywhere of codimension c in U. Then there is a canonical long exact sequence of abelian groups of the form*

$$\ldots \to A_{r+c,i}(Z, X) \to A_{r,i}(U, X) \to A_{r,i}(U - Z, X)$$

$$\to A_{r+c,i-1}(Z, X) \to \ldots.$$

Proof:

1. Since we have natural homomorphisms

$$A_{r,i}(Y, X) \to A_{r+1,i}(Y, X \times \mathbf{A}^1),$$

   Propositions 4.4, 4.5 and resolution of singularities imply that we may assume Y to be smooth and quasi-projective. Then (1) follows from Theorems 8.1, 7.4 and Lemma 4.1.

2. This follows from localization exact sequence (5.12) and (1).

3. This follows immediately from Theorem 8.2 and (2).

4. This follows immediately from Theorem 8.2 and Corollary 5.12.

Remark 8.4 *The "cosuspension" and "suspension" properties of the bivariant theory $A_{*,*}(-,-)$ are particular cases of projective bundle theorems for the first and second variables respectively. For the corresponding general results see [24].*

Let X, X' be schemes of finite type over k and let U be a smooth scheme over k. For any pair of integral closed subschemes $Z \subset X \times U, Z' \subset X' \times U$ equidimensional over U, the fibre product $Z \times_U Z'$ is also equidimensional over U. Thus, sending Z, Z' to the cycle associated to the subscheme $Z \times_U Z' \subset X \times X' \times U$ determines a pairing

$$\times : z_{equi}(X, r) \otimes z_{equi}(X', r') \to z_{equi}(X \times X', r + r')$$

of presheaves.

Using this product pairing, we define the algebraic analogue of the operations introduced in [11] for Lawson homology, formulated in terms of the product pairing in [6], and extended to a bivariant context in [7]. We employ the standard notation $\overset{L}{\otimes}$ to denote the derived tensor product.

Proposition 8.5 *Let k be a field which admits resolution of singularities and let X, Y be schemes of finite type over k. Then there is a natural pairing for any $i \geq 0$*

$$A_{r+1,i}(Y, X) \overset{L}{\otimes} A_{0,j}(Spec(k), \mathbf{P}^1) \to A_{r,i+j}(Y, X).$$

Proof: Our pairing factors through the natural map

$$A_{r+1,i}(Y, X) \overset{L}{\otimes} A_{0,j}(Spec(k), \mathbf{P}^1) \overset{1 \otimes p^*}{\to} A_{r+1,i}(Y, X) \overset{L}{\otimes} A_{0,j}(Y, \mathbf{P}^1)$$

and the projection

$$A_{r+1,i+j}(Y, X \times \mathbf{P}^1) \to A_{r,i+j}(Y, X)$$

associated to the suspension isomorphism of Theorem 8.3(2). Thus, it suffices to exhibit a natural pairing

$$A_{r+1,i}(Y, X) \overset{L}{\otimes} A_{0,j}(Y, \mathbf{P}^1) \to A_{r+1,i+j}(Y, X \times \mathbf{P}^1).$$

This pairing is induced by the product pairing

$$\times : z_{equi}(X, r + 1) \times z_{equi}(\mathbf{P}^1, 0) \to z_{equi}(X \times \mathbf{P}^1, r + 1).$$

We next use duality to provide

$$A_{r,i}(X) = A_{r,i}(Spec(k), X) = \underline{h}_i(z_{equi}(x, r))(Spec(k))$$

with a natural multiplicative pairing for smooth schemes X. This multiplicative structure is inspired by the multiplicative structure on "morphic cohomology" studied in [7].

Proposition 8.6 *Let k be a field which admits resolution of singularities and let X be a smooth scheme of pure dimension n over k. Then there is a natural multiplication*

$$A_{r,i}(X) \overset{L}{\otimes} A_{s,j}(X) \to A_{r+s-n,i+j}(X) , \quad \text{for any } r + s \geq n.$$

Proof: By Theorems 8.3(1), 7.4, we have natural isomorphisms

$$A_{r,i}(Spec(k), X) \simeq A_{r+n,i}(Spec(k), X \times \mathbf{A}^n) \simeq A_{r,i}(X, \mathbf{A}^n).$$

On the other hand, the product pairing

$$\times : z_{equi}(\mathbf{A}^n, r) \otimes z_{equi}(\mathbf{A}^n, s) \to z_{equi}(\mathbf{A}^{2n}, r + s)$$

determines a pairing

$$A_{r,i}(X, \mathbf{A}^n) \overset{L}{\otimes} A_{s,j}(X, \mathbf{A}^n) \to A_{r+s,i+j}(X, \mathbf{A}^{2n}).$$

Using Theorems 8.3(1), 7.4 once again, we have

$$A_{r+s,i+j}(X, \mathbf{A}^{2n}) \simeq A_{r+s+n,i+j}(X \times \mathbf{A}^{2n}) \simeq A_{r+s-n,i+j}(X).$$

Our multiplicative pairing now follows.

We introduce one final pairing on our bivariant cycle cohomology groups. The following "composition pairing" is based on a pairing introduced in [7] and further examined in [5].

Proposition 8.7 *Let k be a field which admits resolution of singularities, let T be a scheme of finite type over k, let U be a smooth scheme, and let X be a projective, smooth scheme. Then there is a natural* composition *pairing of chain complexes*

$$\underline{C}_*(z_{equi}(X,r))(U) \otimes \underline{C}_*(z_{equi}(T,s))(X) \to \underline{C}_*(z_{equi}(T,r+s))(U)$$

which induces the pairing

$$A_{r,i}(U,X) \overset{L}{\otimes} A_{s,j}(X,T) \to A_{r+s,i+j}(U,T).$$

Proof: Given a pair of closed integral subschemes

$$Z \subset X \times U \times \Delta^n \ , \ W \subset T \times X \times \Delta^n,$$

we consider the subscheme

$$W \times_{X \times \Delta^n} Z \subset T \times X \times U \times \Delta^n.$$

We easily verify that if Z is equidimensional over $U \times \Delta^n$ and if W is equidimensional over $X \times \Delta^n$, then $W \times_{X \times \Delta^n} Z$ is equidimensional over $U \times \Delta^n$. We define the asserted pairing of chain complexes by sending (Z, W) to the proper push-forward via the projection $T \times X \times U \times \Delta^n \to T \times U \times \Delta^n$ of the cycle associated to $W \times_{X \times \Delta^n} Z$.

9. Motivic cohomology and homology

In this final section, we introduce four theories defined for schemes of finite type over a field k. They are called correspondingly - motivic homology, motivic cohomology, Borel-Moore motivic homology and motivic cohomology with compact supports. All of them but motivic homology (with compact supports) are closely related to bivariant cycle cohomology.

Since some of the definitions below are rather involved, we want to explain first what motivates them. Let X be a scheme of finite type over k. As shown in [24] one can construct a certain triangulated category $DM_{gm}(k)$ of "geometrical mixed motives over k" and asociated to any such X two objects in this category: $M(X)$ - the "motive" of X and $M_c(X)$ - the "motive with compact supports" of X. The correspondence $X \mapsto M(X)$ is covariantly functorial with respect to all morphisms while the correspondence $X \mapsto M_c(X)$ is covariantly functorial with respect to proper morphisms and contravariantly functorial with respect to equidimensional morphisms of relative dimension zero. For a proper X one has $M_c(X) = M(X)$.

The category $DM_{gm}(k)$ has a tensor structure and a distinguished invertible object $\mathbf{Z}(1)$ called the Tate object. For any object M in $DM_{gm}(k)$ we denote by $M(n)$ the object $M \otimes \mathbf{Z}(1)^{\otimes n}$. In particular $\mathbf{Z}(n) = \mathbf{Z}(1)^{\otimes n}$.

Using this formalism, one could define various motivic theories as follows:

Motivic cohomology:

$$H^j(X, \mathbf{Z}(i)) = Hom_{DM}(M(X), \mathbf{Z}(i)[j])$$

Motivic cohomology with compact supports:

$$H_c^j(X, \mathbf{Z}(i)) = Hom_{DM}(M_c(X), \mathbf{Z}(i)[j])$$

Motivic homology:

$$H_j(X, \mathbf{Z}(i)) = Hom_{DM}(\mathbf{Z}(i)[j], M(X))$$

Borel-Moore motivic homology:

$$H_j^{BM}(X, \mathbf{Z}(i)) = Hom_{DM}(\mathbf{Z}(i)[j], M_c(X))$$

The definitions given below are just "explicit" reformulations of these heuristic ones. There are two main reasons why they are more involved than one might expect. The first which applies mainly to the case of homology is that the object $\mathbf{Z}(i)$ has no geometrical meaning for $i < 0$ and $Hom(\mathbf{Z}(i)[j], M(X))$ is in this case just a formal notation for $Hom(\mathbf{Z}[j], M(X)(-i))$. Thus we have to distinguish the cases $i < 0$ and $i \geq 0$ in the definitions below. The second is that there are several ways to interpret $\mathbf{Z}(i)$ geometrically for $i > 0$. In Definitions 9.2, 9.3 below, we use the fact that $\mathbf{Z}(i) = M_c(\mathbf{A}^i)[-2i]$ and in Definition 9.4 the fact that $M(\mathbf{A}^i - \{0\}) = \mathbf{Z}(i)[2i - 1] \oplus \mathbf{Z}$.

Definition 9.1 *Let X be a scheme of finite type over k. For any $n, r \in \mathbf{Z}$ we define the* Borel-Moore *motivic homology of X as*

$$H_n^{BM}(X, \mathbf{Z}(r)) \equiv \begin{cases} A_{r,n-2r}(X) & for \quad r \geq 0 \\ A_{0,n-2r}(X \times \mathbf{A}^{-r}) & for \quad r \leq 0. \end{cases}$$

Definition 9.2 *Let X be a scheme of finite type over k. For all $m, s \in \mathbf{Z}$, we define the* motivic cohomology *of X as*

$$H^m(X, \mathbf{Z}(s)) \equiv A_{0,2s-m}(X, \mathbf{A}^s)$$

(for $s < 0$ we set $A_{0,2s-m}(X, \mathbf{A}^s) = 0$).

To define motivic cohomology with compact supports we use the construction of cdh-cohomology with compact supports given at the end of Section 3.

Definition 9.3 *Let X be a scheme of finite type over k. For any $m, s \in \mathbf{Z}$ we define the* motivic cohomology of X with compact supports *of X as*

$$H_c^m(X, \mathbf{Z}(s)) \equiv \mathbf{H}_c^{m-2s}(X, \underline{C}_*(z_{equi}(\mathbf{A}^s, 0))).$$

To define the last of our four theories, namely motivic homology (with compact supports) we need to recall first the following definition of algebraic singular homology (see [21]).

For any scheme of finite type X over k, denote by $c_{equi}(X,0)$ the subpresheaf in $z_{equi}(X,0)$ whose values on a smooth scheme U over k the subgroup $c_{equi}(X,0)(U)$ is generated by integral closed subschemes Z in $U \times X$ which are *proper* over U (note that this implies that Z is in fact finite over U). One can easily see that unlike $z_{equi}(X,0)$ the presheaf $c_{equi}(X,0)$ is covariantly functorial with respect to *all* morphisms $X_1 \to X_2$ and contravariantly functorial only with respect to flat *proper* morphisms. For any X the presheaf $c_{equi}(X,0)$ has a structure of a pretheory obtained by restricting of the canonical structure of a pretheory on $z_{equi}(X,0)$. Note also that if X is proper we have $c_{equi}(X,0) = z_{equi}(X,0)$. We define the algebraic singular complex of X as the complex of sheaves $\underline{C}_*(c_{equi}(X,0))$. Its homology sheaves (or more precisely their sections on $Spec(k)$) are algebraic singular homology of X (cf. [21]). They form a homology theory which is a part of our motivic homology of X.

Generalizing algebraic singular homolgy we have:

Definition 9.4 *Let X be a scheme of finite type over k. For any $n, r \in \mathbf{Z}$ we define motivic homology of X as follows:*

$$H_n(X, \mathbf{Z}(r)) \equiv \begin{cases} \mathbf{H}^{2r-n}_{\{0\}}(\mathbf{A}^r, \underline{C}_*(c_{equi}(X,0))) & r \geq 0 \\ h_{n-2r-1}(\underline{C}_*(c_{equi}(X \times (\mathbf{A}^r - \{0\}),0)/c_{equi}(X \times \{1\},0))) & r \leq 0 \end{cases}$$

In what follows, we summarize some basic properties of our four theories in the case when k admits resolution of singularities. All of these properties follow from results of this paper, [23] and the corresponding properties of the motivic category considered in [24].

Motivic cohomology - is a family of contravariant functors $H^i(-, \mathbf{Z}(j))$ from the category Sch/k to the category of abelian groups. These functors have the following basic properties:

  Homotopy invariance. For any scheme X of finite type over k one has canonical isomorphisms $H^i(X \times \mathbf{A}^1, \mathbf{Z}(j)) = H^i(X, \mathbf{Z}(j))$.

  Algebraic suspension property. For any X as above, one has canonical isomorphisms

$$H^i(X \times \mathbf{P}^1, \mathbf{Z}(j)) = H^i(X, \mathbf{Z}(j)) \oplus H^{i-2}(X, \mathbf{Z}(j-1)).$$

  Mayer-Vietoris exact sequence. For any X as above and any open covering $X = U \cup V$ of X, there is a canonical long exact sequence of the form:

$$\ldots \to H^i(X, \mathbf{Z}(j)) \to H^i(U, \mathbf{Z}(j)) \oplus H^i(V, \mathbf{Z}(j)) \to$$
$$\to H^i(U \cap V, \mathbf{Z}(j)) \to H^{i+1}(X, \mathbf{Z}(j)) \to \ldots$$

Blow-up exact sequence. For any schemes X, Y of finite type over k, proper morphism $p : X \to Y$ and closed subscheme $Z \to Y$ in Y such that $p^{-1}(Y - Z) \to Y - Z$ is an isomorphism, there is a long exact sequence of the form:

$$\ldots \to H^i(Y, \mathbf{Z}(j)) \to H^i(X, \mathbf{Z}(j)) \oplus H^i(Z, \mathbf{Z}(j)) \to$$

$$H^i(p^{-1}(Z), \mathbf{Z}(j)) \to H^{i+1}(Y, \mathbf{Z}(j)) \to \ldots$$

Covariant functoriality. For a proper, flat, equidimensional morphism $f : X \to Y$ of relative dimension d, there are homomorphisms

$$H^i(X, \mathbf{Z}(j)) \to H^{i-2d}(X, \mathbf{Z}(j - d)).$$

Motivic homology - is a family of covariant functors $H_i(-, \mathbf{Z}(j))$ from the category Sch/k to the category of abelian groups. These functors have the following basic properties:

Homotopy invariance. For any scheme X of finite type over k one has canonical isomorphisms $H_i(X \times \mathbf{A}^1, \mathbf{Z}(j)) = H_i(X, \mathbf{Z}(j))$.

Algebraic suspension property. For any X as above, one has canonical isomorphisms

$$H_i(X \times \mathbf{P}^1, \mathbf{Z}(j)) = H_i(X, \mathbf{Z}(j)) \oplus H_{i-2}(X, \mathbf{Z}(j - 1)).$$

Mayer-Vietoris exact sequence. For any X as above and any open covering $X = U \cup V$ of X, there is a canonical long exact sequence of the form:

$$\ldots \to H_i(U \cap V, \mathbf{Z}(j)) \to H_i(U, \mathbf{Z}(j)) \oplus H_i(V, \mathbf{Z}(j)) \to H_i(X, \mathbf{Z}(j))$$

$$\to H^{i-1}(U \cap V, \mathbf{Z}(j)) \to \ldots$$

Blow-up exact sequence. For any schemes X, Y of finite type over k, a proper morphism $p : X \to Y$ and a closed subscheme $Z \to Y$ in Y such that $p^{-1}(Y - Z) \to Y - Z$ is an isomorphism there is a long exact sequence of the form:

$$\ldots \to H_i(p^{-1}(Z), \mathbf{Z}(j)) \to H_i(X, \mathbf{Z}(j)) \oplus H_i(Z, \mathbf{Z}(j)) \to$$

$$H^i(Y, \mathbf{Z}(j)) \to H^{i-1}(p^{-1}(Z), \mathbf{Z}(j)) \to \ldots$$

Contravariant functoriality. For a proper flat equidimensional morphism $f : X \to Y$ of relative dimension d, there are homomorphisms

$$H_i(Y, \mathbf{Z}(j)) \to H_{i+2d}(X, \mathbf{Z}(j + d)).$$

Motivic cohomology with compact support - is a family of contravariant functors $H_c^i(-, \mathbf{Z}(j))$ from the category Sch_c/k of schemes and proper morphisms to the category of abelian groups. These functors have the following basic properties:

Homotopy invariance. For any scheme of finite type over k, one has canonical isomorphisms $H_c^i(X \times \mathbf{A}^1, \mathbf{Z}(j)) = H_c^{i-2}(X, \mathbf{Z}(j-1))$.

Localization. For any X as above and any closed subscheme Z in X, one has a canonical exact sequence:

$$\ldots \to H_c^i(X - Z, \mathbf{Z}(j)) \to H_c^i(X, \mathbf{Z}(j)) \to H_c^i(Z, \mathbf{Z}(j))$$
$$\to H_c^{i+1}(Z, \mathbf{Z}(j)) \to \ldots$$

Covariant functoriality. For a flat equidimensional morphism $f : X \to Y$ of relative dimension d, there are homomorphisms

$$H_c^i(X, \mathbf{Z}(j)) \to H_c^{i-2d}(X, \mathbf{Z}(j-d)).$$

Borel-Moore motivic homology - is a family of covariant functors $H_i^{BM}(-, \mathbf{Z}(j))$ from the category Sch_c/k to the category of abelian groups. These functors have the following basic properties:

Homotopy invariance. For any scheme of finite type over k, one has canonical isomorphisms $H_i^{BM}(X \times \mathbf{A}^1, \mathbf{Z}(j)) = H_{i-2}^{BM}(X, \mathbf{Z}(j))$.

Localization. For any X as above and any closed subscheme Z in X, one has a canonical exact sequence:

$$\ldots \to H_i^{BM}(Z, \mathbf{Z}(j)) \to H_i^{BM}(X, \mathbf{Z}(j)) \to H_i^{BM}(X - Z, \mathbf{Z}(j))$$
$$\to H_{i+1}^{BM}(Z, \mathbf{Z}(j)) \to \ldots.$$

Contravariant functoriality. For a flat equidimensional morphism
$f : X \to Y$ of relative dimension d, there are homomorphisms

$$H_i^{BM}c(Y, \mathbf{Z}(j)) \to H_{i+2d}^{BM}(X, \mathbf{Z}(j+d)).$$

The following two main results relate our four theories.

1. If X is a proper scheme of finite type over k, then one has canonical isomorphisms
$$H_c^i(X, \mathbf{Z}(j)) = H^i(X, \mathbf{Z}(j))$$
$$H_i^{BM}(X, \mathbf{Z}(j)) = H_i(X, \mathbf{Z}(j))$$

2. If X is a smooth scheme of finite type over k of pure dimension d and
Z is a closed subscheme in X, then there are long exact sequences:

$$\ldots \to H_i(X - Z, \mathbf{Z}(j)) \to H_i(X, \mathbf{Z}(j)) \to H_c^{2d-i}(Z, \mathbf{Z}(d-j))$$

$$\to H_{i-1}(X - Z, \mathbf{Z}(j)) \to \ldots$$

$$\ldots \to H_{2d-i}^{BM}(Z, \mathbf{Z}(d-j)) \to H^i(X, \mathbf{Z}(j)) \to H^i(X - Z, \mathbf{Z}(j))$$

$$\to H_{2d-i-1}^{BM}(Z, \mathbf{Z}(d-j)) \to \ldots$$

In particular if X is a smooth scheme of finite type over k one has
canonical isomorphisms

$$H_i(X, \mathbf{Z}(j)) = H_c^{2d-i}(Z, \mathbf{Z}(d-j))$$

$$H^i(X, \mathbf{Z}(j)) = H_{2d-i}^{BM}(Z, \mathbf{Z}(d-j)).$$

Remark 9.5 One can also define motivic (co-)homology theories for coefficients other than $\mathbf{Z}$. More precisely, for any commutative ring R one can define theories $H^i(X, R(j))$, $H_i(X, R(j))$ etc. in such a way that the usual universal coefficients theorems hold.

References

1. S. Bloch. Algebraic cycles and higher K-theory. *Adv. in Math.*, 61:267–304, 1986.
2. K.S. Brown and S.M. Gersten. Algebraic K-theory and generalized sheaf cohomology. *Lecture Notes in Math. 341*, pages 266–292, 1973.
3. Eric M. Friedlander. Algebraic cycles, Chow varieties, and Lawson homology. *Compositio Math.*, 77:55–93, 1991.
4. Eric M. Friedlander. Some computations of algebraic cycle homology. *K-theory*, 8:271–285, 1994.
5. Eric M. Friedlander. Filtrations on algebraic cycles and homology. *Ann. Sci. Ec. Norm. Sup.*, 28(3):317–343, 1995.
6. Eric M. Friedlander and O. Gabber. Cycle spaces and intersection theory. *Topological methods in modern mathematics*, pages 325–370, 1993.
7. Eric M. Friedlander and H. Blaine Lawson. A theory of algebraic cocycles. *Ann. of Math.*, 136:361–428, 1992.
8. Eric M. Friedlander and H. Blaine Lawson. Duality relating spaces of algebraic cocycles and cycles. *Topology*, 36(2):533–565, 1997.
9. Eric M. Friedlander and H. Blaine Lawson. Moving algebraic cycles of bounded degree. *Invent. Math.*, 132(1):91–119, 1998.
10. Eric M. Friedlander and B. Mazur. Correspondence homomorphisms for singular varieties. *Annales d'Institut Fourier*, 44:703–727, 1994.
11. Eric M. Friedlander and B. Mazur. *Filtrations on the homology of algebraic varieties*, volume 529 of *Memoir of the AMS*. AMS, Providence, RI, 1994.

12. H. Hironaka. Resolution of singularities of an algebraic variety over a field of characteristic zero. *Ann. of Math.*, 79:109–326, 1964.
13. H. Blaine Lawson. Algebraic cycles and homotopy theory. *Ann. of Math.*, 129:599–632, 1989.
14. M. Nagata. Imbedding of an abstract variety in a complete variety. *J. Math. Kyoto Univ.*, 2:1–10, 1962.
15. E. Nart. The Bloch complex in codimension one and arithmetic duality. *Journal of Number Theory*, 32(3):321–331, 1989.
16. Y. Nisnevich. The completely decomposed topology on schemes and associated descent spectral sequences in algebraic K-theory. In *Algebraic K-theory: connections with geometry and topology*, pages 241–342. Kluwer Acad. Publ., Dordrecht, 1989.
17. M. Raynaud and L. Gruson. Criteres de platitude et de projectivite. *Inv. Math.*, 13:1–89, 1971.
18. J. Roberts. Chow's moving lemma. In *Algebraic geometry*, pages 89–96. Wolters-Noordhof Publ., Oslo, 1970.
19. P. Samuel. *Methodes d'algebre abstraite en geometrie algebrique*. Ergebnisse der Mathematik, N.F.4. Springer-Verlag, Berlin, 1955.
20. A. Suslin. Higher Chow groups and ctale cohomology. *This volume*.
21. Andrei Suslin and Vladimir Voevodsky. Singular homology of abstract algebraic varieties. *Invent. Math.*, 123(1):61–94, 1996.
22. A. Suslin and V. Voevodsky. Relative cycles and Chow sheaves. *This volume*.
23. V. Voevodsky. Cohomological theory of presheaves with transfers. *This volume*.
24. V. Voevodsky. Triangulated categories of motives over a field. *This volume*.
25. V. Voevodsky. Homology of schemes. *Selecta Mathematica, New Series*, 2(1):111–153, 1996.

5

Triangulated Categories of Motives Over a Field

Vladimir Voevodsky

Contents

1 Introduction   **188**

2 Geometrical motives   **189**
   2.1   The triangulated category of geometrical motives 189
   2.2   Summary of main results 194

3 Motivic complexes   **198**
   3.1   Nisnevich sheaves with transfers and the category $DM_-^{eff}(k)$ 198
   3.2   The embedding theorem 205
   3.3   Etale sheaves with transfers 214
   3.4   Motives of varieties of dimension ≤ 1 215
   3.5   Fundamental distinguished triangles in the category of geo-
        metrical motives 218

4 Homology of algebraic cycles and duality   **223**
   4.1   Motives of schemes of finite type 223
   4.2   Bivariant cycle cohomology 228
   4.3   Duality in the triangulated categories of motives 233

1. Introduction

In this paper we construct for any perfect field k a triangulated category $DM_-^{eff}(k)$ which is called the *triangulated category of (effective) motivic complexes* over k (the minus sign indicates that we consider only complexes bounded from the above). This construction provides a natural categorical framework to study different algebraic cycle cohomology theories ([3],[13],[9],[7]) in the same way as the derived category of the etale sheaves provides a categorical framework for the etale cohomology. The first section of the paper may be considered as a long introduction. In §2.1 we give an elementary construction of a triangulated category $DM_{gm}^{eff}(k)$ of *effective geometrical motives* over k which is equivalent to the full triangulated subcategory in $DM_{-}^{eff}(k)$ generated by "motives" of smooth varieties. In §2.2 we give a detailed summary of main results of the paper.

We do not discuss here the relations of our theory to the hypothetical theory of mixed motives or more generally mixed motivic sheaves ([1],[2]) primarily because it would require giving a definition of what the later theory is (or should be) which deserves a separate careful consideration. We would like to mention also that though for rational coefficients the standard motivic conjectures predict that $DM_{gm}^{eff}(k)$ should be equivalent to the derived category of bounded complexes over the abelian category of mixed motives most probably no such description exists for integral or finite coefficients (see 4.3.8).

Most of the proofs in this paper are based on results obtained in [7] and [15]. In particular due to the resolution of singularities restriction in [7] the results of Section 4 are proven at the moment only for fields of characteristic zero. As a general rule the restrictions on the base field which are given at the beginning of each section are assumed throughout this section if the opposite is not explicitly declared.

We would like to mention two other constructions of categories similar to our $DM_{gm}(k)$. One was given by M. Levine in [8]. Another one appeared in [16]. At the end of §4.1 we give a sketch of the proof that with the rational coefficients it is equivalent to the category defined here if the base field has resolution of singularities.

I would like to thank Eric M. Friedlander and A. Suslin for many useful discussions which, I hope, helped to make the paper clearer. I would also like to thank A. Neeman whose explanations of the localization theory for "large" triangulated categories were very useful.

2. Geometrical motives

2.1. The triangulated category of geometrical motives

Let k be a field. We denote by Sm/k the category of smooth schemes over k.

For a pair X, Y of smooth schemes over k denote by $c(X, Y)$ the free abelian group generated by integral closed subschemes W in $X \times Y$ which are finite over X and surjective over a connected component of X. An element of $c(X, Y)$ is called a *finite correspondence*[1] from X to Y.

Let now X_1, X_2, X_3 be a triple of smooth schemes over k, $\phi \in c(X_1, X_2)$ be a finite correspondence from X_1 to X_2 and $\psi \in c(X_2, X_3)$ be a finite correspondence from X_2 to X_3. Consider the product $X_1 \times X_2 \times X_3$ and let $pr_i : X_1 \times X_2 \times X_3 \to X_i$ be the corresponding projections. One can verify easily that the cycles $(pr_1 \times pr_2)^*(\phi)$ and $(pr_2 \times pr_3)^*(\psi)$ on $X_1 \times X_2 \times X_3$ are in general position. Let $\psi * \phi$ be their intersection. We set $\psi \circ \phi = (pr_1 \times pr_3)_*(\psi * \phi)$. Note that the push-forward is well defined since ϕ (resp. ψ) is finite over X_1 (resp. X_2).

For any composable triple of finite correspondences α, β, γ one has

$$(\alpha \circ \beta) \circ \gamma = \alpha \circ (\beta \circ \gamma)$$

and therefore one can define a category $SmCor(k)$ such that objects of $SmCor(k)$ are smooth schemes of finite type over k, morphisms are finite correspondences and compositions of morphisms are compositions of correspondences defined above. We will denote the object of $SmCor(k)$ which corresponds to a smooth scheme X by $[X]$.

For any morphism $f : X \to Y$ its graph Γ_f is a finite correspondence from X to Y. It gives us a functor $[-] : Sm/k \to SmCor(k)$.

One can easily see that the category $SmCor(k)$ is additive and one has $[X \coprod Y] = [X] \oplus [Y]$.

Consider the homotopy category $\mathcal{H}^b(SmCor(k))$ of bounded complexes over $SmCor(k)$. We are going to define the triangulated category of effective geometrical motives over k as a localization of $\mathcal{H}^b(SmCor(k))$. Let T be the class of complexes of the following two forms:

1. For any smooth scheme X over k the complex

$$[X \times \mathbf{A}^1] \overset{[pr_1]}{\to} [X]$$

   belongs to T.

2. For any smooth scheme X over k and an open covering $X = U \cup V$ of X the complex

$$[U \cap V] \overset{[j_U] \oplus [j_V]}{\to} [U] \oplus [V] \overset{[i_U] \oplus (-[i_V])}{\to} [X]$$

   belongs to T (here j_U, j_V, i_U, i_V are the obvious open embeddings).

---

[1] Note that a correspondence belongs to $c(X, Y)$ if and only if it has a well defined pull-back with respect to any morphism $X' \to X$ and a well defined push-forward with respect to any morphism $Y \to Y'$.

Denote by $\bar{T}$ the minimal thick subcategory of $\mathcal{H}^b(SmCor(k))$ which contains T. It would be most natural for our purposes to define the category of geometrical motives as the localization of $\mathcal{H}^b(SmCor(k))$ with respect to $\bar{T}$. Unfortunately this definition makes it difficult to formulate results relating our theory to more classical motivic theories (see Proposition 2.1.4 and discussion of 1-motices in Section 3.4). The problem is that in the classical approach one usually replaces additive categories of geometrical nature by their pseudo-abelian (or Karoubian) envelopes, i.e. one formally adds kernels and cokernels of projectors. Note that this operation takes triangulated categories into triangulated categories and tensor categories into tensor categories. We follow this tradition in the definition below, but we would like to mention again that in our case the only reason to do so is to make comparison statements which involve the "classical motives" to look more elegant.

Definition 2.1.1 *Let k be a field. The triangulated category $DM_{gm}^{eff}(k)$ of effective geometrical motives over k is the pseudo-abelian envelope of the localization of the homotopy category $\mathcal{H}^b(SmCor(k))$ of bounded complexes over $SmCor(k)$ with respect to the thick subcategory $\bar{T}$. We denote the obvious functor $Sm/k \to DM_{gm}^{eff}(k)$ by M_{gm}.*

Note that one has the following simple but useful result.

Lemma 2.1.2 *Let X be a smooth scheme over k and $X = U \cup V$ be a Zariski open covering of X. Then there is a canonical distinguished triangle in DM_{gm}^{eff} of the form*

$$M_{gm}(U \cap V) \to M_{gm}(U) \oplus M_{gm}(V) \to M_{gm}(X) \to M_{gm}(U \cap V)[1].$$

For a pair of smooth schemes X, Y over k we set

$$[X] \otimes [Y] = [X \times Y].$$

For any smooth schemes X_1, Y_1, X_2, Y_2 the external product of cycles defines a homomorphism:

$$c(X_1, Y_1) \otimes c(X_2, Y_2) \to c(X_1 \times X_2, Y_1 \times Y_2)$$

which gives us a definition of tensor product of morphisms in $SmCor(k)$. Together with the obvious commutativity and associativity isomorphisms it gives us a tensor category structure on $SmCor(k)$. This structure defines in the usual way a tensor triangulated category structure on $\mathcal{H}^b(SmCor(k))$ which can be descended to the category $DM_{gm}^{eff}(k)$ by the universal property of localization. We proved the following simple result.

Proposition 2.1.3 *The category $DM_{gm}^{eff}(k)$ has a tensor triangulated category structure such that for any pair X, Y of smooth schemes over k there is a canonical isomorphism $M_{gm}(X \times Y) \cong M_{gm}(X) \otimes M_{gm}(Y)$.*

Note that the unit object of our tensor structure is $M_{gm}(Spec(k))$. We will denote it by $\mathbf{Z}$. For any smooth scheme X over k the morphism $X \to Spec(k)$ gives us a morphism in $DM_{gm}^{eff}(k)$ of the form $M_{gm}(X) \to \mathbf{Z}$. There is a canonical distinguished triangle $\tilde{M}_{gm}(X) \to M_{gm}(X) \to \mathbf{Z} \to \tilde{M}_{gm}(X)[1]$ where $\tilde{M}_{gm}(X)$ is the reduced motive of X represented in $\mathcal{H}^b(SmCor(k))$ by the complex $[X] \to [Spec(k)]$.

We define the Tate object $\mathbf{Z}(1)$ of $DM_{gm}^{eff}(k)$ as $\tilde{M}_{gm}(\mathbf{P}^1)[-2]$. We further define $\mathbf{Z}(n)$ to be the n-th tensor power of $\mathbf{Z}(1)$. For any object A of $DM_{gm}^{eff}(k)$ we denote by $A(n)$ the object $A \otimes \mathbf{Z}(n)$.

Finally we define the triangulated category $DM_{gm}(k)$ of geometrical motives over k as the category obtained from $DM_{gm}^{eff}(k)$ by inverting $\mathbf{Z}(1)$. More precisely, objects of $DM_{gm}(k)$ are pairs of the form (A, n) where A is an object of $DM_{gm}^{eff}(k)$ and $n \in \mathbf{Z}$ and morphisms are defined by the following formula

$$Hom_{DM_{gm}}((A, n), (B, m)) = \lim_{k \geq -n, -m} Hom_{DM_{gm}^{eff}}(A(k+n), B(k+m)).$$

The category $DM_{gm}(k)$ with the obvious shift functor and class of distinguished triangles is clearly a triangulated category. The situation with the tensor structure on DM_{gm} is more subtle. In general it is not possible to get a tensor structure on the category obtained from a tensor additive category by inverting an object Q but it is possible when the permutation involution on $Q \otimes Q$ is the identity morphism. We will show below (2.1.5) that it is true in our case. For a field k which admits resolution of singularities we will also show that the canonical functor from $DM_{gm}^{eff}(k)$ to $DM_{gm}(k)$ is a full embedding, i.e. that the Tate object is quasi-invertible in $DM_{gm}^{eff}(k)$ (4.3.1).

To prove that the permutation involution for the Tate object equals identity we need first to establish a connection between our $DM_{gm}^{eff}(k)$ and Chow motives.

Consider the category $\mathcal{C}$ whose objects are smooth projective schemes over k and morphisms are given by the formula

$$Hom_{\mathcal{C}_0}(X, Y) = \oplus_{X_i} A_{dim(X_i)}(X_i \times Y)$$

where X_i are the connected components of X and $A_d(-)$ is the group of cycles of dimension d modulo rational equivalence. Its pseudo-abelian envelope (i.e. the category obtained from $\mathcal{C}$ by formal addition of cokernels of projectors) is called the category of effective Chow motives over k. Denote this category by $Chow^{eff}(k)$ and let

$$Chow : SmProj/k \to Chow^{eff}(k)$$

be the corresponding functor on the category of smooth projective varieties over k.

Proposition 2.1.4 *There exists a functor* $Chow^{eff}(k) \to DM_{gm}^{eff}(k)$ *such that the following diagram commutes:*

$$
\begin{array}{ccc}
SmProj/k & \longrightarrow & Sm/k \\
\scriptstyle Chow \downarrow & & \downarrow \scriptstyle M_{gm} \\
Chow^{eff}(k) & \longrightarrow & DM_{gm}^{eff}(k).
\end{array}
$$

Proof: It is clearly sufficient to show that for smooth projective varieties X, Y over k there is a canonical homomorphism

$$A_{dim(X)}(X \times Y) \to Hom_{DM_{gm}^{eff}}(M_{gm}(X), M_{gm}(Y)).$$

Denote by $h_0(X, Y)$ the cokernel of the homomorphism

$$c(X \times \mathbf{A}^1, Y) \to c(X, Y)$$

given by the difference of restrictions to $X \times \{0\}$ and $X \times \{1\}$. One can easily see that the obvious homomorphism

$$c(X, Y) \to Hom_{DM_{gm}^{eff}}(M_{gm}(X), M_{gm}(Y))$$

factors through $h_0(X, Y)$. On the other hand by definition of rational equivalence we have a canonical homomorphism

$$h_0(X, Y) \to A_{dim(X)}(X \times Y)$$

which is an isomorphism by [7, Th. 7.1].

Corollary 2.1.5 *The permutation involution on* $\mathbf{Z}(1) \otimes \mathbf{Z}(1)$ *is identity in* DM_{gm}^{eff}.

Proof: It follows from Proposition 2.1.4 and the corresponding well known fact for Chow motives.

Remark: It follows from Corollary 4.2.6 that if k admits resolution of singularities then the functor constructed in Proposition 2.1.4 is a full embedding and any distinguished triangle in $DM_{gm}^{eff}(k)$ with all three vertices being of the form $M_{gm}(X)$ for smooth projective varieties X splits.

2.2. Summary of main results

We will give in this section a summary of main results of the paper. Their proofs are based on the construction given in Section 3. Essentially our main technical tool is an embedding of the category DM_{gm}^{eff} to the derived category of sheaves with some additional structure (transfers) which allow us to apply all the standard machinery of sheaves and their cohomology to our category of motives.

Motives of singular varieties. For a field k which admits resolution of singularities we construct in Section 4.1 an extension of the functor $M_{gm} : Sm/k \to DM_{gm}^{eff}$ to a functor $M_{gm} : Sch/k \to DM_{gm}^{eff}$ from the category of all schemes of finite type over k. This extended functor has the following main properties.

Kunnet formula. For schemes of finite type X, Y over k one has a canonical isomorphism $M_{gm}(X \times Y) = M_{gm}(X) \otimes M_{gm}(Y)$ (4.1.7).

Homotopy invariance For a scheme of finite type X over k the morphism $M_{gm}(X \times \mathbf{A}^1) \to M_{gm}(X)$ is an isomorphism (4.1.8).

Mayer-Vietoris axiom. For a scheme X of finite type over k and an open covering $X = U \cup V$ of X one has a canonical distinguished triangle of the form

$$M_{gm}(U \cap V) \to M_{gm}(U) \oplus M_{gm}(V) \to M_{gm}(X) \to M_{gm}(U \cap V)[1]$$

(4.1.1).

Blow-up distinguished triangle. For a scheme X of finite type over k and a closed subscheme Z in X denote by $p_Z : X_Z \to X$ the blow-up of Z in X. Then there is a canonical distinguished triangle of the form

$$M_{gm}(p_Z^{-1}(Z)) \to M_{gm}(X_Z) \oplus M_{gm}(Z) \to M_{gm}(X) \to M_{gm}(p_Z^{-1}(Z))[1]$$

(4.1.3).

Projective bundle theorem. Let X be a scheme of finite type over k and $\mathcal{E}$ be a vector bundle on X. Denote by $p : \mathbf{P}(\mathcal{E}) \to X$ the projective bundle over X associated with $\mathcal{E}$. Then one has a canonical isomorphism:

$$M_{gm}(\mathbf{P}(\mathcal{E})) = \oplus_{n=0}^{dim\mathcal{E}-1} M_{gm}(X)(n)[2n]$$

(4.1.11).

It should be mentioned that our functor $M_{gm}(-)$ provides a way to extend motivic homology and cohomology type theories to not necessarily smooth varieties. In the case of *homology* this extension is essentially the only one possible. The situation with cohomology is different. For instance the Picard group for a smooth scheme X is canonically isomorphic to the motivic cohomology group $Hom_{DM_{gm}^{eff}}(M_{gm}(X), \mathbf{Z}(1)[2])$. We do not have any "motivic" description for the Picard groups of arbitrary varieties though, since the functor $X \mapsto Pic(X)$ considered on the category of all schemes is not homotopy invariant and does not have the descent property for general blow-ups.

We hope that there is another more subtle approach to "motives" of singular varieties which makes use of some version of "reciprocity functors" introduced by Bruno Kahn instead of homotopy invariant functors considered in this paper which gives "right" answers for all schemes.

Motives with compact support. For any field k which admits resolution of singularities we construct a functor M_{gm}^c from the category of schemes of finite type over k and proper morphisms to the category DM_{gm}^{eff} which has the following properties:

1. For a proper scheme X over k one has a canonical isomorphism

$$M_{gm}^c(X) = M_{gm}(X)$$

2. For a scheme X of finite type over k and a closed subscheme Z in X one has a canonical distinguished triangle

$$M_{gm}^c(Z) \to M_{gm}^c(X) \to M_{gm}^c(X - Z) \to M_{gm}^c(Z)[1]$$

(4.1.5).

3. For a flat equidimensional morphism $f : X \to Y$ of schemes of finite type over k there is a canonical morphism

$$M_{gm}^c(Y)(n)[2n] \to M_{gm}^c(X)$$

where $n = dim(X/Y)$ (4.2.4).

4. For any scheme of finite type X over k one has a canonical isomorphism $M_{gm}^c(X \times \mathbf{A}^1) = M_{gm}^c(X)(1)[2]$ (4.1.8).

Blow-ups of smooth varieties and Gysin distinguished triangles.
Let k be a perfect field, X be a smooth scheme over k and Z be a smooth
closed subscheme in X everywhere of codimension c. Then one has:

Motives of blow-ups. There is a canonical isomorphism

$$M_{gm}(X_Z) = M_{gm}(X) \oplus (\oplus_{n=1}^{c-1} M_{gm}(Z)(n)[2n])$$

(3.5.3).

Gysin distinguished triangle. There is a canonical distinguished triangle

$$M_{gm}(X - Z) \to M_{gm}(X) \to M_{gm}(Z)(c)[2c] \to M_{gm}(X - Z)[1]$$

(3.5.4).

Quasi-invertibility of the Tate object.   Let k be a field which admits
resolution of singularities. Then for any objects A, B of $DM_{gm}^{eff}(k)$ the
obvious morphism

$$Hom(A, B) \to Hom(A(1), B(1))$$

is an isomorphism. In particular the functor

$$DM_{gm}^{eff} \to DM_{gm}$$

is a full embedding (4.3.1).

Duality.   For any field k which admits resolution of singularities the category $DM_{gm}(k)$ is a "rigid tensor triangulated category". More precisely
one has (4.3.7):

1. For any pair of objects A, B in DM_{gm} there exists the internal *Hom*-
   object $\underline{Hom}_{DM}(A, B)$. We set A^* to be $\underline{Hom}_{DM}(A, \mathbf{Z})$.

2. For any object A in $DM_{gm}(k)$ the canonical morphism $A \to (A^*)^*$ is
   an isomorphism.

3. For any pair of objects A, B in DM_{gm} there are canonical isomorphisms

$$\underline{Hom}_{DM}(A, B) = A^* \otimes B$$

$$(A \otimes B)^* = A^* \otimes B^*$$

For any smooth equidimensional scheme X of dimension n over k there is a canonical isomorphism (4.3.2)

$$M_{gm}(X)^* = M^c_{gm}(X)(-n)[-2n].$$

Let X be a smooth equidimensional scheme of dimension n over k and Z be a closed subscheme of X. Applying duality to the localization sequence for M^c_{gm} we get the following generalized Gysin distinguished triangle

$$M_{gm}(X - Z) \to M_{gm}(X) \to M^c_{gm}(Z)^*(n)[2n] \to M_{gm}(X - Z)[1]$$

Relations to the algebraic cycle homology theories. Let again k be a field which admits resolution of singularities and X be a scheme of finite type over k.

Higher Chow groups. If X is quasi-projective and equidimensional of dimension n the groups $Hom(\mathbf{Z}(i)[j], M^c_{gm}(X))$ (i.e. Borel-Moore homology in our theory) are canonically isomorphic to the higher Chow groups $CH^{n-i}(X, j - 2i)$ (Corollary 4.2.9, see also [3], [12], [7]).

Suslin homology. For any X of finite type over a perfect field k the groups (i.e. homology in our theory) $Hom(\mathbf{Z}[j], M_{gm}(X))$ are isomorphic to the Suslin homology groups $h_j(X)$ (3.2.7, see also [9],[13]).

Motivic cohomology. The groups $Hom(M_{gm}(X), \mathbf{Z}(i)[j])$ (i.e. cohomology in our theory) are isomorphic to motivic cohomology groups $H^j_M(X, \mathbf{Z}(i))$ introduced in [7]. In particular one has the desired relations with algebraic K-theory (loc. cit.).

Bivariant cycle cohomology. For any X, Y of finite type over k any $i \geq 0$ and any j the group $Hom(M_{gm}(X)(i)[j], M^c_{gm}(Y))$ is isomorphic to the bivariant cycle cohomology group $A_{i,j-2i}(X, Y)$ (see [7]).

Relations to the Chow motives. For any pair of smooth projective varieties X, Y over a field which admits resolution of singularities the group $Hom_{DM_{gm}(k)}(M_{gm}(X), M_{gm}(Y))$ is canonically isomorphic to the group of cycles of dimension $dim(X)$ on $X \times Y$ modulo rational equivalence (4.2.6(1)). In particular the full additive subcategory in DM_{gm} which is closed under direct summands and generated by objects of the form $M_{gm}(X)(n)[2n]$ for smooth projective X over k and $n \in \mathbf{Z}$ is canonically equivalent as a tensor additive category to the category of Chow motives over k. Moreover any distinguished triangle with all three vertices being in this subcategory splits by 4.2.6(2).

Motivic complexes and the homotopy t-structure. We construct for any perfect field k an embedding of the category $DM_{gm}^{eff}(k)$ to a bigger tensor triangulated category $DM_-^{eff}(k)$ of motivic complexes over k. The image of $DM_{gm}^{eff}(k)$ is "dense" in $DM_-^{eff}(k)$ in the sense that the smallest triangulated subcategory in $DM_-^{eff}(k)$ which is closed with respect to direct sums amd contains the image of $DM_{gm}^{eff}(k)$ coincides with $DM_-^{eff}(k)$. Almost by definition the category $DM_-^{eff}(k)$ has a (non degenerate) t-structure whose heart is the abelian category $HI(k)$ of homotopy invariant Nisnevich sheaves with transfers on Sm/k. Note that this t-structure is not the desired "motivic t-structure" on $DM_-^{eff}(k)$ whose heart is the abelian category of (effective) mixed motives over k. On the other hand it seems to be important to have at least one non-degenerate t-structure on $DM_-^{eff}(k)$ since one may hope then to construct the "correct" one by the gluing technique.

Motivic complexes in the etale and h- topologies. We construct in Section 3.3 a category $DM_{-,et}^{eff}(k)$ of the etale (effective) motivic complexes. There are canonical functors

$$DM_-^{eff}(k) \to DM_{-,et}^{eff}(k) \to DM_h(k)$$

(where the last category is the one constructed in [16]). We show that the second functor is an equivalence if k admits resolution of singularities and the first one becomes an equivalence after tensoring with $\mathbf{Q}$. We also show that the category $DM_{-,et}^{eff}(k, \mathbf{Z}/n\mathbf{Z})$ of the etale motivic complexes with $\mathbf{Z}/n\mathbf{Z}$ coefficients is equivalent to the derived category of complexes bounded from the above over the abelian category of sheaves of $\mathbf{Z}/n\mathbf{Z}$-modules on the small etale site $Spec(k)_{et}$ for n prime to $char(k)$ and $DM_{-,et}^{eff}(k, \mathbf{Z}/p\mathbf{Z}) = 0$ for $p = char(k)$.

3. Motivic complexes

3.1. Nisnevich sheaves with transfers and the category $DM_-^{eff}(k)$

The definition of the category DM_{gm}^{eff} given above being quite geometrical is very inconvenient to work with. In this section we introduce another triangulated category - the category of (effective) motivic complexes. We will denote this category by $DM_-^{eff}(k)$.

Definition 3.1.1 *Let k be a field. A presheaf with transfers on Sm/k is an additive contravariant functor from the category $SmCor(k)$ to the category of abelian groups. It is called a Nisnevich sheaf with transfers if the corresponding presheaf of abelian groups on Sm/k is a sheaf in the Nisnevich topology.*

We denote by $PreShv(SmCor(k))$ the category of presheaves with transfers on Sm/k and by $Shv_{Nis}(SmCor(k))$ its full subcategory consisting of Nisnevich sheaves with transfers.

For a smooth scheme X over k denote by $L(X)$ the corresponding representable presheaf with transfers, i.e. $L(X)(Y) = c(Y, X)$ for any smooth scheme Y over k. Note that this presheaf is a particular case of Chow presheaves considered in [14]. In the notations of that paper $L(X)$ is the Chow presheaf $c_{equi}(X/Spec(k), 0)$.

The following lemma is straight-forward.

Lemma 3.1.2 *For any smooth scheme X over k the presheaf $L(X)$ is a sheaf in the Nisnevich topology.*

The role of the Nisnevich topology becomes clear in the following proposition:

Proposition 3.1.3 *Let X be a scheme of finite type over k and $\{U_i \to X\}$ be a Nisnevich covering of X. Denote the coproduct $\coprod U_i$ by U and consider the complex of presheaves:*

$$\ldots \to L(U \times_X U) \to L(U) \to L(X) \to 0$$

with the differential given by alternating sums of morphisms induced by the projections. Then it is exact as a complex of Nisnevich sheaves.

Proof: Note first that the presheaves $L(-)$ can be extended in the obvious way to presheaves on the category of smooth schemes over k which are not necessarily of finite type. Since points in the Nisnevich topology are henselian local schemes it is sufficient to verify that for any smooth henselian local scheme S over k the sequence of abelian groups

$$\ldots \to L(U \times_X U)(S) \to L(U)(S) \to L(X)(S) \to 0$$

is exact. For a closed subscheme Z in $X \times S$ which is quasi-finite over S denote by $L(Z/S)$ the free abelian group generated by irreducible components of Z which are finite and surjective over S. Clearly, the groups $L(Z/S)$ are covariantly functorial with respect to morphisms of quasi-finite schemes over S. For a closed subscheme Z in $X \times S$ which is finite over S denote by

Z_U the fiber product $Z \times_{X \times S} (U \times S)$. One can easily see that our complex is a filtered inductive limit of complexes K_Z of the form

$$\ldots \to L(Z_U \times_Z Z_U/S) \to L(Z_U/S) \to L(Z/S) \to 0$$

where Z runs through closed subschemes of $X \times S$ which are finite and surjective over S. Since S is henselian any such Z is again henselian and hence the covering $Z_U \to Z$ splits. Let us choose a splitting $s_1 : Z \to Z_U$. We set $s_k : (Z_U)_Z^k \to (Z_U)_Z^{k+1}$ to be the product $s_1 \times_Z Id_{(Z_U)_Z^k}$. The morphisms s_k induce homomorphisms of abelian groups

$$\sigma_k : L((Z_U)_Z^k/S) \to L((Z_U)_Z^{k+1}/S).$$

One can verify easily that these homomorphisms provide a homotopy of the identity morphism of K_Z to zero.

Remarks:

1. Proposition 3.1.3 is false in the Zariski topology even if the corresponding covering $\{U_i \to X\}$ is a Zariski open covering.

2. The only property of the Nisnevich topology which we used in the proof of Proposition 3.1.3 is that a scheme finite over a "Nisnevich point" is again a "Nisnevich point". In particular exactly the same argument works for the etale topology and moreover for any topology where higher direct images for finite equidimensional morphisms vanish.

Theorem 3.1.4 *The category of Nisnevich sheaves with transfers is abelian and the embedding*

$$Shv_{Nis}(SmCor(k)) \to PreShv(SmCor(k))$$

has a left adjoint functor which is exact.

Proof: Note first that the category $PreShv(SmCor(k))$ is abelian by obvious reasons. To prove that $Shv_{Nis}(SmCor(k))$ is abelian it is clearly sufficient to construct an exact left adjoint functor to the canonical embedding. Its existence is an immediate corollary of Lemma 3.1.6 below.

Lemma 3.1.5 *Let $f : [X] \to [Y]$ be a morphism in $SmCor(k)$ and $p : U \to Y$ be a Nisnevich covering. Then there exists a Nisnevich covering $U' \to X$ and a morphism $f' : [U'] \to [U]$ such that $[p] \circ f' = f \circ [p']$.*

Proof: It follows immediately from the fact that a scheme finite over a henselian local scheme is a disjoint union of henselian local schemes.

Lemma 3.1.6 *Let F be a presheaf with transfers on Sm/k. Denote by F_{Nis} the sheaf in the Nisnevich topology associated with the corresponding presheaf of abelian groups on Sm/k. Then there exists a unique Nisnevich sheaf with transfers such that its underlying presheaf is F_{Nis} and the canonical morphism of presheaves $F \to F_{Nis}$ is a morphism of presheaves with transfers.*

Proof: Let us prove the uniqueness part first. Let F_1 and F_2 be two presheaves with transfers given together with isomorphisms of the corresponding presheaves on Sm/k with F_{Nis}. Consider a morphism $f : [X] \to [Y]$ in $SmCor(k)$. We have to show that for any section ϕ of F_{Nis} on Y one has $F_1(f)(\phi) = F_2(f)(\phi)$. Let $p : U \to Y$ be a Nisnevich covering such that $F_{Nis}(p)(\phi)$ belongs to the image of the homomorphism $F(U) \to F_{Nis}(U)$ (such a covering always exist) and $U' \to X$, $f' : [U'] \to [U]$ be as in Lemma 3.1.5. Then we have

$$F_1([p'])F_1(f)(\phi) = F_2(f')F_2([p])(\phi) = F_2([p'])F_2(f)(\phi) = F_1([p'])F_2(f)(\phi)$$

which implies that $F_1(f)(\phi) = F_2(f)(\phi)$ because p' is a covering and F_1, F_2 are Nisnevich sheaves.

To give F_{Nis} the structure of a presheaf with transfers it is sufficient to construct for any section ϕ of F_{Nis} on a smooth scheme X over k a morphism of presheaves $[\phi] : L(X) \to F_{Nis}$ such that $[\phi]$ takes the tautological section of $L(X)$ over X to ϕ. Let $p : U \to X$ be a Nisnevich covering of X such that $F_{Nis}(p)(\phi)$ corresponds to a section ϕ_U of F over U satisfying the condition $F(pr_1)(\phi_U) = F(pr_2)(\phi_U)$ where $pr_i : U \times_X U \to U$ are the projections. Then ϕ_U defines a morphism of sheaves $[\phi_U] : L(U) \to F_{Nis}$ such that $[\phi_U] \circ L(pr_1) = [\phi_U] \circ L(pr_2)$. Applying Proposition 3.1.3 to our covering we conclude that $[\phi_U]$ can be descended to a morphism $[\phi] : L(X) \to F_{Nis}$.

The proof of the following lemma is standard.

Lemma 3.1.7 *The category $Shv_{Nis}(SmCor(k))$ has sufficiently many injective objects.*

Proposition 3.1.8 *Let X be a smooth scheme over a field k and F be a Nisnevich sheaf with transfers. Then for any $i \in \mathbf{Z}$ there is a canonical isomorphism:*

$$Ext^i_{Shv_{Nis}(SmCor(k))}(L(X), F) = H^i_{Nis}(X, F).$$

Proof: Since the category $Shv_{Nis}(SmCor(k))$ has sufficiently many injective objects by Lemma 3.1.7 and for any Nisnevich sheaf with tranfers G one has $Hom(L(X), G) = G(X)$ we only have to show that for any injective Nisnevich sheaf with transfers I one has $H^n_{Nis}(X, I) = 0$ for $n > 0$. It is sufficient to show that the Chech cohomology groups with coefficients in I vanish for all X (see [10, Prop. III.2.11]). Let $\mathcal{U} = \{U_i \to X\}$ be a Nisnevich covering of X and α be a class in $\check{H}^n_{Nis}(\mathcal{U}/X, I)$. Let us set $U = \coprod U_i$. Then α is given by a section a of I over U^{n+1}_X or equivalently by a morphism $L(U^{n+1}_X) \to I$ in the category of sheaves with transfers. In view of Lemma 3.1.3 the fact that a is a cocycle implies that as a morphism it can be factored through a morphism from $ker(L(U^n) \to L(U^{n-1}))$ to I. Since I is an injective object in $Shv_{Nis}(SmCor(k))$ it implies that a can be factored through U^n, i.e. that $\alpha = 0$.

Consider now the derived category $D^-(Shv_{Nis}(SmCor(k)))$ of complexes of Nisnevich sheaves with transfers bounded from the above. We will use the following generalization of Proposition 3.1.8.

Proposition 3.1.9 *Let X be a smooth scheme over a field k and K be a complex of Nisnevich sheaves with transfers bounded from the above. Then for any $i \in \mathbf{Z}$ there is a canonical isomorphism*

$$Hom_{D^-(Shv_{Nis}(SmCor(k)))}(L(X), K[i]) = \mathbf{H}^i_{Nis}(X, K)$$

where the groups on the right hand side are the hypercohomology of X in the Nisnevich topology with coefficients in the complex of sheaves K .

Proof: By Proposition 3.1.8 and the cohomological dimension theorem for the Nisnevich topology ([11]) we conclude that the sheaf with transfers $L(X)$ has finite Ext-dimension. Therefore it is sufficient to prove our proposition for a bounded complex K. Then morphisms in the derived category can be computed using an injective resolution for K and the statement follows from Proposition 3.1.8.

Remarks:

1. The natural way to prove Proposition 3.1.8 would be to show that if I is an injective Nisnevich sheaf with transfers then I is an injective Nisnevich sheaf. To do so one could construct the free sheaf with transfers functor from the category of Nisnevich sheaves to the category of Nisnevich sheaves with transfers which is left adjoint to the forgetfull functor and then use the same argument as in the standard

proof of the fact that an injective sheaf is an injective presheaf. The problem is that while the free sheaf with transfers functor is right exact simple examples show that it is not left exact.

2. Though Proposition 3.1.8 gives us an interpretation of Ext-groups from sheaves $L(X)$ in the category of Nisnevich sheaves with transfers we do not know how to describe the Ext-groups from $L(X)$ in the category of all Nisnevich sheaves.

Definition 3.1.10 *A presheaf with transfers F is called homotopy invariant if for any smooth scheme X over k the projection $X \times \mathbf{A}^1 \to X$ induces isomorphism $F(X) \to F(X \times \mathbf{A}^1)$.*

A Nisnevich sheaf with transfers is called homotopy invariant if it is homotopy invariant as a presheaf with transfers.

The following proposition relates presheaves with transfers to pretheories ([15]). Though the proof given here is fairly long the statement itself is essentially obvious and the only problem is to "unfold" all the definitions involved.

Proposition 3.1.11 *Let k be a field. Then any presheaf with transfers on Sm/k is a pretheory of homological type over k.*

Proof: Let F be a presheaf with transfers on Sm/k, U be a smooth scheme over k and $X \to U$ be a smooth curve over U. Let us remind that we denote in [15] by $c_{equi}(X/U, 0)$ the free abelian group generated by integral closed subschemes of X which are finite over U and surjective over an irreducible component of U. For any morphism of smooth schemes $f : U' \to U$ there is a base change homomorphism

$$cycl(f) : c_{equi}(X/U, 0) \to c_{equi}(X \times_U U'/U', 0)$$

and for any morphism $p : X \to X'$ of smooth curves over U there is a push-forward homomorphism

$$p_* : c_{equi}(X/U, 0) \to c_{equi}(X'/U, 0).$$

To give F a pretheory structure we have to construct a homomorphism

$$\phi_{X/U} : c_{equi}(X/U, 0) \to Hom(F(X), F(U))$$

such that the following two conditions hold ([15, Def. 3.1]):

1. For an element $\mathcal{Z}$ in $c_{equi}(X/U, 0)$ which corresponds to a section $s : U \to X$ of the projection $X \to U$ we have $\phi_{X/U}(\mathcal{Z}) = F(s)$.

2. For any morphism of smooth schemes $f : U' \to U$ an element $\mathcal{Z}$ in $c_{equi}(X/U, 0)$ and an element u in $F(X)$ we have:

$$F(f)(\phi_{X/U}(\mathcal{Z})(u)) = \phi_{X \times_U U'/U'}(cycl(f)(\mathcal{Z}))(F(f_X)(u))$$

where f_X is the projection $X \times_U U' \to X$.

For (F, ϕ) to be a pretheory of homological type we require in addition that for any morphism $p : X \to X'$ of smooth curves over U, any element u in $F(X')$ and any element $\mathcal{Z}$ in $c_{equi}(X/U, 0)$ we have

$$\phi_{X'/U}(p_*(\mathcal{Z}))(u) = \phi_{X/U}(\mathcal{Z})(F(p)(u)).$$

Let $g : X \to U$ be a smooth curve over a smooth scheme U and Z be an integral closed subscheme in X which belongs to $c_{equi}(X/U, 0)$. Consider the closed embedding $Id_X \times g : X \to X \times U$. Then the image of Z under this embedding belongs to $c(U, X) = Hom_{SmCor(k)}([U], [X])$.

This construction defines homomorphisms

$$\alpha_{X/U} : c_{equi}(X/U, 0) \to Hom_{SmCor(k)}([U], [X])$$

and since F is a functor on $SmCor(k)$ we can define $\phi_{X/U}$ as the composition of $\alpha_{X/U}$ with the canonical homomorphism

$$Hom_{SmCor(k)}([U], [X]) \to Hom(F(X), F(U)).$$

One can verify easily that homomorphisms $\alpha_{X/U}$ satisfy the following conditions which implies immediately the statement of the proposition.

1. For a section $s : U \to X$ of the projection $X \to U$ we have $\alpha_{X/U}(s) = [s]$.

2. For a morphism of smooth schemes $f : U' \to U$ and an element $\mathcal{Z}$ in $c_{equi}(X/U, 0)$ we have

$$\alpha_{X/U}(\mathcal{Z}) \circ [f] = [f_X] \circ \alpha_{X \times_U U'/U'}(cycl(f)(\mathcal{Z}))$$

where again f_X is the projection $X \times_U U' \to X$.

3. For a morphism $p : X \to X'$ of smooth curves over U and an element $\mathcal{Z}$ in $c_{equi}(X/U, 0)$ we have

$$\alpha_{X'/U}(p_*(\mathcal{Z})) = [p] \circ \alpha_{X/U}(\mathcal{Z}).$$

The following proposition summarizes some of the main properties of homotopy invariant presheaves with transfers which follow from the corresponding results about homotopy invariant pretheories proven in [15] by Proposition 3.1.11.

Theorem 3.1.12 *Let F be a homotopy invariant presheaf with transfers on Sm/k. Then the Nisnevich sheaf with transfers F_{Nis} associated with F is homotopy invariant. Moreover as a presheaf on Sm/k it coincides with the Zariski sheaf F_{Zar} associated with F. If in addition the field k is perfect one has:*

1. *The presheaves $H^i_{Nis}(-, F_{Nis})$ have canonical structures of homotopy invariant presheaves with transfers.*

2. *For any smooth scheme over k one has*

$$H^i_{Zar}(X, F_{Zar}) = H^i_{Nis}(X, F_{Nis}).$$

Theorem 3.1.12 implies immediately the following result.

Proposition 3.1.13 *For any perfect field k the full subcategory $HI(k)$ of the category $Shv_{Nis}(SmCor(k))$ which consists of homotopy invariant sheaves is abelian and the inclusion functor $HI(k) \to Shv_{Nis}(SmCor(k))$ is exact.*

We denote by $DM^{eff}_-(k)$ the full subcategory of $D^-(Shv_{Nis}(SmCor(k)))$ which consists of complexes with homotopy invariant cohomology sheaves. Proposition 3.1.13 implies that $DM^{eff}_-(k)$ is a triangulated subcategory. Moreover, the standard t-structure on $D^-(Shv_{Nis}(SmCor(k)))$ induces a t-structure on $DM^{eff}_-(k)$ whose heart is equivalent to the category $HI(k)$. We call this t-structure the homotopy t-structure on $DM^{eff}_-(k)$.

The connection between our geometrical category $DM^{eff}_{gm}(k)$ and the category DM^{eff}_- of (effective) motivic complexes will become clear in the next section. We will use the following simple fact about $DM^{eff}_-(k)$.

Lemma 3.1.14 *The category $DM^{eff}_-(k)$ is pseudo-abelian i.e. each projector in $DM^{eff}_-(k)$ has kernel and cokernel.*

3.2. The embedding theorem

We prove in this section our main technical result - the fact that the category $DM^{eff}_{gm}(k)$ admits a natural full embedding as a tensor triangulated category to the category $DM^{eff}_-(k)$. All through this section we assume that k is a perfect field. Since the objects of $DM^{eff}_-(k)$ are essentially some complexes of sheaves on the category Sm/k of smooth schemes over k the existence of this embedding permits us to apply the machinery of sheaves and their cohomology to the category $DM_{gm}^{eff}(k)$.

Let us define first a tensor structure on the category of Nisnevich sheaves with transfers. In view of Lemma 3.1.6 it is sufficient to define tensor products for presheaves with transfers. Note that for two presheaf with transfers their tensor product in the category of presheaves does not have transfers and thus a more sophisticated construction is required.

Let F be a presheaf with transfers. Let A_F be the set of pairs of the form $(X, \phi \in F(X))$ for all smooth schemes X over k. Since a section $\phi \in F(X)$ is the same as a morphism of presheaves with transfers $L(X) \to F$ there is a canonical surjection of presheaves

$$\oplus_{(X,\phi) \in A_F} L(X) \to F.$$

Iterating this construction we get a canonical left resolution $\mathcal{L}(F)$ of F which consists of direct sums of presheaves of the form $L(X)$ for smooth schemes X over k. We set

$$L(X) \otimes L(Y) = L(X \times Y)$$

and for two presheaves with transfers F, G:

$$F \otimes G = \underline{H}_0(\mathcal{L}(F) \otimes \mathcal{L}(G)).$$

One can verify easily that this construction indeed provides us with a tensor structure on the category of presheaves with transfers and thus on the category of Nisnevich sheaves with transfers. Moreover using these canonical "free resolutions" we get immediately a definition of tensor product on the derived category $D^-(Shv_{Nis}(SmCor(k)))$.

Remark: Note that the tensor product of two homotopy invariant presheaves with transfers is almost never a homotopy invariant presheaf. To define tensor structure on $DM_-^{eff}(k)$ we need the description of this category given in Proposition 3.2.3 below.

For presheaves with transfers F, G denote by $\underline{Hom}(F, G)$ the presheaf with transfers of the form:

$$\underline{Hom}(F, G)(X) = Hom(F \otimes L(X), G)$$

where the group on the right hand side is the group of morphisms in the category of presheaves with transfers. One can verify easily that for any three presheaves with transfers F, G, H there is a canonical isomorphism

$$Hom(F, \underline{Hom}(G, H)) \to Hom(F \otimes G, H),$$

i.e. $\underline{Hom}(-, -)$ is the internal Hom-object with respect to our tensor product. Note also that if G is a Nisnevich sheaf with transfers then $\underline{Hom}(F, G)$ is a Nisnevich sheaf with transfers for any presheaf with transfers F.

To construct a functor $DM_{gm}^{eff}(k) \to DM_-^{eff}(k)$ as well as to define the tensor structure on DM_-^{eff} we will need an alternative description of DM_-^{eff} as a localization of the derived category $D^-(Shv_{Nis}(SmCor(k)))$.

Let $\Delta^\bullet$ be the standard cosimplicial object in Sm/k. For any presheaf with transfers F on Sm/k let $\underline{C}_*(F)$ be the complex of presheaves on Sm/k of the form $\underline{C}_n(F)(X) = F(X \times \Delta^n)$ with differentials given by alternated sums of morphisms which correspond to the boundary morphisms of $\Delta^\bullet$ (note that $\underline{C}_n(F) = \underline{Hom}(L(\Delta^n), F)$). This complex is called the singular simplicial complex of F. One can easily see that if F is a presheaf with transfers (resp. a Nisnevich sheaf with transfers) then $\underline{C}_*(F)$ is a complex of presheaves with transfers (resp. Nisnevich sheaves with transfers). We denote the cohomology sheaves $\underline{H}^{-i}(\underline{C}_*(F))$ by $\underline{h}_i^{Nis}(F)$. In the case when F is of the form $L(X)$ for a smooth scheme X over k we will abbreviate the notation $\underline{C}_*(L(X))$ (resp. $\underline{h}_i^{Nis}(L(X))$) to $\underline{C}_*(X)$ (resp. $\underline{h}_i^{Nis}(X)$). The complex $\underline{C}_*(X)$ is called the Suslin complex of X and its homology groups over $Spec(k)$ are called the Suslin homology of X (see [13], [15]).

Lemma 3.2.1 *For any presheaf with transfers F over k the sheaves $\underline{h}_i^{Nis}(F)$ are homotopy invariant.*

Proof: The cohomology presheaves $\underline{h}_i(F)$ of the complex $\underline{C}_*(F)$ are homotopy invariant for any presheaf F (see [15, Prop. 3.6]). The fact that the associated Nisnevich sheaves are homotopy invariant follows from Theorem 3.1.12.

We say that two morphisms of presheaves with transfers $f_0, f_1 : F \to G$ are homotopic if there is a morphism $h : F \otimes L(\mathbf{A}^1) \to G$ such that

$$h \circ (Id_F \otimes L(i_0)) = f_0$$

$$h \circ (Id_F \otimes L(i_1)) = f_1$$

where $i_0, i_1 : Spec(k) \to \mathbf{A}^1$ are the points 0 and 1 respectively. A morphism of presheaves with transfers $f : F \to G$ is a (strict) homotopy equivalence if there is a morphism $g : G \to F$ such that $f \circ g$ and $g \circ f$ are homotopic to the identity morphisms of G and F respectively. One can verify easily that the composition of two homotopy equivalences is a homotopy equivalence.

Consider homomorphisms

$$\eta_n : F \to \underline{C}_n(F)$$

which take a section of F on a smooth scheme X to its preimage on $X \times \Delta^n$ under the projection $X \times \Delta^n \to X$. We will need the following lemma.

Lemma 3.2.2 *The morphisms η_n are strict homotopy equivalences.*

Proof: Since Δ^n is (noncanonically) isomorphic to $\mathbf{A}^n$ we have for any $n > 0$:

$$\underline{C}_n(F) = \underline{C}_1(\underline{C}_{n-1}(F))$$

and therefore it is sufficient to show that η_1 is a homotopy equivalence. Let $\alpha : \underline{C}_1(F) \to F$ be the morphism which sends a section of F on $X \times \mathbf{A}^1$ to its restriction to $X \times \{0\}$. Then $\alpha \circ \eta_1 = Id$ and it remains to show that there is a morphism $h : \underline{C}_1(F) \otimes L(\mathbf{A}^1) \to \underline{C}_1(F)$ such that

$$h \circ (Id \otimes L(i_1)) = Id$$

$$h \circ (Id \otimes L(i_0)) = \eta_1 \circ \alpha.$$

We set h to be the morphism adjoint to the morphism

$$\underline{C}_1(F) \to \underline{Hom}(L(\mathbf{A}^1), \underline{C}_1(F)) = \underline{C}_2(F)$$

which sends a section of F on $X \times \mathbf{A}^1$ to its preimage on $X \times \mathbf{A}^2$ under the morphism $\mathbf{A}^2 \to \mathbf{A}^1$ given by the multiplication of functions.

Lemma 3.2.1 implies that $\underline{C}_*(-)$ is a functor from the category of Nisnevich sheaves with transfers on Sm/k to $DM_-^{eff}(k)$. The following proposition shows that it can be extended to a functor from the corresponding derived category which provides us with the alternative description of $DM_-^{eff}(k)$ mentioned above.

Proposition 3.2.3 *The functor $\underline{C}_*(-)$ can be extended to a functor*

$$\mathbf{R}C : D^-(Shv_{Nis}(SmCor(k))) \to DM_-^{eff}(k)$$

which is left adjoint to the natural embedding. The functor $\mathbf{R}C$ identifies $DM_-^{eff}(k)$ with localization of $D^-(Shv_{Nis}(SmCor(k)))$ with respect to the localizing subcategory generated by complexes of the form

$$L(X \times \mathbf{A}^1) \overset{L(pr_1)}{\to} L(X)$$

for smooth schemes X over k.

Proof: Denote by A the class of objects in $D^-(Shv_{Nis}(SmCor(k)))$ of the form $L(X \times \mathbf{A}^1) \xrightarrow{L(pr_1)} L(X)$ for smooth schemes X over k and let $\mathcal{A}$ be the localizing subcategory generated by A, i.e. the minimal triangulated subcategory in $D^-(Shv_{Nis}(SmCor(k)))$ which contains A and is closed under direct sums and direct summands. Consider the localization $D^-(Shv_{Nis}(SmCor(k)))/\mathcal{A}$ of $D^-(Shv_{Nis}(SmCor(k)))$ with respect to the class of morphisms whose cones are in $\mathcal{A}$. Our proposition asserts that the restriction of the canonical projection

$$D^-(Shv_{Nis}(SmCor(k))) \to D^-(Shv_{Nis}(SmCor(k)))/\mathcal{A}$$

to the subcategory $DM_-^{eff}(k)$ is an equivalence and that the composition of this projection with the inverse equivalence coincides on sheaves with transfers with the functor $\underline{C}_*(F)$.

To prove this assertion it is sufficient to show that the following two statements hold:

1. For any sheaf with transfers F on Sm/k the canonical morphism $F \to \underline{C}_*(F)$ is an isomorphism in $D^-(Shv_{Nis}(SmCor(k)))/\mathcal{A}$.

2. For any object T of $DM_-^{eff}(k)$ and any object B of $\mathcal{A}$ one has $Hom(B,T) = 0$.

To prove the second statement we may assume (by definition of $\mathcal{A}$) that B is of the form $L(X \times \mathbf{A}^1) \to L(X)$ for a smooth scheme X over k. By Proposition 3.1.9 it is sufficient to show that the projection $X \times \mathbf{A}^1 \to X$ induces isomorphisms on the hypercohomology groups

$$\mathbf{H}^*(X,T) \to \mathbf{H}^*(X \times \mathbf{A}^1, T).$$

which follows by the hypercohomology spectral sequence from Theorem 3.1.12.

To prove the first statement we need the following two lemmas.

Lemma 3.2.4 *For any object T of $D^-(Shv_{Nis}(SmCor(k)))$ and an object S of $\mathcal{A}$ the object $T \otimes S$ belongs to $\mathcal{A}$.*

Proof: Since $\mathcal{A}$ is the localizing subcategory generated by A it is sufficient to consider the case of T being of the form $L(X)$ and S being in A. Then it follows from the fact that $L(X) \otimes L(Y) = L(X \times Y)$.

Lemma 3.2.5 *Let $f : F \to G$ be a homotopy equivalence of presheaves with transfers. Then the cone of f belongs to $\mathcal{A}$.*

Proof: We have to show that f becomes an isomorphism after localization with respect to $\mathcal{A}$. It is sufficient to show that a morphism of sheaves with transfers which is homotopic to the identity equals the identity morphism in $D^-(Shv_{Nis}(SmCor(k)))/\mathcal{A}$. By definition of homotopy we have only to show that for any F the morphisms

$$Id \otimes L(i_0) : F \to F \otimes L(\mathbf{A}^1)$$

$$Id \otimes L(i_1) : F \to F \otimes L(\mathbf{A}^1)$$

are equal. Denote by I^1 the kernel of the morphism $L(\mathbf{A}^1) \to L(pt)$. Then the difference $Id \otimes L(i_0) - Id \otimes L(i_1)$ can be factored through $F \otimes I^1$ which is zero in $D^-(Shv_{Nis}(SmCor(k)))/\mathcal{A}$ by Lemma 3.2.4 since I^1 belongs to $\mathcal{A}$.

Let F be a sheaf with transfers. Denote by $\underline{C}_{\geq 1}(F)$ the cokernel of the obvious morphism of complexes of sheaves $F \to \underline{C}_*(F)$. To finish the proof of the proposition we have to show that $\underline{C}_{\geq 1}(F)$ belongs to $\mathcal{A}$. Let $\tilde{\underline{C}}_n(F) = coker(\eta_n)$ where η_n is the morphism defined right before Lemma 3.2.2. Since the differential in $\underline{C}_*(F)$ takes $Im(\eta_n)$ to $Im(\eta_{n-1})$ the sheaves $\tilde{\underline{C}}_n(F)$ form a quotient complex of $\underline{C}_*(F)$ which is clearly quasi-isomorphic to $\underline{C}_{\geq 1}(F)$. Since $\mathcal{A}$ is a localizing subcategory it is sufficient now to note that for each n $\tilde{\underline{C}}_n(F)$ belongs to $\mathcal{A}$ by Lemmas 3.2.5, 3.2.2. Proposition is proven.

We define the tensor structure on $DM_-^{eff}(k)$ as the descent of the tensor structure on $D^-(Shv_{Nis}(SmCor(k)))$ with respect to the projection $\mathbf{R}C$. Note that such a descent exists by the universal property of localization and Lemma 3.2.4.

Remark: Note that the inclusion functor

$$DM_-^{eff}(k) \to D^-(Shv_{Nis}(SmCor(k)))$$

does not preserve tensor structures. In fact many hard conjectures in motivic theory (Beilinson's vanishing conjectures and Quillen-Lichtenbaum conjectures in the first place) can be formulated as statements about the behavior of the tensor structures on these two categories with respect to the inclusion functor.

The following theorem is the main technical result we use to study the category $DM_{gm}^{eff}(k)$.

Theorem 3.2.6 *Let k be a perfect field. Then there is a commutative diagram of tensor triangulated functors of the form*

$$\begin{array}{ccc}
\mathcal{H}^b(SmCor(k)) & \xrightarrow{L} & D^-(Shv_{Nis}(SmCor(k))) \\
\downarrow & & \downarrow \mathbf{R}C \\
DM_{gm}^{eff}(k) & \xrightarrow{i} & DM_-^{eff}(k)
\end{array}$$

such that the following conditions hold:

1. *The functor i is a full embedding with a dense image.*

2. *For any smooth scheme X over k the object $\mathbf{R}C(L(X))$ is canonically isomorphic to the Suslin complex $\underline{C}_*(X)$*

Proof: The only statement which requires a proof is that the functor i exists and is a full embedding. The fact that it is a tensor triangulated functor follows then immediately from the corresponding property of the composition $\mathbf{R}C \circ L$ and the fact that DM_{gm}^{eff} is a localization of $\mathcal{H}^b(SmCor(k))$. In view of Lemma 3.1.14 we may replace $DM_{gm}^{eff}(k)$ by the localization of the category $\mathcal{H}^b(SmCor(k))$ with respect to the thick subcategory generated by objects of the following two types:

1. Complexes of the form $L(X \times \mathbf{A}^1) \to L(X)$ for smooth schemes X over k.

2. Complexes of the form $L(U \cap V) \to L(U) \oplus L(V) \to L(X)$ for Zariski open coverings of the form $X = U \cup V$.

We denote the class of objects of the first type by T_{hom} and the class of objects of the second type by T_{MV}.

To prove the existence of i it is clearly sufficient to show that $\mathbf{R}C$ takes T_{hom} and T_{MV} to zero. The fact that $\mathbf{R}C(T_{hom}) = 0$ follows immediately from the definition of this functor (see Proposition 3.2.3). Note that $\mathbf{R}C(L(Y)) = \underline{C}_*(Y)$ for any smooth scheme Y over k. Thus to prove that $\mathbf{R}C(T_{MV}) = 0$ we have to show that for an open covering $X = U \cup V$ of a smooth scheme X the total complex of the bicomplex

$$0 \to \underline{C}_*(U \cap V) \to \underline{C}_*(U) \oplus \underline{C}_*(V) \to \underline{C}_*(X) \to 0$$

is exact in the Nisnevich topology.

Note that this sequence of complexes of sheaves is left exact and the cokernel of the last arrow is isomorphic to the singular simplicial complex $\underline{C}_*(L(X)/(L(U) + L(V)))$ of the quotient presheaf $L(X)/(L(U) + L(V))$. The sheaf in the Nisnevich topology associated with this presheaf is zero by Proposition 3.1.3 and therefore $\underline{C}_*(L(X)/(L(U) + L(V)))$ is quasi-isomorphic to zero in the Nisnevich topology by [15, Theorem 5.9]. This proves the existence of the functor i.

To prove that i is a full embedding we proceed as follows. Consider the category $D^-(PreShv(SmCor(k)))$. One can verify easily that $D^-(Shv_{Nis}(SmCor(k)))$ is the localization of $D^-(PreShv(SmCor(k)))$ with respect to the localizing subcategory generated by presheaves with transfers F such that $F_{Nis} = 0$. The functor L from our diagram lifts to a functor

$$L_0 : \mathcal{H}^b(SmCor(k)) \to D^-(PreShv(SmCor(k)))$$

which is clearly a full embedding. Moreover the localizing subcategory generated by the image of L_0 coincides with $D^-(PreShv(SmCor(k)))$.

Let $T = T_{hom} \cup T_{MV}$ and let $\mathcal{T}$ be the localizing subcategory in $D^-(PreShv(SmCor(k)))$ generated by T. The general theory of localization of triangulated categories implies now that it is sufficient to show that for any presheaf with transfers F such that $F_{Nis} = 0$ we have $F \in \mathcal{T}$.

Consider the family of functors $H^i : Sm/k \to Ab$ of the form

$$H^i(X) = Hom_{D^-(PreShv(SmCor(k)))/\mathcal{T}}(L(X), F[i]).$$

It clearly suffice to show that $H^i = 0$ for all i. Note that since $\mathcal{T}$ contains T_{MV} our family has the Mayer-Vietoris long exact sequences for Zariski open coverings. Thus by [5, Theorem 1'] we have only to show that the sheaves in the Zariski topology associated with H^i's are zero. By the construction H^i's are presheaves with transfers and since T_{hom} belongs to $\mathcal{T}$ they are homotopy invariant. Thus $(H^i)_{Zar} = (H^i)_{Nis}$ by Theorem 3.1.12(2).

A morphism from $L(X)$ to $F[i]$ in $D^-(PreShv(SmCor(k)))/\mathcal{T}$ can be represented by a diagram of the form

$$
\begin{array}{ccc}
L(X) & & F[i] \\
 f \searrow & & \swarrow g \\
 & K &
\end{array}
$$

such that the cone of g belongs to $\mathcal{T}$. Since $\underline{C}_*(-)$ is an exact functor from the category of presheaves with transfers to the category of complexes of presheaves with transfers it can be extended to $D^-(PreShv(SmCor(k)))$ and as was established in the proof of Proposition 3.2.3 for any object K of $D^-(PreShv(SmCor(k)))$ the canonical morphism $K \to \underline{C}_*(K)$ is an isomorphism in $D^-(PreShv(SmCor(k)))/\mathcal{T}$. Thus it is sufficient to show that there exists a Nisnevich covering $U \to X$ of X such that the composition:

$$L(U) \to L(X) \xrightarrow{f} K \to \underline{C}_*(K)$$

is zero. It follows from the fact that $(\underline{C}_*(K))_{Nis} = 0$.

The following corollary gives an "explicit" description of morphisms in the category $DM_{gm}^{eff}(k)$ in terms of certain hypercohomology groups. It will be used extensively in further sections to provide "motivic" interpretations for different algebraic cycle homology type theories.

Corollary 3.2.7 *Let k be a perfect field. Then for any smooth schemes X, Y over k and any $j \in \mathbf{Z}$ one has a canonical isomorphism*

$$Hom_{DM_{gm}^{eff}}(M_{gm}(X), M_{gm}(Y)[j]) = \mathbf{H}_{Nis}^j(X, \underline{C}_*(Y)) = \mathbf{H}_{Zar}^j(X, \underline{C}_*(Y)).$$

In particular for any smooth scheme X over k the groups $Hom_{DM_{gm}^{eff}}(\mathbf{Z}[j], M_{gm}(X))$ are isomorphic to the Suslin homology of X (see [13],[9]).

Proof: The first isomorphism follows from Theorem 3.2.6 and Proposition 3.1.9. The second one follows from Theorem 3.1.12 and Lemma 3.2.1.

One of the important advantages of the category $DM_-^{eff}(k)$ is that it has internal Hom-objects for morphisms from objects of DM_{gm}^{eff}.

Proposition 3.2.8 *Let A, B be objects of $DM_-^{eff}(k)$ and assume that A belongs to the image of $DM_{gm}^{eff}(k)$. Then there exists the internal Hom-object $\underline{Hom}_{DM_-^{eff}}(A, B)$. If $A = \underline{C}_*(X)$ for a smooth scheme X and $p_X : X \to Spec(k)$ is the canonical morphism then*

$$\underline{Hom}_{DM_-^{eff}}(A, B) = \mathbf{R}(p_X)_*((p_X)^*(B)).$$

Proof: We may consider the internal $RHom$-object $\underline{RHom}(A, B)$ in the derived category of unbounded complexes over $Shv_{Nis}(SmCor(k))$. It is clearly sufficient to verify that it belongs to $DM_-^{eff}(k)$. The fact that its cohomology sheaves are homotopy invariant follows from Theorem 3.1.12(1). The fact that $H^i(\underline{RHom}(A, B)) = 0$ for sufficiently large i follows from the lemma below.

Lemma 3.2.9 *Let X be a smooth scheme over a perfect field k and F be a homotopy invariant Nisnevich sheaf with transfers over k. Denote by $p : X \to Spec(k)$ the canonical morphism. Then the sheaves $R^i p_*(p^*(F))$ on Sm/k are zero for $i > dim(X)$.*

Proof: By Theorem 3.1.12 these sheaves are homotopy invariant Nisnevich sheaves with transfers. Therefore, by [15, Cor. 4.19] for any smooth scheme Y over k and any nonempty open subset U of Y the homomorphisms

$$R^i p_*(p^*(F))(Y) \to R^i p_*(p^*(F))(U)$$

are injective and our result follows from the cohomological dimension theorem for the Nisnevich cohomology ([11]).

3.3. Etale sheaves with transfers

To avoid unpleasant technical difficulties we assume in this section that k has finite etale cohomological dimension. This section is very sketchy mainly because of Propositions 3.3.2, 3.3.3 below which show that with rational coefficients the etale topology gives the same motivic answers as the Nisnevich topology and with finite coefficients everything degenerates to the usual etale cohomology.

Denote by $Shv_{et}(SmCor(k))$ the category of presheaves with transfers on Sm/k which are etale sheaves. One can easily see that the arguments of Section 3.1 work for the etale topology as well as for the Nisnevich topology. In particular one has the following result.

Proposition 3.3.1 *For any field k the category $Shv_{et}(SmCor(k))$ is abelian and there exists the associated sheaf functor*

$$Shv_{Nis}(SmCor(k)) \to Shv_{et}(SmCor(k))$$

which is exact. Denote by $D^-(Shv_{et}(SmCor(k)))$ the derived category of complexes bounded from the above over $Shv_{et}(SmCor(k))$. Then for any object A of this category and any smooth scheme X over k one has a canonical isomorphism

$$Hom_{D^-(Shv_{et})}(L(X), A) = \mathbf{H}^0_{et}(X, A).$$

We denote by $DM^{eff}_{-,et}(k)$ the full subcategory of $D^-(Shv_{et}(SmCor(k)))$ which consists of complexes with homotopy invariant cohomology sheaves. Using results of [15] we see immediately that this is a triangulated subcategory, the analog of Proposition 3.2.3 holds in the etale case and the associated etale sheaf functor gives us a functor $DM^{eff}_-(k) \to DM^{eff}_{-,et}(k)$.

Proposition 3.3.2 *The functor*

$$DM^{eff}_-(k) \otimes \mathbf{Q} \to DM^{eff}_{-,et}(k) \otimes \mathbf{Q}$$

is an equivalence of triangulated categories.

Proof: It follows immediately from Proposition 3.3.1 and the comparison Theorem [15, Prop. 5.28].

Denote by $DM^{eff}_{-,et}(k, \mathbf{Z}/n\mathbf{Z})$ the category constructed in the same way as $DM^{eff}_{-,et}(k)$ from the abelian category $Shv_{et}(SmCor(k), \mathbf{Z}/n\mathbf{Z})$ of etale sheaves of $\mathbf{Z}/n\mathbf{Z}$-modules with transfers.

Proposition 3.3.3 *Denote by p the exponential characteristic of the field k. Then one has:*

1. *Let $n \geq 0$ be an integer prime to p. Then the functor*

$$DM^{eff}_{-,et}(k, \mathbf{Z}/n\mathbf{Z}) \to D^-(Shv(Spec(k)_{et}, \mathbf{Z}/n\mathbf{Z}))$$

   where $Shv(Spec(k)_{et}, \mathbf{Z}/n\mathbf{Z})$ is the abelian category of sheaves of $\mathbf{Z}/n\mathbf{Z}$-modules on the small etale site $Spec(k)_{et}$ which takes a complex of sheaves on Sm/k to its restriction to $Spec(k)_{et}$ is an equivalence of triangulated categories.

2. *For any $n \geq 0$ the category $DM^{eff}_{-,et}(k, \mathbf{Z}/p^n\mathbf{Z})$ is equivalent to the zero category.*

Proof: The first statement follows from the rigidity theorem [15, Th. 5.25]. The second one follows from the fact that $\mathbf{Z}/p\mathbf{Z} = 0$ in $DM^{eff}_{-,et}(k)$ which was proven (in slightly different form) in [16].

3.4. Motives of varieties of dimension ≤ 1

Consider the thick subcategory $d_{\leq n}DM^{eff}_{gm}(k)$ in the category $DM^{eff}_{gm}(k)$ generated by objects of the form $M_{gm}(X)$ for smooth schemes of dimension $\leq n$ over k. Similarly let $d_{\leq n}DM^{eff}_-(k)$ be the localizing subcategory in $DM^{eff}_-(k)$ generated by objects of the form $\underline{C}_*(X)$ for smooth schemes X of dimension $\leq n$ over k. These categories are called the category of (effective) geometrical n-motives and the category of (effective) n-motivic complexes respectively. One can observe easily that the inclusion functors

$$d_{\leq n}DM^{eff}_-(k) \to d_{\leq n+1}DM^{eff}_-(k)$$

have right adjoints given on the level of sheaves with transfers by taking the canonical free resolution of the restriction of a sheaf to the category of smooth schemes of dimension $\leq n$. Unfortunately, the corresponding left adjoint functors most probably do not exist for $n \geq 2$. In fact it can be shown that if the standard motivic assumptions hold the existence of such an adjoint for $n = 2$ would imply that the group of 1-cycles modulo algebraic equivalence on a variety of dimension three is either finitely generated or is not countable, which is known to be wrong.

In this section we will describe "explicitly" the category $d_{\leq 0}DM_{gm}^{eff}(k)$ of geometrical 0-motives and give some partial description of the category $d_{\leq 1}DM_{gm}^{eff}(k)$ of geometrical 1-motives. In particular we obtain a description of the motivic cohomology of weight one which will be used in the next section to construct the standard distinguished triangles in $DM_{gm}^{eff}(k)$.

We start with the category $d_{\leq 0}DM_{gm}^{eff}(k)$ of zero motives. As always we assume that k is a perfect field. Choose an algebraic closure $\bar{k}$ of k and let $G_k = Gal(\bar{k}/k)$ be the Galois group of $\bar{k}$ over k. Denote by $Perm(G_k)$ the full additive subcategory of the category of $\mathbf{Z}[G_k]$-modules which consists of permutational representations (i.e. representations which are formal linear envelopes of finite G_k-sets). The tensor structure on the category of $\mathbf{Z}[G_k]$ modules gives us a tensor structure on $Perm(k)$ such that $Perm(k)$ becomes a rigid tensor additive category.

Let $Shv(Perm(G_k))$ be the category of additive contravariant functors from $Perm(k)$ to the category of abelian groups. Clearly the category $Shv(Perm(G_k))$ is abelian and we have a full embedding

$$Perm(k) \to Shv(Perm(G_k))$$

which takes an object to the corresponding representable functor. We denote this functor by $L_G(-)$. Note that objects of the form $L_G(-)$ are projective objects in $Shv(Perm(G_k))$ and therefore the corresponding functor

$$\mathcal{H}^b(Perm(k)) \to D^b(Shv(Perm(k)))$$

is also a full embedding.

Denote by $\mathcal{H}^b(Perm(k))_\oplus$ the pseudo-abelian envelope of $\mathcal{H}^b(Perm(k))$. Note that the embedding

$$\mathcal{H}^b(Perm(k)) \to D^b(Shv(Perm(k)))$$

has a canonical extension to an embedding

$$\mathcal{H}^b(Perm(k))_\oplus \to D^b(Shv(Perm(k)))$$

since the right hand side category is pseudo-abelian.

The following proposition which gives an explicit description of the categories $d_{\leq 0}DM_{gm}^{eff}(k)$ and $d_{\leq 0}DM_-^{eff}(k)$ is an easy corollary of Theorem 3.2.6 and the elementary Galois theory.

Proposition 3.4.1 *There is a commutative diagram of the form*

$$\begin{array}{ccc} \mathcal{H}^b(Perm(k))_\oplus & \to & D^-(Shv(Perm(k))) \\ \downarrow & & \downarrow \\ d_{\leq 0}DM_{gm}^{eff}(k) & \to & d_{\leq 0}DM_-^{eff}(k) \end{array}$$

with vertical arrows being equivalences of tensor triangulated categories.

Remarks:

1. Note that while the category $D^-(Shv(Perm(k)))$ (and thus the category $d_{\leq 0}DM_-^{eff}(k)$) has an obvious t-structure it is not clear how to construct any nondegenerate t-structure on the category $\mathcal{H}^b(Perm(k))_\oplus$. In particular, even for zero motives there seems to be no reasonable abelian theory underlying the triangulated theory which we consider.

2. The rational coefficients analog of $\mathcal{H}^b(Perm(k))_\oplus$ is canonically equivalent to the derived category of bounded complexes over the abelian category of G_k-representations over **Q**. Thus with rational coefficients our theory for zero dimensional varieties gives the usual Artin motives.

The following result which was proven in different forms in [9] and [13] (see also [6]) contains essentially all the information one needs to describe the category $d_{\leq 1}DM_{gm}^{eff}(k)$ of 1-motives.

Theorem 3.4.2 *Let $p: C \to Spec(k)$ be a smooth connected curve over a field k. Denote by $Alb(C)$ the Albanese variety of C and let $\underline{Alb}(C)$ be the sheaf of abelian groups on Sm/k represented by $Alb(X)$. Then one has:*

1. $\underline{h}_i^{Nis}(C) = 0$ *for $i \neq 0, 1$.*

2. $\underline{h}_1^{Nis}(C) = p_*\mathbf{G}_m$ *if C is proper and $\underline{h}_1^{Nis}(C) = 0$ otherwise. In particular $\underline{h}_1^{Nis}(C) = \mathbf{G}_m$ if and only if C is proper and geometrically connected.*

3. *The kernel of the canonical homomorphism $\underline{h}_0^{Nis}(C) \to \mathbf{Z}$ is canonically isomorphic to a subsheaf $\underline{Alb}(C)$.*

Corollary 3.4.3 *Let X be a smooth scheme over k. Then one has:*

$$Hom_{DM_{gm}^{eff}}(M_{gm}(X), \mathbf{Z}(1)[j]) = H_{Zar}^{j-1}(X, \mathbf{G}_m).$$

Proof: It follows from the definition of $\mathbf{Z}(1)$, Corollary 3.2.7 and the fact that

$$\underline{h}_i(\mathbf{P}^1) = \begin{cases} \mathbf{Z} & \text{for } i = 0 \\ \mathbf{G}_m^* & \text{for } i = 1 \\ 0 & \text{for } i \neq 0, 1. \end{cases}$$

by Theorem 3.4.2.

Unfortunately we are unable to give a reasonable description of the category $d_{\leq 1} DM_{gm}^{eff}(k)$ due to the fact that with the integral coefficients it most probably has no "reasonable" t-structure (see 4.3.8 for a precise statement in the case of the whole category $DM_{gm}^{eff}(k)$). The rational coefficients analog $d_{\leq 1} DM_{gm}^{eff}(k, \mathbf{Q})$ can be easily described in "classical" terms as the derived category of bounded complexes over the abelian category of the Deligne 1-motives over k with rational coefficients. We do not consider this description here mainly because it will not play any role in the rest of the paper.

3.5. Fundamental distinguished triangles in the category of geometrical motives

In this section we will construct several canonical distinguished triangles in $DM_{gm}^{eff}(k)$ which correspond to the standard exact sequences in the cohomology of algebraic varieties. We will also prove the standard decomposition results for motives of projective bundles and blow-ups.

Proposition 3.5.1 *Let X be a smooth scheme over k and $\mathcal{E}$ be a vector bundle over X. Denote by $p : \mathbf{P}(\mathcal{E}) \to X$ the projective bundle over X associated with $\mathcal{E}$. Then one has a canonical isomorphism in $DM_{gm}^{eff}(k)$ of the form:*

$$M_{gm}(\mathbf{P}(\mathcal{E})) = \oplus_{n=0}^{dim\mathcal{E}-1} M_{gm}(X)(n)[2n].$$

Proof: We may assume that $d = dim(\mathcal{E}) > 0$. Let $\mathcal{O}(1)$ be the standard line bundle on $\mathbf{P}(\mathcal{E})$. By Corollary 3.4.3 it defines a morphism $\tau_1 : M_{gm}(\mathbf{P}(\mathcal{E})) \to \mathbf{Z}(1)[2]$. For any $n \geq 0$ we set τ_n to be the composition

$$M_{gm}(\mathbf{P}(\mathcal{E})) \overset{M_{gm}(\Delta)}{\to} M_{gm}(\mathbf{P}(\mathcal{E})^n) = M_{gm}(\mathbf{P}(\mathcal{E}))^{\otimes n} \overset{\tau_1^{\otimes n}}{\to} \mathbf{Z}(n)[2n].$$

Let further σ_n be the composition

$$M_{gm}(\mathbf{P}(\mathcal{E})) \overset{M_{gm}(\Delta)}{\to} M_{gm}(\mathbf{P}(\mathcal{E})) \otimes M_{gm}(\mathbf{P}(\mathcal{E})) \overset{Id_{M_{gm}(\mathbf{P})} \otimes \tau_n}{\to} M_{gm}(X)(n)[2n].$$

We have a morphism

$$\Sigma = \oplus_{n=0}^{d-1}\sigma_n : M_{gm}(\mathbf{P}(\mathcal{E})) \to \oplus_{n=0}^{dim\mathcal{E}-1} M_{gm}(X)(n)[2n].$$

Let us show that it is an isomorphism. Note first that Σ is natural with respect to X. Using the induction on the number of open subsets in a trivializing covering for $\mathcal{E}$ and the distinguished triangles from Lemma 2.1.2 we may assume that $\mathcal{E}$ is trivial. In this case Σ is the tensor product of the corresponding morphism for the trivial vector bundle over $Spec(k)$ and $Id_{M_{gm}(X)}$. It means that we have only to consider the case $X = Spec(k)$. Then our proof goes exactly as the proof of the similar result in [16].

Proposition 3.5.2 *Let X be a smooth scheme over k and $Z \subset X$ be a smooth closed subscheme in X. Denote by $p : X_Z \to X$ the blow-up of Z in X. Then one has a canonical distinguished triangle of the form:*

$$M_{gm}(p^{-1}(Z)) \to M_{gm}(Z) \oplus M_{gm}(X_Z) \to M_{gm}(X) \to M_{gm}(p^{-1}(Z))[1].$$

Proof: Consider the complex $[p^{-1}(Z)] \to [Z] \oplus [X_Z] \to [X]$ in SmCor(k). We have to show that the corresponding object of $DM_{gm}^{eff}(k)$ is zero. Consider the complex Φ of sheaves with transfers of the form

$$L(p^{-1}(Z)) \to L(Z) \oplus L(X_Z) \to L(X).$$

By Theorem 3.2.6 we have only to show that for any homotopy invariant sheaf with transfers F on Sm/k and any $i \in \mathbf{Z}$ we have

$$Hom_{D^-(Shv_{Nis}(SmCor(k)))}(\Phi, F[i]) = 0.$$

Let Φ_0 be the complex of Nisnevich sheaves of the form

$$\mathbf{Z}_{Nis}(p^{-1}(Z)) \to \mathbf{Z}_{Nis}(Z) \oplus \mathbf{Z}_{Nis}(X_Z) \to \mathbf{Z}_{Nis}(X)$$

where $\mathbf{Z}_{Nis}(-)$ denote the freely generated Nisnevich sheaf. By Proposition 3.1.8 the canonical homomorphisms:

$$Hom_{D^-(Shv_{Nis}(SmCor(k)))}(\Phi, F[i]) \to Hom_{D^-(Shv_{Nis}(Sm/k))}(\Phi_0, F[i])$$

are isomorphisms for all $i \in \mathbf{Z}$. The complex Φ_0 is clearly left exact and

$$coker(\mathbf{Z}_{Nis}(Z) \oplus \mathbf{Z}_{Nis}(X_Z) \to \mathbf{Z}_{Nis}(X)) = coker(\mathbf{Z}_{Nis}(X_Z) \overset{\mathbf{Z}_{Nis}(p)}{\to} \mathbf{Z}_{Nis}(X))$$

since $p^{-1}(Z) \to Z$ is the projective bundle of a vector bundle and thus $\mathbf{Z}_{Nis}(p^{-1}(Z)) \to \mathbf{Z}_{Nis}(Z)$ is a surjection in the Nisnevich topology. Therefore we have canonical isomorphisms:

$$Hom_{D^-(Shv_{Nis}(Sm/k))}(\Phi_0, F[i]) = Ext^i_{Shv_{Nis}(Sm/k)}(coker(\mathbf{Z}_{Nis}(p)), F).$$

The last groups are zero by [15, Prop. 5.21].

Proposition 3.5.3 *Let X be a smooth scheme over k and Z be a smooth closed subscheme in X everywhere of codimension c. Denote by $p : X_Z \to X$ the blow-up of Z in X. Then there is a canonical isomorphism*

$$M_{gm}(X_Z) = M_{gm}(X) \oplus (\oplus_{n=1}^{c-1} M_{gm}(Z)(n)[2n]).$$

Proof: In view of Propositions 3.5.2 and 3.5.1 we have only to show that the morphism $M_{gm}(X_Z) \to M_{gm}(X)$ has a canonical splitting. Consider the following diagram (morphisms are numbered for convenience):

$$
\begin{array}{ccc}
M_{gm}(p^{-1}(Z)) & \xrightarrow{\ 1\ } & M_{gm}(q^{-1}(Z \times \{0\})) \\
{\scriptstyle 3}\downarrow & & \downarrow{\scriptstyle 4} \\
M_{gm}(Z) \oplus M_{gm}(X_Z) & \xrightarrow{\ 2\ } & M_{gm}(Z \times \{0\}) \oplus M_{gm}((X \times \mathbf{A}^1)_{Z \times \{0\}}) \\
{\scriptstyle 5}\downarrow & & \downarrow{\scriptstyle 6} \\
M_{gm}(X) & \xrightarrow{M_{gm}(Id \times \{0\})} & M_{gm}(X \times \mathbf{A}^1) \\
{\scriptstyle 7}\downarrow & & \downarrow{\scriptstyle 8} \\
M_{gm}(p^{-1}(Z))[1] & \xrightarrow{1[1]} & M_{gm}(q^{-1}(Z \times \{0\}))[1].
\end{array}
$$

where $q : (X \times \mathbf{A}^1)_{Z \times \{0\}} \to X \times \mathbf{A}^1$ is the blow-up of $Z \times \{0\}$ in $X \times \mathbf{A}^1$.

Note that the morphism $M_{gm}(Id \times \{0\})$ is an isomorphism equal to the isomorphism $M_{gm}(Id \times \{1\})$. Since the later morphism obviously has a lifting to $M_{gm}(Z \times \{0\}) \oplus M_{gm}((X \times \mathbf{A}^1)_{Z \times \{0\}})$ we conclude that the morphism marked (6) has a canonical splitting. Since the vertical triangles are distinguished by Proposition 4.1.3 it gives us a canonical splitting of the morphism (4).

The morphism $p^{-1}(Z) \to q^{-1}(Z \times \{0\})$ is the canonical embedding of the projective bundle $\mathbf{P}(N(Z))$ over Z (where $N(Z)$ is the normal bundle to Z in X) to the projective bundle $\mathbf{P}(N(Z) \oplus \mathcal{O})$ over Z (where $\mathcal{O}$ is the trivial line bundle on Z). Thus by Proposition 3.5.1 the morphism (1) is a splitting monomorphism with a canonical splitting. Thus the composition of the morphism (2) with the canonical splitting of the morphism (4) with the canonical splitting of the morphism (1) gives us a canonical splitting of the morphism (3). Using the fact that the left vertical triangle is distinguished we conclude that the morphism (5) also has a canonical splitting.

The diagram from the proof of Proposition 3.5.3 has another important application. Let X be a smooth scheme over k and Z be a smooth closed subscheme in X everywhere of codimension c.

Using this diagram we can construct a morphism

$$g_Z : M_{gm}(X) \to M_{gm}(Z)(c)[2c]$$

as follows. Consider two morphisms from $M_{gm}(X)$ to $M_{gm}(Z \times \{0\}) \oplus M_{gm}((X \times \mathbf{A}^1)_{Z \times \{0\}})$. One is the composition of the canonical splitting of the morphism (5) with the morphism (2). Another one is the canonical lifting of the morphism $Id \times \{1\}$. Let f be their difference. Composition of f with the morphism (6) is zero. Thus f has a canonical lifting to a morphism $M_{gm}(X) \to M_{gm}(q^{-1}(Z \times \{0\}))$. To get the morphism g_Z we note that $q^{-1}(Z \times \{0\})$ is the projective bundle associated with a vector bundle of dimension $c + 1$ over Z and use Proposition 3.5.1.

The composition of this morphism with the canonical morphism

$$M_{gm}(Z)(c)[2c] \to \mathbf{Z}(c)[2c]$$

which is induced by the projection $Z \to Spec(k)$ is the class of Z in the motivic cohomology group $H^{2c}_M(X, \mathbf{Z}(c))$.

The following proposition shows that g_Z fits into a canonical distinguished triangle which leads to the Gysin exact sequences in cohomology theories.

Proposition 3.5.4 *Let X be a smooth scheme over k and Z be a smooth closed subscheme in X everywhere of codimension c. Then there is a canonical distinguished triangle in $DM^{eff}_{gm}(k)$ of the form*

$$M_{gm}(X - Z) \overset{M_{gm}(j)}{\to} M_{gm}(X) \overset{g_Z}{\to} M_{gm}(Z)(c)[2c] \to M_{gm}(X - Z)[1]$$

(here j is the open embedding $X - Z \to X$).

Proof: Denote by $M_{gm}(X/(X - Z))$ the object in $DM^{eff}_{gm}(k)$ which corresponds to the complex $[X - Z] \to [X]$. Note that its image in $DM^{eff}_-(k)$ is canonically isomorphic to $\underline{C}_*(L(X)/L(X - Z))$ and that there is a distinguished triangle of the form:

$$M_{gm}(X - Z) \to M_{gm}(X) \to M_{gm}(X/(X - Z)) \to M_{gm}(X - Z)[1].$$

To prove the proposition it is sufficient to show that there is an isomorphism $M_{gm}(X/(X - Z)) \to M_{gm}(Z)(c)[2c]$.

To do it we consider again the diagram from the proof of Proposition 3.5.3. Exactly the same arguments as before show that there is a canonical isomorphism

$$M_{gm}(X_Z/(X_Z - p^{-1}(Z))) = M_{gm}(X/(X - Z)) \oplus (\oplus_{n=1}^{c-1} M_{gm}(Z)(n)[2n])$$

which is compatible in the obvious sense with the distinguished triangles for $M_{gm}(X_Z/(X_Z - p^{-1}(Z)))$ and $M_{gm}(X/(X - Z))$. Proceeding as in the construction of the morphism g_Z we see that it can in fact be factored through a canonical morphism

$$\alpha_{(X,Z)} : M_{gm}(X/(X-Z)) \to M_{gm}(Z)(c)[2c].$$

The word "canonical" here means that the following conditions hold:

1. Let $f : X' \to X$ be a smooth morphism. Denote $f^{-1}(Z)$ by Z'. Then the diagram

$$
\begin{array}{ccc}
M_{gm}(X'/(X'-Z')) & \overset{\alpha_{(X',Z')}}{\to} & M_{gm}(Z')(c)[2c] \\
\downarrow & & \downarrow \\
M_{gm}(X/(X-Z)) & \overset{\alpha_{(X,Z)}}{\to} & M_{gm}(Z)(c)[2c]
\end{array}
$$

   commutes.

2. For any smooth scheme Y over k we have

$$\alpha_{(X\times Y, Z\times Y)} = \alpha_{(X,Z)} \otimes Id_{M_{gm}(Y)}.$$

Consider an open covering $X = U \cup V$ of X. Let

$$Z_U = Z \cap U$$

$$Z_V = Z \cap V.$$

One can easily see that there is a canonical distinguished triangle of the form

$$M_{gm}(U \cap V/(U \cap V - Z_U \cap Z_V)) \to M_{gm}(U/(U-Z_U)) \oplus M_{gm}(V/(V-Z_V)) \to$$

$$\to M_{gm}(X/(X-Z)) \to M_{gm}(U \cap V/(U \cap V - Z_U \cap Z_V))[1]$$

and morphisms $\alpha_{-,-}$ map it to the corresponding Mayer-Vietoris distinguished triangle for the open covering $Z = Z_U \cup Z_V$ of Z.

Thus to prove that $\alpha_{(X,Z)}$ is an isomorphism it is sufficient to show that X has an open covering $X = \cup U_i$ such that $\alpha_{(V,Z_V)}$ is an isomorphism for any open subset V which lies in one of the U_i's. In particular we may assume that there exists an etale morphism $f : X \to \mathbf{A}^d$ such that $Z = f^{-1}(\mathbf{A}^{d-c})$. Consider the Cartesian square:

$$
\begin{array}{ccc}
Y & \longrightarrow & Z \times \mathbf{A}^c \\
\downarrow & & \downarrow {\scriptstyle f_{|Z} \times Id} \\
X & \overset{f}{\longrightarrow} & \mathbf{A}^d.
\end{array}
$$

Since f (and therefore $f_{|Z}$) is an etale morphism the diagonal is a connected component of $Z \times_{\mathbf{A}^{d-c}} Z$ and we may consider the open subscheme $X' = Y - (Z \times_{\mathbf{A}^{d-c}} Z - Z)$ of Y. Let Z' be the image of Z in Y. Then pr_1 maps Z' isomorphically to $Z \times \{0\} \subset Z \times \mathbf{A}^c$ and pr_2 maps Z' isomorphically to $Z \subset X$. Moreover

$$pr_1^{-1}(Z \times \{0\}) = Z'$$

and

$$pr_2^{-1}(Z) = Z'.$$

Thus by [15, Prop. 5.18] the obvious morphisms of Nisnevich sheaves

$$\mathbf{Z}_{Nis}(X')/\mathbf{Z}_{Nis}(X' - Z') \to \mathbf{Z}_{Nis}(X)/\mathbf{Z}_{Nis}(X - Z)$$

$$\mathbf{Z}_{Nis}(X')/\mathbf{Z}_{Nis}(X' - Z') \to \mathbf{Z}_{Nis}(Z \times \mathbf{A}^c)/\mathbf{Z}_{Nis}(Z \times \mathbf{A}^c - Z \times \{0\})$$

are isomorphisms. It follows immediately from Propositoon 3.1.8 that the same holds for the sheaves $L(-)/L(-)$ and therefore by Theorem 3.2.6 the morphisms

$$M_{gm}(X'/(X' - Z')) \to M_{gm}(X/(X - Z))$$

$$M_{gm}(X'/(X' - Z')) \to M_{gm}(Z \times \mathbf{A}^c/(Z \times \mathbf{A}^c - Z \times \{0\}))$$

are isomorphisms. Due to the naturality properties of the morphisms $\alpha_{(-,-)}$ stated above it remains to show that $\alpha_{(\mathbf{A}^c,\{0\})}$ or equivalently $\alpha_{(\mathbf{P}^c,pt)}$ is an isomorphism. It follows easily from the construction of $\alpha_{(-,-)}$ (see [16]).

Corollary 3.5.5 *Let k be a field which admits resolution of singularities. Then DM_{gm}^{eff} is generated as a triangulated category by direct summands of objects of the form $M_{gm}(X)$ for smooth projective varieties X over k.*

4. Homology of algebraic cycles and duality

4.1. Motives of schemes of finite type

Let X be a scheme of finite type over a field k. For any smooth scheme U over k consider the following two abelian groups

1. The group $L(X)(U)$ is the free abelian group generated by closed integral subschemes Z of $X \times U$ such that Z is finite over U and dominant over an irreducible (=connected) component of U,

2. The group $L^c(X)(U)$ is the free abelian group generated by closed integral subschemes Z of $X \times U$ such that Z is quasi-finite over U and dominant over an irreducible (=connected) component of U.

One can define easily Nisnevich sheaves with transfers $L(X), L^c(X)$ on Sm/k such that for any smooth U over k the groups $L(X)(U), L^c(X)(U)$ are the groups described above. In the case when X is smooth over k our notation agrees with the notation $L(X)$ for the presheaf with transfers represented by X. The presheaves $L(X)$ (resp. $L^c(X)$) are covariantly functorial with respect to X (resp. with respect to proper morphisms $X \to X'$) which gives us two functors:

$$L(-) : Sch/k \to PreShv(SmCor(k))$$

$$L^c(-) : Sch^{prop}/k \to PreShv(SmCor(k))$$

(here Sch^{prop}/k is the category of schemes of finite type over k and proper morphisms). Note that the functor $L(-)$ from Sch/k extends the functor $L(-)$ from Sm/k which we considered before.

For a scheme of finite type X over k we write $\underline{C}_*(X)$ (resp. $\underline{C}_*^c(X)$) instead of $\underline{C}_*(L(X))$ (resp. $\underline{C}_*(L^c(X))$). By Lemma 3.2.1 these complexes of Nisnevich sheaves with transfers are objects of the category $DM_-^{eff}(k)$. It provides us with two functors:

$$\underline{C}_*(-) : Sch/k \to DM_-^{eff}(k)$$

$$\underline{C}_*^c(-) : Sch^{prop}/k \to DM_-^{eff}(k).$$

We will show later in this section that if k admits resolution of singularities they can be factored through the canonical embedding $DM_{gm}^{eff}(k) \to DM_-^{eff}(k)$ and therefore define "motives" and "motives with compact support" for schemes which are not necessarily smooth over k.

Proposition 4.1.1 *Let X be a scheme of finite type over k and $X = U \cup V$ be an open covering of X. Then there is a canonical distinguished triangle in DM_-^{eff} of the form*

$$\underline{C}_*(U \cap V) \to \underline{C}_*(U) \oplus \underline{C}_*(V) \to \underline{C}_*(X) \to \underline{C}_*(U \cap V)[1].$$

Proof: It is sufficient to notice that the sequence of Nisnevich sheaves

$$0 \to L(U \cap V) \to L(U) \oplus L(V) \to L(X) \to 0$$

is exact.

The main technical result which allow us to work effectively with objects of the form $L(X), L^c(X)$ for singular varieties is the theorem below which is a particular case of [7, Th. 5.5(2)].

Theorem 4.1.2 *Let k be a field which admits resolution of singularities and F be a presheaf with transfers on Sm/k such that for any smooth scheme X over k and a section $\phi \in F(X)$ there is a proper birational morphism $p : X' \to X$ with $F(p)(\phi) = 0$. Then the complex $\underline{C}_*(F)$ is quasi-isomorphic to zero.*

The first application of this theorem is the following blow-up distinguished triangle.

Proposition 4.1.3 *Consider a Cartesian square of morphisms of schemes of finite type over k of the form*

$$
\begin{array}{ccc}
p^{-1}(Z) & \longrightarrow & X_Z \\
\downarrow & & \downarrow p \\
Z & \longrightarrow & X
\end{array}
$$

such that the following conditions hold:

1. *The morphism $p : X_Z \to X$ is proper and the morphism $Z \to X$ is a closed embedding.*

2. *The morphism $p^{-1}(X - Z) \to X$ is an isomorphism.*

Then there is a canonical distinguished triangle in $DM_-^{eff}(k)$ of the form

$$\underline{C}_*(p^{-1}(Z)) \to \underline{C}_*(Z) \oplus \underline{C}_*(X_Z) \to \underline{C}_*(X) \to \underline{C}_*(p^{-1}(Z))[1].$$

Proof: It is sufficient to notice that the sequence of presheaves

$$0 \to L(p^{-1}(Z)) \to L(X_Z) \oplus L(Z) \to L(X)$$

is exact and the quotient presheaf $L(X)/(L(X_Z) \oplus L(Z))$ satisfies the condition of Theorem 4.1.2.

Corollary 4.1.4 *Let k be a field which admits resolution of singularities. Then for any scheme X of finite type over k the object $\underline{C}_*(X)$ belongs to $DM_{gm}^{eff}(k)$.*

The following proposition explains why $\underline{C}_*^c(X)$ is called the motivic complex with compact support.

Proposition 4.1.5 *Let k be a field which admits resolution of singularities, X be a scheme of finite type over k and Z be a closed subscheme of X. Then there is a canonical distinguished triangle of the form*

$$\underline{C}_*^c(Z) \to \underline{C}_*^c(X) \to \underline{C}_*^c(X - Z) \to \underline{C}_*^c(Z)[1].$$

If X is proper than there is a canonical isomorphism $\underline{C}_*^c(X) = \underline{C}_*(X)$.

Proof: The second statement is obvious. To prove the first one it is again sufficient to notice that the sequence of presheaves

$$0 \to L^c(Z) \to L^c(X) \to L^c(X - Z)$$

is exact and the cokernel $L^c(X - Z)/L^c(X)$ satisfies the condition of Theorem 4.1.2.

Corollary 4.1.6 *Let k be a field which admits resolution of singularities. Then for any scheme X of finite type over k the object $\underline{C}_*^c(X)$ belongs to $DM_{gm}^{eff}(k)$.*

Proposition 4.1.7 *Let X, Y be schemes of finite type over k. Then there are canonical isomorphisms:*

$$\underline{C}_*(X \times Y) = \underline{C}_*(X) \otimes \underline{C}_*(Y)$$

$$\underline{C}_*^c(X \times Y) = \underline{C}_*^c(X) \otimes \underline{C}_*^c(Y)$$

Proof: One can construct easily natural morphisms

$$\underline{C}_*(X) \otimes \underline{C}_*(Y) \to \underline{C}_*(X \times Y)$$

$$\underline{C}_*^c(X) \otimes \underline{C}_*^c(Y) \to \underline{C}_*^c(X \times Y).$$

For smooth projective X, Y they are isomorphisms in DM^{eff} by the definition of the tensor structure on this category. The general case follows now formally from Propositions 4.1.3 and 4.1.5.

Corollary 4.1.8 *For any scheme of finite type X over k one has canonical isomorphisms:*

$$\underline{C}_*(X \times \mathbf{A}^1) = \underline{C}_*(X)$$

$$\underline{C}_*^c(X \times \mathbf{A}^1) = \underline{C}_*^c(X)(1)[2].$$

In particular we have:

$$\underline{C}_*^c(\mathbf{A}^n) = \mathbf{Z}(n)[2n].$$

Proof: In view of Proposition 4.1.7 it is sufficient to show that

$$\underline{C}_*(\mathbf{A}^1) = \mathbf{Z}$$

$$\underline{C}_*^c(\mathbf{A}^1) = \mathbf{Z}(1)[2].$$

The first fact follows immediately from our definitions. The second follows from the definition of the Tate object and Proposition 4.1.5.

We want to describe now morphisms of the form $Hom(\underline{C}_*(X), \underline{C}_*(F))$ for schemes of finite type X over k and Nisnevich sheaves with transfers F. The first guess that this group is isomorphic to $\mathbf{H}_{Nis}(X, \underline{C}_*(F))$ as it was proved in 3.2.7 for smooth schemes X turns out to be wrong.

To get the correct answer we have to consider the cdh-topology on the category Sch/k of schemes of finite type over k.

Definition 4.1.9 *The cdh-topology on Sch/k is the minimal Grothendieck topology on this category such that the following two types of coverings are cdh-coverings.*

1. *the Nisnevich coverings.*

2. *Coverings of the form $X' \coprod Z \overset{p \coprod i}{\to} X$ such that p is a proper morphism, i is a closed embedding and the morphism $p^{-1}(X - i(Z)) \to X - i(Z)$ is an isomorphism.*

Denote by $\pi : (Sch/k)_{cdh} \to (Sm/k)_{Nis}$ the obvious morphism of sites. Note that the definition of $\underline{C}_*(F)$ given above for presheaves on Sm/k also works for presheaves on Sch/k. The theorem below follows formally from Theorem 4.1.2.

Theorem 4.1.10 *Let X be a scheme of finite type over k and F be a presheaf with transfers on Sm/k. Then for any $i \geq 0$ there are canonical isomorphisms*

$$Hom(\underline{C}_*(X), \underline{C}_*(F)[i]) = \mathbf{H}^i_{cdh}(X, \underline{C}_*(\pi^*(F))) = \mathbf{H}^i_{cdh}(X, \pi^*(\underline{C}_*(F))).$$

In particular if X is smooth one has

$$\mathbf{H}^i_{cdh}(X, \underline{C}_*(\pi^*(F))) = \mathbf{H}^i_{cdh}(X, \pi^*(\underline{C}_*(F))) = \mathbf{H}^i_{Nis}(X, \underline{C}_*(F)).$$

Corollary 4.1.11 *Let k be a field which admits resolution of singularities and X be a scheme of finite type over k. Let further $\mathcal{E}$ be a vector bundle over X. Denote by $p : \mathbf{P}(\mathcal{E}) \to X$ the projective bundle over X associated with $\mathcal{E}$. Then one has a canonical isomorphism in $DM^{eff}_-(k)$ of the form:*

$$\underline{C}_*(\mathbf{P}(\mathcal{E})) = \oplus_{n=0}^{dim\mathcal{E}-1} \underline{C}_*(X)(n)[2n].$$

Proof: Theorem 4.1.10 implies that we can construct a natural morphism

$$\underline{C}_*(\mathbf{P}(\mathcal{E})) \to \oplus_{n=0}^{dim\mathcal{E}-1} \underline{C}_*(X)(n)[2n]$$

in exactly the same way as in Proposition 3.5.1. The fact that it is an isomorphism follows now formally from Proposition 3.5.1, Proposition 4.1.3 and resolution of singularities.

Let us recall that in [16] a triangulated category $DM_h(k)$ was defined as a localization of the derived category $D^-(Shv_h(Sch/k))$ of complexes bounded from the above over the category of sheaves of abelian groups in the h-topology on Sch/k. One can easily see that there is a canonical functor

$$DM^{eff}_{-,et}(k) \to DM_h(k).$$

Using the comparison results of [16] (for sheaves of $\mathbf{Q}$-vector spaces) together with the technique described in this section one can verify easily that this functor is an equivalence after tensoring with $\mathbf{Q}$. Moreover, using the description of the category $DM^{eff}_{-,et}(k, \mathbf{Z}/n\mathbf{Z})$ given in Section 3.3 and comparison results for torsion sheaves from [16] one can also show that it is an equivalence for finite coefficients. Combining these two results we obtain the following theorem.

Theorem 4.1.12 *Let k be a field which admits resolution of singularities. Then the functor*

$$DM^{eff}_{-,et}(k) \to DM_h(k)$$

is an equivalence of triangulated categories. In particular, the categories $DM^{eff}_-(k) \otimes \mathbf{Q}$ and $DM_h(k) \otimes \mathbf{Q}$ are equivalent.

4.2. Bivariant cycle cohomology

Let us recall the definition of the bivariant cycle cohomology given in [7]. For any scheme of finite type X over k and any $r \geq 0$ we denote by $z_{equi}(X, r)$ the presheaf on the category of smooth schemes over k which takes a smooth scheme Y to the free abelian group generated by closed integral subschemes Z of $Y \times X$ which are equidimensional of relative dimension r over Y (note that it means in particular that Z dominates an irreducible component of Y). One can verify easily that $z_{equi}(X, r)$ is a sheaf in the Nisnevich topology and moreover that it has a canonical structure of a presheaf with transfers. The presheaves with transfers $z_{equi}(X, r)$ are covariantly functorial with respect to proper morphisms of X by means of the usual proper push-forward of cycles and contravariantly functorial with an appropriate dimension shift with respect to flat equidimensional morphisms. For $r = 0$ the presheaf $z_{equi}(X, r)$ is isomorphic to the presheaf $L^c(X)$ defined in the previous section.

We will also use below the notation $z_{equi}(Y, X, r)$ introduced in [15] for the presheaf which takes a smooth scheme U to $z_{equi}(X, r)(U \times Y)$.

Definition 4.2.1 *Let X, Y be schemes of finite type over k and $r \geq 0$ be an integer. The bivariant cycle cohomology $A_{r,i}(Y, X)$ of Y with coefficients in r-cycles on X are the hypercohomology $\mathbf{H}^{-i}_{cdh}(Y, \pi^*(\underline{C}_*(z_{equi}(X, r))))$.*

It follows immediately from this definition and Theorem 4.1.10 that we have

$$A_{r,i}(Y, X) = Hom_{DM^{eff}_-}(\underline{C}_*(Y)[i], \underline{C}_*(z_{equi}(X, r))).$$

We will show below (4.2.3) that in fact these groups admit more explicit interpretation in terms of the category DM^{eff}_-.

The following theorem summarizes the most important for us properties of bivariant cycle cohomology proven in [7].

Theorem 4.2.2 *Let k be a field which admits resolution of singularities, X, Y be schemes of finite type over k and $d \geq 0$ be an integer.*

1. *For any $i, r \geq 0$ there are canonical isomorphisms of the form*

$$A_{r,i}(Y \times \mathbf{P}^1, X) = A_{r,i}(Y, X) \oplus A_{r+1,i}(Y, X)$$

   where the projection $A_{r,i}(Y \times \mathbf{P}^1, X) \to A_{r,i}(Y, X)$ (resp. embedding $A_{r,i}(Y, X) \to A_{r,i}(Y \times \mathbf{P}^1, X)$) is given by the embedding of Y to $Y \times \mathbf{P}^1$ by means of a point of $\mathbf{P}^1$ (resp. by the projection $Y \times \mathbf{P}^1 \to Y$) ([7, Th. 8.3(3)]).

2. *For any $i, r \geq 0$ there are canonical isomorphisms of the form*

$$A_{r+1,i}(Y, X \times \mathbf{A}^1) = A_{r,i}(Y, X)$$

   ([7, Th. 8.3(1)]).

3. *If Y is smooth and equidimensional of dimension n there are canonical isomorphisms*

$$A_{r,i}(Y, X) = A_{r+n,i}(Spec(k), Y \times X)$$

   ([7, Th. 8.2]).

4. *For any smooth quasi-projective Y the obvious homomorphisms*

$$\underline{h}_i(\underline{C}_*(z_{equi}(X, r))(Y)) \to A_{r,i}(Y, X)$$

   are isomorphisms ([7, Th. 8.1]).

Proposition 4.2.3 *Let X, Y be schemes of finite type over k. Then for any $r \geq 0$ there are canonical isomorphisms*

$$Hom_{DM^{eff}_-}(\underline{C}_*(Y)(r)[2r + i], \underline{C}^c_*(X)) = A_{r,i}(Y, X).$$

Proof: For $r = 0$ it follows immediately from Definition 4.2.1 and Theorem 4.1.10 since $z_{equi}(X, 0) = L^c(X)$. We proceed by the induction on r. By definition we have $\underline{C}_*(Y)(1)[2] = \underline{C}_*(Y) \otimes \mathbf{Z}(1)[2]$ and $\mathbf{Z}(1)[2]$ is canonically isomorphic to the direct summand of $\underline{C}_*(\mathbf{P}^1)$ which is the kernel of the projector induced by the composition of morphisms $\mathbf{P}^1 \to Spec(k) \to \mathbf{P}^1$. Thus by Proposition 4.1.7 $\underline{C}_*(Y)(r)[2r]$ is canonically isomorphic to the kernel of the corresponding projector on $\underline{C}_*(Y \times \mathbf{P}^1)(r-1)[2r-2]$. By the inductive assumption we have a canonical isomorphism

$$Hom_{DM_-^{eff}}(\underline{C}_*(Y \times \mathbf{P}^1)(r-1)[2r-2+i], \underline{C}_*^c(X)) = A_{r-1,i}(Y \times \mathbf{P}^1, X)$$

and thus by Theorem 4.2.2(1) we have

$$Hom_{DM_-^{eff}}(\underline{C}_*(Y)(r)[2r+i], \underline{C}_*^c(X)) = A_{r,i}(Y, X).$$

Corollary 4.2.4 *Let $f : X \to Y$ be a flat equidimensional morphism of relative dimension n of schemes of finite type over k. Then there is a canonical morphism in DM_-^{eff} of the form:*

$$f^* : \underline{C}_*^c(Y)(n)[2n] \to \underline{C}_*^c(X)$$

and these morphisms satisfy all the standard properties of the contravariant functoriality of algebraic cycles.

Proof: Let $\Gamma_f \subset X \times Y$ be the graph of f. Considered as a cycle on $X \times Y$ it clearly belongs to $z_{equi}(X, n)(Y)$ and our statement follows from Proposition 4.2.3.

Corollary 4.2.5 *Let X be a smooth scheme over k. Denote by $A^i(X)$ the group of cycles of codimension i on X modulo rational equivalence. Then there is a canonical isomorphism*

$$A^i(X) = Hom_{DM_{gm}^{eff}}(M_{gm}(X), \mathbf{Z}(i)[2i]).$$

Proof: It follows immediately from the fact that $\mathbf{Z}(i)[2i] = \underline{C}_*^c(\mathbf{A}^i)$ (Lemma 4.1.8), Proposition 4.2.3 and Theorem 4.2.2(2).

Corollary 4.2.6 *Let X, Y be smooth proper schemes over k. Then one has:*

$$Hom_{DM_{gm}^{eff}}(M_{gm}(X), M_{gm}(Y)) = A_{dim(X)}(X \times Y)$$

$$Hom_{DM_{gm}^{eff}}(M_{gm}(X), M_{gm}(Y)[i]) = 0 \text{ for } i > 0.$$

Proof: It follows immediately from Proposition 4.2.3 and Theorem 4.2.2(3).

Corollary 4.2.7 *Let X be a smooth scheme over k and Y be any scheme of finite type over k. Then there is a canonical isomorphism in $DM_-^{eff}(k)$ of the form:*

$$\underline{Hom}_{DM_-^{eff}}(\underline{C}_*(X), \underline{C}_*^c(Y)) = \underline{C}_*(z_{equi}(X, Y, 0))$$

where $z_{equi}(X, Y, n)$ is the sheaf of the form

$$z_{equi}(X, Y, r)(-) = z_{equi}(Y, r)(- \times X).$$

Proposition 4.2.3 has the following "global" reformulation.

Proposition 4.2.8 *Let X be a scheme of finite type over a field k which admits resolution of singularities. Then for any $r \geq 0$ there is a canonical isomorphism in $DM_-^{eff}(k)$ of the form:*

$$\underline{C}_*(z_{equi}(X, r)) = \underline{Hom}_{DM_-^{eff}}(\mathbf{Z}(r)[2r], \underline{C}_*^c(X)).$$

Proof: It is sufficient to show that the isomorphisms of Proposition 4.2.3 are induced by a morphism

$$\underline{Hom}_{DM_-^{eff}}(\mathbf{Z}(r)[2r], \underline{C}_*^c(X)) \to \underline{C}_*(z_{equi}(X, r))$$

in DM_-^{eff}.

Denote by p_r the formal linear combination of endomorphisms of $(\mathbf{P}_k^1)^r$ which gives in DM_-^{eff} the projector from $\underline{C}_*((\mathbf{P}_k^1)^r)$ to $\mathbf{Z}(r)[2r]$ and by p the canonical morphism $(\mathbf{P}_k^1)^r \to Spec(k)$. By Proposition 3.2.8 the object $\underline{Hom}_{DM_-^{eff}}(\mathbf{Z}(r)[2r], \underline{C}_*^c(X))$ is the direct summand of the complex of sheaves $\mathbf{R}p_*(p^*(\underline{C}_*(z_{equi}(X, 0))))$ defined by p_r. Consider the obvious morphism of complexes:

$$\underline{C}_*(z_{equi}(\mathbf{P}_k^1)^r, X, 0) = \underline{C}_*(p_*(p^*(z_{equi}(X, 0)))) \to \mathbf{R}p_*(p^*(\underline{C}_*(z_{equi}(X, 0)))).$$

By Theorem 4.2.2(4) this morphism is a quasi-isomorphism of complexes of sheaves on Sm/k and thus the same holds for the direct summands of both complexes defined by p_r. It remains to construct a morphism

$$z_{equi}((\mathbf{P}_k^1)^r, X, 0) \to z_{equi}(X, r).$$

Let U be a smooth scheme over k. The group $z_{equi}((\mathbf{P}_k^1)^r, X, 0)(U)$ is by definition the subgroup of cycles on $(\mathbf{P}_k^1)^r \times U \times X$ which consists of cycles equidimensional of relative dimension 0 over $(\mathbf{P}_k^1)^r \times U$. Pushing them forward with respect to the projection $(\mathbf{P}_k^1)^r \times U \times X \to U \times X$ we get the required homomorphism.

The following proposition is essentially a reformulation of a result proven by A. Suslin in [12].

Proposition 4.2.9 *Let X be a quasi-projective equidimensional scheme over k of dimension n. Then for all $i, j \in \mathbf{Z}$ there are canonical isomorphisms:*

$$CH^{n-i}(X, j - 2i) = \begin{cases} Hom_{DM_-^{eff}}(\mathbf{Z}(i)[j], \underline{C}_*^c(X)) & for \quad i \geq 0 \\ Hom_{DM_-^{eff}}(\mathbf{Z}, \underline{C}_*^c(X)(-i)[-j]) & for \quad i \leq 0. \end{cases}$$

which commute with the boundary maps in the localization long exact sequences.

Proof: Consider first the case $i \geq 0$. By Proposition 4.2.3 the left hand side group is isomorphic to the group $A_{i,j-2i}(X)$ which is by definition the $(j - 2i)$-th homology group of the complex $C_*(z_{equi}(X, i))$. Consider the Bloch's complex $\mathcal{Z}^{n-i}(X, *)$ which computes the higher Chow groups ([3]). The group $C_k(z_{equi}(X, i))$ is the group of cycles on $X \times \Delta^k$ which are equidimensional of relative dimension i over Δ^k while the group $\mathcal{Z}^{n-i}(X, k)$ consists of cycles of codimension $n - i$ on $X \times \Delta^k$ which itersect all faces of Δ^k properly. One can easily see that since $i \geq 0$ we have an inclusion $C_k(z_{equi}(X, i)) \subset \mathcal{Z}^{n-i}(X, k)$. The differencials in both complexes are defined by the intersection with the faces which implies that for any X we have a monomorphism of complexes of abelian groups of the form:

$$\psi : C_*(z_{equi}(X, i)) \to \mathcal{Z}^{n-i}(X, *)$$

which is clearly canonical with respect to both flat (the contravariant functoriality) and proper (the covariant functoriality) morphisms in X.

Let Z be a closed subscheme in X of pure codimension m. To define the boundary homomorphism in the localization long exact sequences forthe bivariant cycle cohomology and the higher Chow groups one considers the exact sequences of complexes:

$$0 \to C_*(z_{equi}(Z, i)) \to C_*(z_{equi}(X, i)) \to C_*(z_{equi}(X - Z, i)) \to coker_1 \to 0$$

$$0 \to \mathcal{Z}^{n-i-m}(Z, *) \to \mathcal{Z}^{n-i}(X, *) \to \mathcal{Z}^{n-i}(X - Z, *) \to coker_2 \to 0$$

and then shows that both cokernels are quasi-isomorphic to zero (see [7] for $coker_1$ and [4] for $coker_2$). Our inclusion ψ gives a morphism of these

exact sequences which implies immediately that the corresponding homomorphisms on (co-)homology groups commute with the boundary homomorphisms.

By [12] the monomorphism ψ is a quasi-isomorphism of complexes for
any *affine* scheme X. By induction on $dim(X)$ and the five isomorphisms
lemma applied to the localization long exact sequences in both theories it
implies that ψ is a quasi-isomorphism for all X.

Suppose now that $i < 0$. In this case the right hand side group is isomorphic to $Hom(\mathbf{Z}[j - 2i], \underline{C}_*^c(X)(-i)[-2i])$. By Lemma 4.1.8 we have
$\underline{C}_*^c(X)(-i)[-2i]) = \underline{C}_*^c(X \times \mathbf{A}^{-i})$ and by Proposition 4.2.3 we conclude
that the group in question is isomorphic to the group $A_{0,j-2i}(X \times \mathbf{A}^{-i})$.
By the reasoning given above it is canonically isomorphic to the group
$CH^{n-i}(X \times \mathbf{A}^{-i}, j - 2i)$. Finally applying the homotopy invariance theorem for higher Chow groups we get the assertion of the proposition for
$i < 0$.

4.3. Duality in the triangulated categories of motives

One of the most important corollaries of the results of the previous section
is the following theorem.

Theorem 4.3.1 *For any objects* A, B *in* $DM_{gm}^{eff}(k)$ *the natural map*

$$- \otimes Id_{\mathbf{Z}(1)} : Hom(A, B) \to Hom(A(1), B(1))$$

is an isomorphism. Thus the canonical functor

$$DM_{gm}^{eff}(k) \to DM_{gm}(k)$$

is a full embedding.

Proof: By Corollary 3.5.5 we may assume that $A = M_{gm}(X)[i]$ and
$B = M_{gm}(Y)$ for some smooth projective varieties X, Y over k and some
$i \in \mathbf{Z}$. By Corollary 4.1.8 we have $\underline{C}_*(Y)(1)[2] = \underline{C}_*^c(Y \times \mathbf{A}^1)$ and therefore
the right hand side group is isomorphic by Proposition 4.2.3 to the group
$A_{1,i}(X, Y \times \mathbf{A}^1)$. By Theorem 4.2.2(2) it is isomorphic to $A_{0,i}(X, Y)$ which
is (again by Proposition 4.2.3) isomorphic to $Hom_{DM^{eff}}(\underline{C}_*(X), \underline{C}_*(Y))$.
We have shown that there is a canonical isomorphism

$$Hom_{DM_{gm}^{eff}}(M_{gm}(X)[i], M_{gm}(Y))$$
$$= Hom_{DM_{gm}^{eff}}(M_{gm}(X)(1)[i], M_{gm}(Y)(1)).$$

The fact that it coincides with the morphism induced by the tensor multiplication with $Id_{\mathbf{Z}(1)}$ follows easily from the explicit form of morphisms which we used to construct it.

We will use Theorem 4.3.1 to identify DM_{gm}^{eff} with its image in DM_{gm}.

Let us show now that the properties of the bivariant cycle cohomology given by Theorem 4.2.2 imply in particular that $DM_{gm}(k)$ is a "rigid tensor triangulated category".

Theorem 4.3.2 *Let X be a smooth proper scheme of dimension n over k. Consider the morphism*

$$M_{gm}(X) \otimes M_{gm}(X) \to \mathbf{Z}(n)[2n]$$

which corresponds to the diagonal $X \to X \times X$ by Corollary 4.2.5. Then the corresponding morphism

$$M_{gm}(X) \to \underline{Hom}_{DM_-^{eff}}(M_{gm}(X), \mathbf{Z}(n)[2n])$$

is an isomorphism.

Proof: We have to show that for any object A of DM_{gm}^{eff} the morphism

$$Hom(A, M_{gm}(X)) \to Hom(A \otimes M_{gm}(X), \mathbf{Z}(n)[2n])$$

given by the diagonal is an isomorphism. Since these morphisms are natural with respect to A we may assume that $A = M_{gm}(U)[i]$ for a smooth scheme U over k. In this case the statement follows from Theorem 4.2.2(3) and Proposition 4.2.3.

Proposition 4.3.3 *Let X be a scheme of dimension n over k. Then for any $m \geq n$ the morphism*

$$\underline{Hom}_{DM_-^{eff}}(\underline{C}_*(X), \mathbf{Z}(n))(m - n) \to \underline{Hom}_{DM_-^{eff}}(\underline{C}_*(X), \mathbf{Z}(m))$$

is an isomorphism.

Proof: Since our morphisms are natural we may assume using resolution of singularities that X is smooth and projective. Then the statement follows in the standard way from Proposition 4.2.3 and Theorem 4.2.2(3).

Corollary 4.3.4 *For any scheme X of dimension n over k and any $m \geq n$ the object $\underline{Hom}_{DM_-^{eff}}(\underline{C}_*(X), \mathbf{Z}(m))$ belongs to DM_{gm}^{eff}.*

Corollary 4.3.5 *For any pair of objects A, B in $DM_{gm}(k)$ there exists the internal Hom-object $\underline{Hom}_{DM_{gm}}(A, B)$ in DM_{gm}.*

Corollary 4.3.6 *Let X be a variety of dimension n. Then for any $m \geq n$ the object $\underline{Hom}_{DM_{gm}}(M_{gm}(X), \mathbf{Z}(m))$ belongs to $DM_{gm}^{eff}(k)$ and its image in $DM_-^{eff}(k)$ is canonically isomorphic to $\underline{Hom}_{DM_-^{eff}}(\underline{C}_*(X), \mathbf{Z}(m))$.*

Remark: Note that even if the internal Hom-object $\underline{Hom}(\underline{C}_*(X), \underline{C}_*(Y))$ in DM_-^{eff} belongs to DM_{gm}^{eff} its image in DM_{gm} does not in general coincide with $\underline{Hom}_{DM_{gm}}(M_{gm}(X), M_{gm}(Y))$. Consider for example $X = \mathbf{P}^1$ and $Y = Spec(k)$. Then

$$\underline{Hom}_{DM_-^{eff}}(\underline{C}_*(X), \underline{C}_*(Y)) = \mathbf{Z}$$

while

$$\underline{Hom}_{DM_{gm}}(M_{gm}(X), M_{gm}(Y)) = \mathbf{Z} \oplus \mathbf{Z}(-1)[-2].$$

For any object A of $DM_{gm}(k)$ we define A^* as the internal Hom-object $\underline{Hom}_{DM_{gm}}(A, \mathbf{Z})$. The following theorem can be stated by saying that $DM_{gm}(k)$ is a rigid tensor triangulated category.

Theorem 4.3.7 *Let k be a field which admits resolution of singularities. Then one has.*

1. *For any object A in $DM_{gm}(k)$ the canonical morphism $A \to (A^*)^*$ is an isomorphism.*

2. *For any pair of objects A, B of $DM_{gm}(k)$ there are canonical isomorphisms:*

$$(A \otimes B)^* = A^* \otimes B^*$$

$$\underline{Hom}(A, B) = A^* \otimes B$$

3. *For a smooth scheme X of pure dimension n over k one has canonical isomorphisms*

$$M_{gm}(X)^* = M_{gm}^c(X)(-n)[-2n]$$

$$M_{gm}^c(X)^* = M_{gm}(X)(-n)[-2n].$$

Proof: To prove statements (1),(2) we may assume by Corollary 3.5.5 that $A = M_{gm}(X)$, $B = M_{gm}(Y)[i]$ for smooth projective varieties X, Y over k and some $i \in \mathbf{Z}$. Then everything follows easily from Theorem 4.3.2.

Let us show that the second statement holds. Since $(-)^{**} = (-)$ it is sufficient by Corollary 4.3.6 to construct a canonical isomorphism

$$\underline{Hom}_{DM^{eff}}(\underline{C}_*(X), \mathbf{Z}(n)[2n]) \to \underline{C}_*^c(X).$$

By Corollaries 4.2.7 and 4.1.8 the left hand side object is canonically isomorphic to $\underline{C}_*(z_{equi}(X, \mathbf{A}^n, 0))$. By [7, Th. 7.4] there is a canonical morphism of sheaves with transfers

$$z_{equi}(X, \mathbf{A}^n, 0) \to z_{equi}(X \times \mathbf{A}^n, n)$$

which induces a quasi-isomorphism on the corresponding complexes $\underline{C}_*(-)$. Finally the flat pull-back morphism

$$z_{equi}(X, 0) \to z_{equi}(X \times \mathbf{A}^n, n)$$

induces a quasi-isomorphism of $\underline{C}_*(-)$ complexes by Theorem 4.2.2(2).

Finally we are going to formulate in this section a simple result which shows that integrally $DM_{gm}^{eff}(k)$ does not have a "reasonable" t-structure. More precisely let us say that a t-structure $\tau = (\mathcal{D}^{\leq 0}, \mathcal{D}^{\geq 0})$ on $DM_{gm}^{eff}(k)$ is reasonable if the following conditions hold:

1. τ is compatible with the Tate twist, i.e. an object M belongs to $\mathcal{D}^{\leq 0}$ (resp. $\mathcal{D}^{\geq 0}$) if and only if $M(1)$ does.

2. For a smooth affine scheme X of dimension n one has:

$$H_i^\tau(M_{gm}(X)) = 0 \text{ for } i < 0 \text{ or } i > n$$

$$H_i^\tau(M_{gm}^c(X)) = 0 \text{ for } i < n \text{ or } i > 2n$$

where H_i^τ are the cohomology objects with respect to τ.

Note that in view of Theorem 4.3.7 the last two conditions are dual to each other.

Proposition 4.3.8 *Let k be a field such that there exists a conic X over k with no k-rational points. Then $DM_{gm}^{eff}(k)$ has no reasonable t-structure.*

Proof: One can easily see that our conditions on τ imply that for any smooth plane curve X we have

$$H_i^\tau(M_{gm}(X)) = \begin{cases} 0 & \text{for} \quad i \neq 0, 1, 2 \\ \mathbf{Z} & \text{for} \quad i = 0 \\ \mathbf{Z}(1) & \text{for} \quad i = 2. \end{cases}$$

and for a smooth hypersurface Y in $\mathbf{P}^3$ we have $H_1^\tau(M_{gm}(Y)) = 0$.

Let now X be a conic over k with no rational points. The diagonal gives an embedding of $X \times X$ in $\mathbf{P}^3$ and since $M_{gm}(X)$ is clearly a direct summand of $M_{gm}(X \times X)$ for any X we conclude that

$$
H_i^\tau(M_{gm}(X)) = \begin{cases} 0 & \text{for} \quad i \neq 0, 2 \\ \mathbf{Z} & \text{for} \quad i = 0 \\ \mathbf{Z}(1) & \text{for} \quad i = 2. \end{cases}
$$

Therefore we have a distinguished triangle of the form

$$
\mathbf{Z}(1)[2] \to M_{gm}(X) \to \mathbf{Z} \to \mathbf{Z}(1)[3].
$$

By Corollary 3.4.3 we have:

$$
Hom_{DM_{gm}^{eff}}(\mathbf{Z}, \mathbf{Z}(1)[3]) = H_{Zar}^2(Spec(k), \mathbf{G}_m) = 0.
$$

Thus our triangle splits and we have an isomorphism $M_{gm}(X) = \mathbf{Z} \oplus \mathbf{Z}(1)[2]$. It implies that the canonical morphism

$$
Hom_{DM_{gm}^{eff}}(\mathbf{Z}, M_{gm}(X)) \to \mathbf{Z}
$$

is surjective which contradicts our assumption on X since the left hand side group is $A_0(X)$ and X has no zero cycles of degree one.

References

1. A. Beilinson. Height pairing between algebraic cycles. In *K-theory, Arithmetic and Geometry.*, volume 1289 of *Lecture Notes in Math.*, pages 1–26. Springer-Verlag, 1987.

2. A. Beilinson, R. MacPherson, and V. Schechtman. Notes on motivic cohomology. *Duke Math. J.*, pages 679–710, 1987.

3. S. Bloch. Algebraic cycles and higher K-theory. *Adv. in Math.*, 61:267–304, 1986.

4. S. Bloch. The moving lemma for higher Chow groups. *J. Algebr. Geom.*, 3(3):537–568, Feb. 1994.

5. K.S. Brown and S.M. Gersten. Algebraic K-theory and generalized sheaf cohomology. *Lecture Notes in Math. 341*, pages 266–292, 1973.

6. Eric M. Friedlander. Some computations of algebraic cycle homology. *K-theory*, 8:271–285, 1994.

7. Eric M. Friedlander and V. Voevodsky. Bivariant cycle cohomology. *This volume*.

8. Marc Levine. *Mixed motives.* American Mathematical Society, Providence, RI, 1998.

9. Stephen Lichtenbaum. Suslin homology and Deligne 1-motives. In *Algebraic K-theory and algebraic topology (Lake Louise, AB, 1991)*, pages 189–196. Kluwer Acad. Publ., Dordrecht, 1993.

10. J.S. Milne. *Etale Cohomology*. Princeton Univ. Press, Princeton, NJ, 1980.
11. Y. Nisnevich. The completely decomposed topology on schemes and associated descent spectral sequences in algebraic K-theory. In *Algebraic K-theory: connections with geometry and topology*, pages 241–342. Kluwer Acad. Publ., Dordrecht, 1989.
12. A. Suslin. Higher Chow groups and etale cohomology. *This volume*.
13. Andrei Suslin and Vladimir Voevodsky. Singular homology of abstract algebraic varieties. *Invent. Math.*, 123(1):61–94, 1996.
14. A. Suslin and V. Voevodsky. Relative cycles and Chow sheaves. *This volume*.
15. V. Voevodsky. Cohomological theory of presheaves with transfers. *This volume*.
16. V. Voevodsky. Homology of schemes. *Selecta Mathematica, New Series*, 2(1):111–153, 1996.

6

Higher Chow Groups and Etale Cohomology

Andrei A. Suslin[*][†]

Introduction

The main purpose of the present paper is to relate the higher Chow groups of varieties over an algebraically closed field introduced by S. Bloch [B1] to etale cohomology. We follow the approach suggested by the author in 1987 during the Lumini conference on algebraic K-theory. The first and most important step in this direction was done in [SV1], where singular cohomology of any qfh-sheaf were computed in terms of Ext-groups. The difficulty in the application of the results of [SV1] to higher Chow groups lies in the fact that a priori higher Chow are not defined as singular homology of a sheaf. To overcome this difficulty we prove that for an affine variety X higher Chow groups $CH^i(X, n)$ of codimension $i \leq dim\, X$ may be computed using equidimensional cycles only (this is done in the first two sections of the paper). In section 3 we generalize this result to all quasiprojective varieties over a field of characteristic zero. Together with the main theorem of [SV1] this result shows that for an equidimensional quasiprojective variety X over an algebraically closed field F of characteristic zero the higher Chow groups of codimension $d = dim\, X$ with finite coefficients $\mathbf{Z}/m$ are dual to $Ext_{qfh}(z_0(X), \mathbf{Z}/m)$, where $z_0(X)$ is the qfh-sheaf of equidimensional cycles of relative dimension zero, introduced and studied in [SV2]. The above Ext-groups are easy to compute and in this way we come to the following main result.

[*]St. Petersburg Branch of the Steklov Mathematical Institute (POMI), Fontanka 27, St. Petersburg, 191011, Russia and Department of Mathematics, Northwestern University, Evanston, IL 60208, USA.

[†]The research was supported in part by a grant from the International Science Foundation.

Theorem (Theorem 4.2 and Corollary 4.3). *Let X be an equidimensional quasiprojective scheme over an algebraically closed field F of characteristic zero. Assume that $i \geq d = \dim X$. Then*

$$CH^i(X, n; \mathbf{Z}/m) = H_c^{2(d-i)+n}(x, \mathbf{Z}/m(d-i))^{\#},$$

where H_c stands for etale cohomology with compact supports. If the scheme X is smooth then this formula simplifies to $CH^i(X, n; \mathbf{Z}/m) = H_{et}^{2i-n}(X, \mathbf{Z}/m(i))$.

Throughout the paper we work with schemes of finite type over a field F. We denote the category of these schemes by Sch/F. We use the term variety as a synonym for integral scheme. We denote by p the exponential characteristic of F.

1. Generic equidimensionality

Theorem 1.1. *Assume that S is an affine scheme, V is a closed subscheme in $\mathbf{A}^n \times S$ and t is a nonnegative integer such that $\dim V \leq n + t$. Assume further that Z is an effective divisor in $\mathbf{A}^n$ and $\varphi : Z \times S \to \mathbf{A}^n \times S$ is any S-morphism. Then there exists an S-morphism $\Phi : \mathbf{A}^n \times S \to \mathbf{A}^n \times S$ such that*

1. *$\Phi|_{Z \times S} = \varphi$*

2. *Fibers of the projection $\Phi^{-1}(V) \to \mathbf{A}^n$ over points of $\mathbf{A}^n - Z$ have dimension $\leq t$*

Proof: Notice first that it suffices to treat the special case $S = \mathbf{A}^m$. In fact, if this case is already settled then we can proceed as follows. Choose a closed embedding $S \hookrightarrow \mathbf{A}^m$. To give an S-morphism $\varphi : Z \times S \to \mathbf{A}^n \times S$ is the same as to give a morphism $\psi = pr_1 \circ \varphi : Z \times S \to \mathbf{A}^n$. Extend ψ to a morphism $\psi' : Z \times \mathbf{A}^m \to \mathbf{A}^n$ and denote by φ' the corresponding $\mathbf{A}^m$-morphism $Z \times \mathbf{A}^m \to \mathbf{A}^n \times \mathbf{A}^m$. Using the theorem for $\mathbf{A}^m$ extend φ' to an $\mathbf{A}^m$-morphism $\Phi' : \mathbf{A}^n \times \mathbf{A}^m \to \mathbf{A}^n \times \mathbf{A}^m$ such that fibers of the projection $(\Phi')^{-1}(V) \to \mathbf{A}^n$ over points of $\mathbf{A}^n - Z$ are of dimension $\leq t$ and finally set $\Phi = \Phi'|_{\mathbf{A}^n \times S}$.

From now on we assume that $S = \mathbf{A}^m$, we denote the coordinate functions in $\mathbf{A}^n$ by $Y_1, \ldots, Y_n$ and by $X_1, \ldots, X_m$ we denote the coordinate functions in $\mathbf{A}^m$. Let $h \in F[Y_1, \ldots, Y_n]$ be the equation of the divisor Z. The morphism $\varphi : Z \times \mathbf{A}^m \to \mathbf{A}^n \times \mathbf{A}^m$ is given by the formula

$\varphi(y, x) = (\varphi_1(y, x), \ldots, \varphi_n(y, x), x)$, where $\varphi_i \in F[Y, X]$ are certain polynomials (uniquely defined mod h). Define $\Phi : \mathbf{A}^n \times \mathbf{A}^m \to \mathbf{A}^n \times \mathbf{A}^m$ by means of the formula

$$\Phi(y, x) = (\varphi_1(y, x) + h(y)p_1(x), \ldots, \varphi_n(y, x) + h(y)p_n(x), x)$$

where $p_i \in F[X]$ are forms of degree N. We will show that if N is sufficiently big and p_i are sufficiently generic then the corresponding Φ has the desired property. To prove this we need some auxilliary facts.

(1.2) Let R be a finitely generated F-algebra and let $X = \mathbf{Spec}\, R$ be the corresponding affine scheme. Consider the polynomial ring $R[Y_1, \ldots, Y_n]$. For any $f \in R[Y_1, \ldots, Y_n]$ we denote by l.f.(f) the leading form of f, we also set $^h f = (Y_0)^{\deg f} f(Y_1/Y_0, \ldots, Y_n/Y_0)$. Thus l.f.$(f)$ is a form of degree $\deg f$ in variables $Y_1, \ldots, Y_n$, while $^h f$ is a form of degree $\deg f$ in variables $Y_0, \ldots, Y_n$, moreover l.f.$(f) = {}^h f(0, Y_1, \ldots, Y_n)$ and $f = {}^h f(1, Y_1, \ldots, Y_n)$. If $\mathfrak{A}$ is an ideal in $R[Y_1, \ldots, Y_n]$ then leading forms of polynomials from $\mathfrak{A}$ give a homogeneous ideal in $R[Y_1, \ldots, Y_n]$ which we denote l.f.$(\mathfrak{A})$. In the same way forms $Y_0^k \cdot {}^h f$ $(k \geq 0, f \in \mathfrak{A})$ give a homogeneous ideal in $R[Y_0, \ldots, Y_n]$ which we denote $^h \mathfrak{A}$ (compare [10, ch. 7, §5]). Note the following evident relations between ideals $\mathfrak{A}$, l.f.$(\mathfrak{A})$, $^h \mathfrak{A}$:

(1.2.1) Let $y \in R[Y_0, \ldots, Y_n]$ be a form, then

$$g \in {}^h \mathfrak{A} \Longleftrightarrow g(1, Y_1, \ldots, Y_n) \in \mathfrak{A}.$$

(1.2.2) l.f.$(\mathfrak{A})$ coincides with the image of $^h \mathfrak{A}$ under a surjective homomorphism of graded rings $s : R[Y_0, \ldots, Y_n] \to R[Y_1, \ldots, Y_n]$ sending Y_0 to zero.

Identify $\mathbf{A}^n = \mathbf{Spec}\, F[Y_1, \ldots, Y_n]$ with an open subscheme $\mathbf{P}^n_{Y_0}$ of $\mathbf{P}^n = \mathbf{Proj}\, F[Y_0, \ldots, Y_n]$ by means of an open embedding $j : \mathbf{A}^n \hookrightarrow \mathbf{P}^n$ given by the formula $j(y_1, \ldots, y_n) = [1 : y_1 \ldots : y_n]$. In other words j is defined by a trivial linear bundle $\mathcal{O}_{\mathbf{A}^n}$ over $\mathbf{A}^n$ and $(n+1)$-tuple of its sections (without common zeros) $(1, Y_1, \ldots, Y_n)$.

Lemma 1.3. *In the above notations denote the closed subscheme of* $X \times \mathbf{A}^n = \mathbf{Spec}\, R[Y_1, \ldots, Y_n]$ *defined by the ideal* $\mathfrak{A}$ *by* V. *Then the closed subscheme of* $\mathbf{Proj}\, R[Y_0, \ldots, Y_n] = X \times \mathbf{P}^n$ *defined by the homogeneous ideal* $^h \mathfrak{A}$ *coincides with the closure* $\overline{V}$ *of* V *in* $X \times \mathbf{P}^n$ *and the closed subscheme of* $\mathbf{Proj}\, R[Y_1, \ldots, Y_n] = X \times \mathbf{P}^{n-1}$ *defined by the homogeneous ideal* l.f.$(\mathfrak{A})$ *coincides with* $V_\infty = \overline{V} \cap (X \times \mathbf{P}^{n-1})$.

Proof: Denote the sheaf of ideals defining $\overline{V}$ by I and let $\mathfrak{P}$ be the homogeneous ideal of $R[Y_0, \ldots, Y_n]$ corresponding to I. Tensoring the exact sequence of sheaves

$$0 \to I \to \mathcal{O}_{X \times \mathbf{P}^n} \to j_*(\mathcal{O}_V)$$

by $\mathcal{O}(q)$ and then taking global sections we get exact sequences of R-modules

$$0 \to \mathfrak{P}_q = \Gamma(X \times \mathbf{P}^n, I(q)) \to R[Y_0, \ldots, Y_n]_q$$
$$= \Gamma(X \times \mathbf{P}^n, \mathcal{O}(q)) \to \Gamma(V, \mathcal{O}_V(q))$$
$$= \Gamma(V, \mathcal{O}_V) = R[Y_1, \ldots, Y_n]/\mathfrak{A}$$

The homomorphism $R[Y_0, \ldots, Y_n]_q \to R[Y_1, \ldots, Y_n]/\mathfrak{A}$ sends the form g to $g(1, Y_1, \ldots, Y_n) \bmod \mathfrak{A}$. This shows that $\mathfrak{P}_q = \{g \in R[Y_0, \ldots, Y_n]_q : g(1, Y_1, \ldots, Y_n) \in \mathfrak{A}\} = ({}^h\mathfrak{A})_q$.

To prove the second assertion denote the closed embedding $X \times \mathbf{P}^{n-1} \to X \times \mathbf{P}^n$ by i. This embedding corresponds to the epimorphism of graded rings $s : R[Y_0, \ldots, Y_n] \to R[Y_1, \ldots, Y_n]$ considered in (**1.2.2**). In these notations we get:

$$\mathcal{O}_{V_\infty} = i^*(\mathcal{O}_{\overline{V}}) = i^*((R[Y_0 \ldots, Y_n]/{}^h\mathfrak{A})^\sim)$$
$$= (R[Y_1, \ldots, Y_n] \otimes_{R[Y_0, \ldots, Y_n]} (R[Y_0, \ldots, Y_n]/{}^h\mathfrak{A}))^\sim$$
$$= (R[Y_1, \ldots, Y_n]/\text{l.f.}(\mathfrak{A}))^\sim.$$

Thus V_∞ is defined by the homogenous ideal l.f.$(\mathfrak{A})$.

(**1.4**) In the situation of (**1.2**) assume further that R is a graded F-algebra such that $R_0 = F$ and R is generated over F by R_1. If a polynomial $f \in R[Y_1, \ldots, Y_n]$ is R-homogenous then l.f.(f) and ${}^h f$ are bihomogenous polynomials. In the same way if $\mathfrak{A}$ is an R-homogenous ideal in $R[Y_1, \ldots, Y_n]$ then the ideals l.f.$(\mathfrak{A})$ and ${}^h\mathfrak{A}$ are bihomogenous. Repeating the arguments used in the proof of Lemma 1.3 we get easily the following fact (where now X stands for $\mathbf{Proj}\, R$).

Lemma 1.4.1. *Let $\mathfrak{A}$ be an R-homogenous ideal in $R[Y_1, \ldots, Y_n]$ and let $V \subset \mathbf{Proj}\,(R[Y_1, \ldots, Y_n]) = X \times \mathbf{A}^n$ be the corresponding closed subscheme. Then the closed subscheme of $X \times \mathbf{P}^n$ defined by the bihomogenous ideal ${}^h\mathfrak{A}$ coincides with the closure $\overline{V}$ of V in $X \times \mathbf{P}^n$. Moreover the closed subscheme of $X \times \mathbf{P}^{n-1}$ defined by the bihomogenous ideal l.f.$(\mathfrak{A})$ coincides with $V_\infty = \overline{V} \cap (X \times \mathbf{P}^{n-1})$.*

(1.5) Consider the bigraded ring $F[Y, X] = F[Y_1, \ldots, Y_n, X_1, \ldots, X_m]$. For any $f \in F[Y, X]$ denote by $deg_Y f, deg_X f, \mathrm{l.f}_Y(f), \mathrm{l.f}_X(f)$, the degree and the leading form of f with respect to the corresponding variables. Introduce the lexicographical order on the set of bidegrees and define $deg f$ as the highest bidegree of nonzero bihomogenous summands of f. Finally define $\mathrm{l.f.}(f)$ to be the bihomogenous summand of f of bidegree $deg f$. The following relations are obvious from the definitions:

(1.5.1) $deg f = (deg_Y f, deg_X(\mathrm{l.f}_Y f))$.

(1.5.2) $\mathrm{l.f.}(f) = \mathrm{l.f.}_X(\mathrm{l.f.}_Y(f))$.

If $\mathfrak{A}$ is any ideal in $F[Y, X]$ then leading forms of polynomials from $\mathfrak{A}$ give a bihomogenous ideal $\mathrm{l.f.}(\mathfrak{A})$ and in view of (1.5.2) we have:

(1.5.3) $\mathrm{l.f.}(\mathfrak{A}) = \mathrm{l.f.}_X(\mathrm{l.f.}_Y(\mathfrak{A}))$.

Using (1.3.1) and (1.4.1) we get now the following lemma.

Lemma 1.5.4. *Denote by V the closed subscheme of $\mathbf{A}^n \times \mathbf{A}^m$ defined by the ideal $\mathfrak{A}$, let $\overline{V}$ denote the closure of V in $\mathbf{P}^n \times \mathbf{A}^m$ and set $V_\infty = \overline{V} \cap (\mathbf{P}^{n-1} \times \mathbf{A}^m)$. Next denote $\overline{V_\infty}$ the closure of V_∞ in $\mathbf{P}^{n-1} \times \mathbf{P}^m$ and set $V_{\infty,\infty} = \overline{V_\infty} \cap (\mathbf{P}^{n-1} \times \mathbf{P}^{m-1})$. Then $V_{\infty,\infty}$ coincides with the closed subscheme of $\mathbf{P}^{n-1} \times \mathbf{P}^{m-1}$ defined by the bihomogenous ideal $\mathrm{l.f.}(\mathfrak{A})$.*

Corollary 1.5.5. *Dimension of the closed subscheme of $\mathbf{P}^{n-1} \times \mathbf{P}^{m-1}$ defined by the bihomogenous ideal $\mathrm{l.f.}(\mathfrak{A})$ is not more than $\dim V - 2$.*

(1.6) Continuation of the proof of theorem 1.1

Denote by $\mathfrak{A}$ the ideal of $F[Y, X]$ defining the closed subscheme $V \subset \mathbf{A}^n \times \mathbf{A}^m$ and choose $f_1, \ldots, f_s \in \mathfrak{A}$ such that the bihomogenous forms $\tilde{f}_j = \mathrm{l.f.}(f_j)$ generate the ideal $\mathrm{l.f.}(\mathfrak{A})$. Assume in the sequel that $N > \max(deg_X \varphi_i, deg_X f_j)$. Then one cheks easily that for any choice of N-forms $p_i(X)$ we have:

1. $deg_X f_j(\varphi_i + hp_1, \ldots, \varphi_n + hp_n, X) \le N \, deg_Y \tilde{f}_j + deg_X \tilde{f}_j$
2. The X-homogenous summand of $f_j(\varphi + hp, X)$ of degree $N \, deg_Y \tilde{f}_j + deg_X \tilde{f}_j$ is equal to $h^{deg_Y f_j} \cdot \tilde{f}_j(p_1, \ldots, p_n, X)$

Thus denoting by $\mathfrak{P}$ the ideal defining the closed subscheme $\Phi^{-1}(V)$ and denoting by $\mathfrak{P}_h \subset F[Y]_h[X]$ the corresponding localization we come to the following conclusion

Lemma 1.6.1. *Define a morphism $\Phi : \mathbf{A}^n \times \mathbf{A}^m \to \mathbf{A}^n \times \mathbf{A}^m$ by the formula $\Phi(y, x) = (\varphi_1(y, x) + h(y)p_1(x), \ldots, \varphi_n(y, x) + h(y)p_n(x), x)$, where p_i are forms of degree $N > \max(deg_X \varphi_i, deg_X f_j)$. Then the ideal $\mathrm{l.f.}_X(\mathfrak{P}_h)$ contains the forms $\tilde{f}_j(p_1, \ldots, p_n, X)$*

Consider further the closed subscheme $\overline{pr_2(V)} \subset \mathbf{A}^m$. It's defined by the ideal $\mathfrak{Q} = \mathfrak{A} \cap F[X]$. It's clear that for any $\mathbf{A}^m$-morphism Φ the ideal $\mathfrak{P}$ contains $\mathfrak{Q}$ and hence $\mathrm{l.f.}_X(\mathfrak{P}_h) \supset \mathrm{l.f.}_X(\mathfrak{Q})$. Denote by W the closed subscheme of $\mathbf{P}^{m-1}$ defined by the homogenous ideal $\mathrm{l.f.}_X(\mathfrak{Q})$. Since $dim\,\overline{pr_2(V)} \leq dim\,V \leq n + t$ we conclude from lemma 1.3.1 that $dim\,W \leq n + t - 1$. To finish the proof of theorem 1.1 we need the following fact.

Proposition 1.7. *Let $\tilde{f}_1(Y, X), \ldots, \tilde{f}_s(Y, X)$ be bihomogenous forms such that the subscheme $C \subset \mathbf{P}^{n-1} \times \mathbf{P}^{m-1}$ of their common zeros is of dimension $\leq n + t - 2$. Assume further that we are given a closed subscheme $W \subset \mathbf{P}^{m-1}$ of dimension $\leq n + t - 1$. Then for any $N \geq 0$ we can find forms $p_1, \ldots, p_n \in F[X]$ of degree N such that the dimension of the closed subscheme of W given by the equations $\tilde{f}_j(p_1, \ldots, p_n, X_1, \ldots, X_m) = 0$ is not more than $t - 1$.*

Proof: Identifying each form with the family of its coefficients we get a bijection between the set of forms and set of rational points of an affine space $\mathbf{A}^M$ $\left(M = \binom{N+m-1}{m-1}\right)$. In the same way n-tuples of N-forms are in one to one correspondence with rational points of the affine space $\mathbf{A}^{Mn}$. Set $T_1 = \{(w, p) \in W \times \mathbf{A}^{Mn} : p_1(w) = \ldots = p_n(w) = 0\}$. It's clear that the set T_1 is closed and $dim\,T_1 = dim\,W + n(M - 1) \leq nM + t - 1$. Consider now the following morphism $g : W \times \mathbf{A}^{Mn} - T_1 \to W \times \mathbf{P}^{n-1}$, $(w, p) \mapsto (w, [p_1(w) : \ldots : p_n(w)])$ and set $T_2 = g^{-1}(C \cap (W \times \mathbf{P}^{n-1}))$. The set T_2 is closed in $W \times \mathbf{A}^{Mn} - T_1$ and hence $T = T_1 \cup T_2$ is closed in $W \times \mathbf{A}^{Mn}$. One checks easily that fibers of g are of dimension $Mn - (n-1)$ and hence $dim\,T_2 \leq dim\,C + Mn - (n - 1) \leq nM + t - 1$. This shows that $dim\,T \leq nM + t - 1$. Finally consider the projection $q : T \to \mathbf{A}^{Mn}$. According to the theorem on the dimension of fibers of a morphism there exists a nonempty open set $U \subset \mathbf{A}^{Mn}$ such that for any $p \in U$ $dim\,q^{-1}(p) \leq t - 1$. Now we can take $p = (p_1, \ldots, p_n)$ to be any rational point of U.

(1.8) End of the proof of theorem 1.1

Note first that we are in the situation considered in (1.7). In fact, according to the choice of f_j, the closed subscheme of $\mathbf{P}^{n-1} \times \mathbf{P}^{m-1}$ defined by the equations $\tilde{f}_j = 0$ coincides with the one defined by the bihomogenous ideal $\mathrm{l.f.}(\mathfrak{A})$ and hence is of dimension $\leq dim\,V - 2 \leq n + t - 2$ according to (1.5.5). Moreover $dim\,W \leq dim\,V - 1 \leq n + t - 1$ as was noted already in (1.6). According to the proposition 1.7 we can choose forms $p_1, \ldots, p_n$ of degree $N > \max(deg_X \varphi_i, deg_X f_j)$ so that the dimension of $W_0 = W \cap \{\tilde{f}_1(p, x) = \ldots = \tilde{f}_s(p, x) = 0\}$ is not more than $t - 1$. It suffices to note now that according to (1.6.1) the infinite part of the fiber of

$\Phi^{-1}(V)$ over any point $y \in \mathbf{A}^n - Z$ is contained in W_0 and hence is of dimension $\leq t - 1$. This implies that the dimension of the fiber itself is not more than t.

2. Higher Chow groups and equidimensional cycles

Denote by Δ^n the linear subvariety of $\mathbf{A}^{n+1}$ given by the equation $t_0 + \cdots + t_n = 1$. The points $p_i = (0, \ldots, \underset{i}{1}, \ldots, 0)$ $(0 \leq i \leq n)$ are called the vertices of the "simplex" Δ^n. They have the following evident property: if $X \subset \mathbf{A}^m$ is any linear subvariety (i.e. X is defined by a sistem of linear equations on coordinates) and if $x_0, \ldots, x_n$ is any $(n+1)$-tuple of rational points of X then there exists a unique linear morphism $\Delta^n \longrightarrow X$, taking p_i to x_i. This shows in particular that any nondecreasing map $\phi : \{0, \ldots, n\} \to \{0, \ldots, m\}$ defines a canonical morphism $\Delta^n \to \Delta^m$ (taking p_i to $p_{\phi(i)}$), which we will denote by the same letter ϕ. In this way $\Delta^{\cdot}$ becomes a cosimplicial scheme. The morphisms $\Delta^n \to \Delta^m$ corresponding to strictly increasing maps $\{1, \ldots, n\} \to \{1, \ldots, m\}$ are called cofaces (of codimension $m - n$). Each coface is a closed embedding, the corresponding closed subscheme of Δ^m is clearly a linear subvariety and will be called a face of Δ^m (of codimension $m - n$). As always, there exist $m + 1$ cofaces of codimension one, they are denoted $\delta_i : \Delta^{m-1} \to \Delta^m$ $(0 \leq i \leq m)$.

Assume that $X \in Sch/F$ is any equidimensional scheme, S and T are smooth absolutely irreducible varieties and $f : X \times S \to X \times T$ is any X-morphism. In this situation the schemes $X \times S$ and $X \times T$ are equidimensional of dimension $dim X + dim S$ and $dim X + dim T$ respectively. Consider the graph decomposition of f

$$
\begin{array}{ccc}
X \times S & \overset{i}{\longrightarrow} & X \times S \times T \\
f \downarrow & & \downarrow pr_{1,3} \\
X \times T & \overset{=}{\longrightarrow} & X \times T
\end{array}
$$

Since T is smooth the embedding i is regular (of codimension $dim T$). This shows that for any subvariety V of $X \times T$ and any component W of $f^{-1}(V) = i^{-1}(V \times S)$ we have the usual inequality $codim_{X \times S}(W) \leq codim_{X \times S \times T}(V \times S) = codim_{X \times T}(V)$. When all components of $f^{-1}(V)$ have correct codimension we can use the Tor-formula to define the cycle $f^*([V])$ (since the morphism $pr_{1,2}$ is flat and i is a regular embedding we conclude that f is of finite Tor-dimension). Alternatively we could say that f is a local complete intersection morphism and use the machinery developed in [4, ch. 6].

Denote by $z^i(X,n)$ the free abelian group generated by codimension i subvarieties $V \subset X \times \Delta^n$, which intersect properly $X \times \Delta^m$ for any face $\Delta^m \hookrightarrow \Delta^n$. It's clear from the above definition that if $[V]$ is a generator of $z^i(X,n)$ then cycles $\partial_j([V]) = (1_X \times \delta_j)^*([V])$ are defined and lie in $z^i(X, n-1)$. More generally, if $\phi : \Delta^m \to \Delta^n$ is any structure morphism of the cosimplicial scheme Δ^* then the cycle $(1_X \times \phi)^*([V])$ is defined and lies in $z^i(X,m)$. Thus $z^i(X,-)$ is a simplicial abelian group. The homotopy groups of this simplicial abelian group are denoted $CH^i(X,n)$—see [B1]. In other words $CH^i(X,n)$ is the n-th homology group of the complex

$$z^i(X,0) \overset{d}{\leftarrow} z^i(X,1) \overset{d}{\leftarrow} \ldots$$

where, as always, d is the alternating sum of face operators.

A morphism of varieties $f : X \to Y$ is called equidimensional of relative dimension t if it is dominant and each fiber of f has pure dimension t. In this situation $dim\,V$ is equal to $t + dim\,X$.

Returning to the definition of the higher Chow groups assume that $i \leq d = dim\,X$ and set $t = d - i$. Let V be a closed subvariety in $X \times \Delta^n$ and assume that the projection $V \to \Delta^n$ is an equidimensional morphism of relative dimension t. This implies that $codim_{X \times \Delta^n}(V) = i$. Moreover, if $\phi : \Delta^m \to \Delta^n$ is any morphism and if W is any component of $(1_X \times \phi)^{-1}(V)$ then $codim_{X \times \Delta^m}(W) \leq codim_{X \times \Delta^n}(V)$, i.e. $dim\,W \geq m + t$. On the other hand each fiber of the projection $W \to \Delta^m$ is contained in the corresponding fiber of $V \to \Delta^n$ and hence has dimension $\leq t$. Since each component of each fiber is of dimension $\geq dim\,W - m$ we conclude that $dim\,W = m + t$ (i.e. $codim_{X \times \Delta^m}(W) = codim_{X \times \Delta^n}(V)$) and moreover the projection $W \to \Delta^m$ is equidimensional of relative dimension t. Thus, denoting by $z^i_{eq}(X,n)$ the free abelian group generated by closed subvarieties $V \subset X \times \Delta^n$ for which the projection $W \to \Delta^n$ is equidimensional of relative dimension t, we see that $z^i_{equi}(X,-)$ is a simplicial abelian subgroup in $z^i(X,n)$. The main purpose of this section is to prove the following result

Theorem 2.1. *Assume that X is an affine equidimensional scheme and $i \leq d = dim\,X$. Then the embedding of complexes $z^i_{equi}(X,-) \hookrightarrow z^i(X,-)$ is a quasiisomorphism.*

The proof is based on a certain auxilliary construction.

(2.2) Let N be a positive integer. Assume that for every n $(0 \leq n \leq N)$ we are given an X-morphism $\varphi_n : X \times \Delta^n \to X \times \Delta^n$ such that the following diagrams commute for all $0 \leq j \leq n \leq N$

$$
\begin{array}{ccc}
X \times \Delta^{n-1} & \xrightarrow{\varphi_{n-1}} & X \times \Delta^{n-1} \\
{\scriptstyle 1_X \times \delta_j} \downarrow & & \downarrow {\scriptstyle 1_X \times \delta_j} \\
X \times \Delta^n & \xrightarrow{\varphi_n} & X \times \Delta^n
\end{array}
$$

(2.2.1)

Define $_\varphi z^i(X, n)$ to be a free abelian group generated by close subvarieties $V \subset X \times \Delta^n$ such that

a) $[V] \in z^i(X, n)$

b) The cycle $(\varphi_n)^*([V])$ is defined and lies in $z^i(X, n)$

The commutativity of (2.2.1) shows that for any generator $[V]$ of $_\varphi z^i(X, n)$ and any j ($0 \leq j \leq n$) the cycle $\partial_j([V]) = (1_X \times \delta_j)^*([V])$ lies in $_\varphi z^i(X, n-1)$, so that setting $_\varphi z^i(X, n) = 0$ for $n > N$ we get a subcomplex $_\varphi z^i(X, -)$ of the complex $z^i(X, -)$. Moreover the commutativity of (2.2.1) implies that the assignment $[V] \mapsto \varphi_n^*([V])$ defines a homomorphism $\varphi^* : {}_\varphi z^i(X, -) \to z^i(X-)$.

We will say that two homomorphisms $C_* \rightrightarrows D_*$ of nonnegative complexes are weakly homotopic if their restrictions to any finitely generated subcomplex of C_* are homotopic. It's clear that weakly homotopic homomorphisms induce the same maps on homology.

Proposition 2.3. *The homomorphisms* $_\varphi z^i(X, -) \overset{\varphi^*}{\underset{\text{inc}}{\rightrightarrows}} z^i(X, -)$ *are weakly homotopic.*

Proof: Assume that for each n ($0 \leq n \leq N$) we have fixed a finite number of closed subvarieties $V_k^n \subset X \times \Delta^n$ such that $[V_k^n] \in {}_\varphi z^i(X, n)$ and the family $\{V_*^{n-1}\}$ contains all components of cycles $\partial_j([V_k^n])$. Denote by C_n the free abelian group generated by $[V_k^n]$. It's clear that C_* is a finitely generated subcomplex in $_\varphi z^i(X, -)$ and that subcomplexes of this type are cofinal among all finitely generated subcomplexes of $_\varphi z^i(X, -)$. Thus it suffices to show that $(\varphi^* - \text{inc})|_{C_*}$ is homotopic to zero. To prove this we construct by induction on n ($0 \leq n \leq N$) X-morphisms $\Phi_n : X \times \Delta^n \times \mathbf{A}^1 \to X \times \Delta^n \times \mathbf{A}^1$ with the following properties

(2.3.1) The following diagram commutes (here i_0 and i_1 denote closed embeddings defined by points $0, 1 \in \mathbf{A}^1$

$$
\begin{array}{ccc}
X \times \Delta^n & \xrightarrow{\text{id}} & X \times \Delta^n \\
{\scriptstyle i_0} \downarrow & & \downarrow {\scriptstyle i_0} \\
X \times \Delta^n \times \mathbf{A}^1 & \xrightarrow{\Phi_n} & X \times \Delta^n \times \mathbf{A}^1
\end{array}
$$

(2.3.1.1)

$$\begin{array}{ccc} X \times \Delta^n & \xrightarrow{\varphi_n} & X \times \Delta^n \\ {\scriptstyle i_1}\downarrow & & \downarrow{\scriptstyle i_1} \\ X \times \Delta^n \times \mathbf{A}^1 & \xrightarrow{\Phi_n} & X \times \Delta^n \times \mathbf{A}^1 \end{array}$$

(2.3.1.2)

$$\begin{array}{ccc} X \times \Delta^{n-1} \times \mathbf{A}^1 & \xrightarrow{\Phi_{n-1}} & X \times \Delta^{n-1} \times \mathbf{A}^1 \\ {\scriptstyle 1_X \times \delta_j \times 1_{\mathbf{A}^1}}\downarrow & & \downarrow{\scriptstyle 1_X \times \delta_j \times 1_{\mathbf{A}^1}} \\ X \times \Delta^n \times \mathbf{A}^1 & \xrightarrow{\Phi_n} & X \times \Delta^n \times \mathbf{A}^1 \end{array}$$

(2.3.1.3)

(2.3.2) The fibers of the projection $\Phi_n^{-1}(\bigcup_k V_k^n \times \mathbf{A}^1) \to \Delta^n \times \mathbf{A}^1$ over points not lying on the divisor $Z = (\Delta^n \times 0) + (\Delta^n \times 1) + \sum_{j=0}^{n} \Delta_j^{n-1} \times \mathbf{A}^1$ (here Δ_j^{n-1} is the j-th face of Δ^n) are of dimension $\leq t$.

Note that the commutativity of diagrams (2.2.1) together with the induction hypothesis implies that the requirements (2.3.1) are compatible with each other and define the morphism Φ_n on $X \times Z$. The existence of extension of this morphism to $X \times \Delta^n \times \mathbf{A}^1$ satisfying the additional requirement (2.3.2) is guaranteed by the theorem 1.1.

The vertices of $\Delta^n \times \mathbf{A}^1$ are the points $p_i \times 0$ and $p_i \times 1$. We introduce the partial ordering on these vertices, taking the standard order on p_i's ($p_i \leq p_j \Leftrightarrow i \leq j$) and taking the standard order on the set $\{0,1\}$. For any nondecreasing system of vertices $q_0 \leq q_1 \leq \ldots \leq q_m$ denote by θ_q the linear morphism $\Delta^m \to \Delta^n \times \mathbf{A}^1$, taking p_i to q_i.

Lemma 2.4. *Let $q_0 < q_1 < \ldots < q_m$ be a strictly increasing sequence of vertices of $\Delta^n \times \mathbf{A}^1$ (thus $m \leq n+1$). Then*

$$dim[\Phi_n \circ (1_X \times \theta_q)]^{-1}\left(\bigcup_k V_k^n \times \mathbf{A}^1\right) \leq m+t.$$

Proof: We proceed by induction on n. Consider first the case $n = m-1$. In this case θ_q is an isomorphism and our statement reduces to the following formula: $dim\, \Phi_n^{-1}(\bigcup_k V_k^n \times \mathbf{A}^1) \leq n+1+t$. Using the induction hypothesis, the properties of the varieties V_k^n, and (2.3.1) we prove immediately the following statement

(2.4.1) The dimension of the part of $\Phi_n^{-1}(\bigcup V_k^n \times \mathbf{A}^1$ lying over Z does not exceed $n+t$.

The fibers of the projection $\Phi_n^{-1}(\bigcup V_k^n \times \mathbf{A}^1) \to \Delta^n \times \mathbf{A}^1$ over the points not lying in Z are of dimension $\leq t$. This implies that the dimension of the part of $\Phi_n^{-1}(\bigcup V_k^n \times \mathbf{A}^1)$ lying over $\Delta^n \times \mathbf{A}^1 - Z$ is not more than $n+1+t$. Finally $dim\, \Phi_n^{-1}(\bigcup V_k^n \times \mathbf{A}^1) \le \max(n+t, n+1+t) = n+1+t$.

Consider next the case $n = m$. Note that θ_q is always a closed embedding. Denote by Z' the inverse image of Z under θ_q. Using (2.4.1) we conclude immediately that the dimension of the part of $[\Phi_n \circ (1_X \times \theta_q)]^{-1}(\bigcup V_k^n \times \mathbf{A}^1) = (1_X \circ \theta_q)^{-1}\Phi_n^{-1}(\bigcup_k V_k^n \times \mathbf{A}^1)$ lying over Z' does not exceed $n+t$. The fibers of the projection $(1_X \times \theta_q)^{-1}\Phi_n^{-1}(\bigcup V_k^n \times \mathbf{A}^1) \to \Delta^n$ over the points of $\Delta^n - Z'$ are of dimension not more than t and hence the dimension of the part of $[\Phi_n \circ (1_X \times \theta_q)]^{-1}(\bigcup V_k^n \times \mathbf{A}^1)$ which lies over $\Delta^n - Z'$ does not exceed $n+t$. Finally $dim([\Phi_n \circ (1_X \times \theta_q)]^{-1}(\bigcup V_k^n \times \mathbf{A}^1)) \le \max(n+t, n+t) = n+t$.

Finally suppose that $m < n$. In this case one can find a strictly increasing sequence of vertices $q_0' < q_1' < \ldots < q_m'$ in $\Delta^{n-1} \times \mathbf{A}^1$ and an index j $(0 \le j \le n)$ such that $\theta_q = (\delta_j \times 1_{\mathbf{A}^1}) \circ \theta_{q'}$ and hence

$$[\Phi_n \circ (1_X \times \theta_q)]^{-1}\left(\bigcup V_k^n \times \mathbf{A}^1\right)$$

$$= [(1_X \times \delta_j \times 1_{\mathbf{A}^1}) \circ \Phi_{n-1} \circ (1_X \times \theta_{q'})]^{-1}\left(\bigcup V_k^n \times \mathbf{A}^1\right)$$

$$\subset [\Phi_{n-1} \circ (1_X \times \theta_{q'})]^{-1}\left(\bigcup V_*^{n-1} \times \mathbf{A}^1\right)$$

so that the induction hypothesis applies.

(2.5) End of the proof of the proposition 2.3

Denote by θ_j $(0 \le j \le n)$ the morphisms defined by the strictly increasing sequences of vertices $p_0 \times 0 < \ldots < p_j \times 0 < p_j \times 1 < \ldots < p_n \times 1$. Lemma 2.4 shows that the cycles $[\Phi_n \circ (1_X \times \theta_j)]^*([V_k^n \times \mathbf{A}^1])$ are defined and lie in $z^i(X, n+1)$. The homotopy we are seeking may be defined now by means of the formula $\sigma([V_k^n]) = \sum_{j=0}^n (-1)^j [\Phi_n \circ (1_X \times \theta_j)]^*([V_k^n \times \mathbf{A}^1])$ (cf. [B1, §2]).

We also need a version of (2.3) for the complex $z_{equi}^i(X, -)$. In the above situation denote by $_\varphi z_{equi}^i(X, n)$ the free abelian group generated by subvarieties $V \subset X \times \Delta^n$ such that

1. $[V] \in z_{equi}^i(X, n)$
2. The cycle $\varphi_n^*([V])$ is defined and lies in $z_{equi}^i(X, n)$

In the same way as above we see that $_\varphi z_{equi}^i(X, -)$ is a subcomplex in $z_{equi}^i(X, -)$ and we have a canonical homomorphism $\varphi^* : {_\varphi z_{equi}^i(X, -)} \to z_{equi}^i(X, -)$.

Proposition 2.6. *The homomorphisms* $_\varphi z_{equi}^i(X, -) \underset{\text{inc}}{\overset{\varphi^*}{\rightrightarrows}} z_{equi}^i(X, -)$ *are weekly homotopic.*

Proof: Repeat the construction used in the proof of proposition 2.3. Lemma 2.4 is replaced by the following fact (which is proved by a straightforward induction on n).

Lemma 2.7. *All fibers of the projection $\Phi_n^{-1}(\bigcup_k V_k^n \times \mathbf{A}^1) \to \Delta^n \times \mathbf{A}^1$ are of dimension $\leq t$.*

This lemma shows that the cycles $[\Phi_n \circ (1_X \times \theta_j)]^*([V_k^n \times \mathbf{A}^1])$ are defined and lie in $z_{equi}^i(X, n+1)$ so that we can define the homotopy by the same formula as above.

Proposition 2.8. *Let C_* be a finitely generated subcomplex in $z^i(X, -)$. Then we can find N and φ as above such that $C_* \subset {}_\varphi z^i(X, -)$ and $\varphi^*(C_*) \subset z_{equi}^i(X, -)$.*

Proof: We may suppose that C_* is generated by a system of subvarieties $V_k^n \subset X \times \Delta^n$ ($0 \leq n \leq N$) with the corresponding properties (see the proof of (2.3)). Using induction on n we construct X-morphisms $\varphi_n : X \times \Delta^n \to X \times \Delta^n$ with the following properties:

1. The diagrames (2.2.1) commute

2. The fibers of the projection $\varphi_n^{-1}(\bigcup_k V_k^n) \to \Delta^n$ over the points not lying on the divisor $Z = \sum_{j=0}^n \Delta_j^{n-1}$ are of dimension $\leq t$.

The induction step in the construction of φ's is guaranteed again by theorem 1.1. Another induction on n shows immediately that all fibers of the projection $\varphi_n^{-1}(\bigcup V_k^n) \to \Delta^n$ are of dimension $\leq t$. Thus the cycles $\varphi_n^*([V_k^n])$ are defined and lie in $z_{equi}^i(X, -)$.

(2.9) Proof of the theorem 2.1

Let v be any n-dimensional cycle of the complex $z^i(X, -)$. Using proposition 2.8 find φ such that $v \in {}_\varphi z^i(X, n)$ and $\varphi^*(v) \in z_{equi}^i(X, n)$. According to proposition 2.3 $v - \varphi^*(v)$ is a boundary. This shows that the homomorphism $H_n(z_{equi}^i(X, -)) \to H_n(z^i(X, -))$ is surjective.

Assume now that v is an n-dimensional cycle in $z_{equi}^i(X, -)$ such that $v = d(w)$ for a certain $w \in z^i(X, n+1)$. Using proposition 2.8 find φ such that $v, w \in {}_\varphi z^i(X, -)$ and $\varphi^*(v), \varphi^*(w) \in z_{equi}^i(X, -)$. Since $\varphi^*(v) = d(\varphi^*(w))$ we see that $\varphi^*(v)$ is a boundary in $z_{equi}^i(X, -)$. On the other hand $v - \varphi^*(v)$ is also a boundary in $z_{equi}^i(X, -)$ according to (2.6). Thus v is a boundary in $z_{equi}^i(X, -)$, i.e. the homomorphism $H_n(z_{equi}^i(X, -)) \to H_n(z^i(X, -))$ is injective.

3. Sheaves of equidimensional cycles

Denote by p the exponential characteristic of F. Recall from [SV2] that for any scheme $X \in Sch/F$ and any $t \geq 0$ there exists a qfh-sheaf $z_t(X)$ which is characterized by the following property:

For any normal connected scheme $S \in Sch/F$ the group of sections of $z_t(X)$ over S is equal to the free $\mathbf{Z}[1/p]$-module generated by closed subvarieties $Z \subset X \times S$ which are equidimensional of relative dimension t over S.

For any presheaf $\mathcal{F}$ on the category Sch/F we will denote by $C_*(\mathcal{F})$ its singular complex—see [SV1], i.e. $C_n(\mathcal{F}) = \mathcal{F}(\Delta^n)$ and the differential $d : C^n(\mathcal{F}) \to C_{n-1}(\mathcal{F})$ is the alternating sum of face operators. We will use the notation $H_*^{sing}(\mathcal{F}), H_*^{sing}(\mathcal{F}, \mathbf{Z}/n), H_{sing}^*(\mathcal{F}, \mathbf{Z}/n)$ for the homology groups of the complexes $C_*(\mathcal{F}), C_*(\mathcal{F}) \otimes^L \mathbf{Z}/n, RHom(C_*(\mathcal{F}), \mathbf{Z}/n)$. The functor C_* is an exact functor from the category of presheaves to the category of nonnegative complexes of presheaves. Furthermore we have the following important theorem

Theorem 3.1 ([V1]). *Assume that $char(F) = 0$ and let $\mathcal{F}$ be a presheaf with transfers on the category Sch/F. If the h-sheaf associated with $\mathcal{F}$ is trivial then $H_*^{sing}(\mathcal{F}) = 0$.*

Using this theorem of Voevodsky and the localization exact sequence for higher Chow groups (see [B2]) we can generalize theorem 2.1 to arbitrary quasiprojective schemes.

Theorem 3.2. *Assume that $char(F) = 0$ then for any equidimensional quasiprojective scheme $X \in Sch/F$ and any $i \leq d = dim\, X$ the canonical embedding of complexes*

$$C_*(z_{d-i}(X)) = z_{equi}^i(X, -) \hookrightarrow z^i(X, -)$$

is a quasiisomorphism.

Proof: We proceed by induction on $d = dim\, X$. The case $d = 0$ is trivial, so assume that $d > 0$ and for schemes of dimension $d - 1$ the theorem is true. We can easily find an effective Cartier divisor $Y \subset X$ such that the open subscheme $U = X - Y$ is affine. The sequence of presheaves

$$0 \longrightarrow z_{d-i}(Y) \longrightarrow z_{d-i}(X) \longrightarrow z_{d-i}(U)$$

is exact and the h-sheaf associated with the presheaf $z_{d-i}(U)/z_{d-i}(X)$ is trivial. Thus we get a long exact localization sequence for homology of equidimensional cycles. Comparing it with the localization sequence for higher Chow groups [B2] and using the induction hypothesis and theorem 2.1 we conclude that $H_n^{sing}(z_{d-i}(X) \to CH^i(X, n)$ is an isomorphism.

4. Higher Chow groups and etale cohomology

In this section we assume F to be an algebraically closed field of characteristic zero. Let $X \in Sch/F$ be an equidimensional quasiprojective scheme of dimension d. Theorem 3.2 shows that for any $i \leq d$ the group $CH^i(X, n; \mathbf{Z}/m) \stackrel{\text{def}}{=} H_n(z^i(X, -) \otimes \mathbf{Z}/m)$ coincides with $H_n^{sing}(z_{d-i}(X), \mathbf{Z}/m)$. For any periodical abelian group A we denote by $A^{\#}$ its dual, i.e. $A^{\#} = Hom(A, \mathbf{Q}/\mathbf{Z})$. Thus

$$CH^i(X, n; \mathbf{Z}/m)^{\#} = H_n^{sing}(z_{d-i}(X), \mathbf{Z}/m)^{\#} = H_{sing}^n(z_{d-i}(X), \mathbf{Z}/m)$$

and the last group according to the main result of [SV1] is canonically isomorphic to $Ext_{qfh}^n(z_{d-i}(X), \mathbf{Z}/m)$. Consider now special case $i = d$. In this case we may use the following result.

Proposition 4.1. *Let X be a separated scheme of finite type over an arbitrary field F (of exponential characteristic p). Then for any m prime to p we have canonical isomorphisms*

$$Ext_{qfh}^n(z_0(X), \mathbf{Z}/m) = Ext_h^n(z_0(X)_h^{\sim}, \mathbf{Z}/m) = H_c^n(X, \mathbf{Z}/m)$$

where H_c^n stands for etale cohomology with compact supports.

Proof: The first equality is proved in [SV1, §10]. To prove the second one choose an open embedding $j : X \hookrightarrow \overline{X}$, where $\overline{X}$ is a complete separated scheme, set $Y = \overline{X} - X$ and let i denote the closed embedding $Y \hookrightarrow \overline{X}$. According to [SV2] we have an exact sequence of h-sheaves

$$0 \to z_0(Y)_h^{\sim} \to z_0(\overline{X})_h^{\sim} \to z_0(X)_h^{\sim} \to 0.$$

For any complete separated scheme Z the sheaf $z_0(Z)$ coincides with the free qfh-sheaf $\mathbf{Z}[\frac{1}{p}]_{qfh}(Z)$ generated by Z (see [SV1, theorem 6.3]). Taking associated h-sheaves we see that $z_0(Z)_h^{\sim} = \mathbf{Z}[\frac{1}{p}]_h(Z)$. Thus the above sequence takes the form $0 \to \mathbf{Z}[\frac{1}{p}]_h(Y) \to \mathbf{Z}[\frac{1}{p}]_h(\overline{X}) \to z_0(X)_h^{\sim} \to 0$ This exact sequence shows that for any h-sheaf of $\mathbf{Z}[\frac{1}{p}]$-modules $\mathcal{F}$ we have the formula

$$Hom(z_0(X)_h^{\sim}, \mathcal{F}) = Ker(\Gamma(\overline{X}, \mathcal{F}) \to \Gamma(Y, \mathcal{F}))$$

Choose an injective resolution I^* of the h-sheaf $\mathbf{Z}/m$, consisting of $\mathbf{Z}[\frac{1}{p}]$-modules. The previous remarks show that $Ext_h^n(z_0(X)_h^{\sim}, \mathbf{Z}/m) =$

$H^n(Ker(\Gamma(\overline{X}, I^*) \to \Gamma(Y, I^*)))$. Moreover the homomorphism of complexes $\Gamma(\overline{X}, I^*) \to \Gamma(Y, I^*)$ is surjective. Consider the natural morphism of sites $\gamma : (Sch/F)_h \to (Sch/F)_{et}$. The comparison theorem for h-cohomology [SV1, §10] shows that $R^i\gamma_*(\mathbf{Z}/m) = 0$ for $i > 0$. This means that I^* is an injective resolution of $\mathbf{Z}/m$ not only in h-topology, but in etale topology as well. Note further that to give a sheaf $\mathcal{F}$ on big etale site $(Sch/F)_{et}$ is the same as to give a sheaf $\mathcal{F}_Z$ on small etale site Z_{et} for each $Z \in (Sch/F)$ and to give homomorphisms $\mathcal{F}_Z \to f_*(\mathcal{F}_T)$ defined for each morphism $f : T \to Z$ and satisfying evident compatibility properties. It's clear that functors $Shv(Sch/F)_{et} \to Shv(Z_{et})$ $(\mathcal{F} \mapsto \mathcal{F}_Z)$ are exact and preserve injectivity. Consider the bicomplex of sheaves on $\overline{X}_{et}$

$$I_{\overline{X}}^* \to i_*(I_Y^*)$$

This bicomplex has only one nonzero homology (in degree zero) equal to $Ker(\mathbf{Z}/m \to i_*(\mathbf{Z}/m)) = j_!(\mathbf{Z}/m)$. Hence

$$
\begin{aligned}
H_c^n(X, \mathbf{Z}/m) &= H^n(\overline{X}, j_!(\mathbf{Z}/m)) \\
&= H^n(\Gamma(\overline{X}, (I_{\overline{X}}^* \to i_*(I_Y^*)))) \\
&= H^n((\Gamma(\overline{X}, I^*) \to \Gamma(Y, I^*))) \\
&= H^n(Ker(\Gamma(\overline{X}, I^*) \to \Gamma(Y, I^*))) \\
&= Ext_h^n(z_0(X)_h^\sim, \mathbf{Z}/m).
\end{aligned}
$$

Theorem 4.2. *Let X be an equidimensional quasiprojective scheme over an algebraically closed field F of characteristic zero. Assume that $i \geq d = \dim X$. Then for any $m > 0$*

$$CH^i(X, n; \mathbf{Z}/m) = H_c^{2(d-i)+n}(X, \mathbf{Z}/m(d-i))^{\#}.$$

Proof: Assume first that $i = d$. In this case according to proposition 4.2 we have:

$$CH^d(X, n; \mathbf{Z}/m)^{\#} = Ext_{qfh}^n(z_0(X), \mathbf{Z}/m) = H_c^n(X, \mathbf{Z}/m)$$

Because this group is finite we conclude that $CH^d(X, n; \mathbf{Z}/m)$ is also finite and hence $CH^d(X, n; \mathbf{Z}/m) = [CH^d(X, n; \mathbf{Z}/m)^{\#}]^{\#} = H_c^n(X, \mathbf{Z}/m)^{\#}$. The general case follows now from the homotopy invariance of higher Chow groups—see [B1, §2]: $CH^i(X, n; \mathbf{Z}/m) = CH^i(X \times \mathbf{A}^{i-d}, n; \mathbf{Z}/m) = H_c^n(X \times \mathbf{A}^{i-d}, \mathbf{Z}/m)^{\#} = H_c^{n-2(i-d)}(X, \mathbf{Z}/m(d-i))^{\#}$.

Corollary 4.3. *In assumptions and notations of theorem 4.2 assume further that X is smooth. Then $CH^i(X, n; \mathbf{Z}/m) = H_{et}^{2i-n}(X, \mathbf{Z}/m(i))$.*

Proof: This follows immediately from theorem 4.2 and the Poincaré duality theorem—see [M].

References

[B1]   S. Bloch. Algebraic cycles and higher K-theory. *Adv. in Math.*, 61:267–304, 1986.

[B2]   ———. The moving lemma for higher Chow groups. *Journal of algebraic geometry* (to appear).

[Bl]   S. Bloch and S. Lichtenbaum. A spectral sequence for motivic cohomology. *Preprint*, 1994.

[M]   J. S. Milne. *Etale cohomology*, Princeton Mathematical Series, Vol. 33. Princeton University Press, 1980.

[SV1] A. Suslin and V. Voevodsky. Singular homology of abstract algebraic varieties. *Invent. Math.* (to appear).

[SV2] ———. Relative cycles and Chow sheaves. This volume, pages 10–86.

[V1]   V. Voevodsky. Cohomological theory of presheaves with transfers. This volume, pages 87–137.